partial differential equations and applications

LECTURE NOTES IN PURE AND APPLIED MATHEMATICS

1. *N. Jacobson,* Exceptional Lie Algebras
2. *L.-Å. Lindahl and F. Poulsen,* Thin Sets in Harmonic Analysis
3. *I. Satake,* Classification Theory of Semi-Simple Algebraic Groups
4. *F. Hirzebruch, W. D. Newmann, and S. S. Koh, Differentiable Manifolds and Quadratic Forms*
5. *I. Chavel,* Riemannian Symmetric Spaces of Rank One
6. *R. B. Burckel,* Characterization of C(X) Among Its Subalgebras
7. *B. R. McDonald, A. R. Magid, and K. C. Smith,* Ring Theory: Proceedings of the Oklahoma Conference
8. *Y.-T. Siu,* Techniques of Extension on Analytic Objects
9. *S. R. Caradus, W. E. Pfaffenberger, and B. Yood,* Calkin Algebras and Algebras of Operators on Banach Spaces
10. *E. O. Roxin, P.-T. Liu, and R. L. Sternberg,* Differential Games and Control Theory
11. *M. Orzech and C. Small,* The Brauer Group of Commutative Rings
12. *S. Thomier,* Topology and Its Applications
13. *J. M. Lopez and K. A. Ross,* Sidon Sets
14. *W. W. Comfort and S. Negrepontis,* Continuous Pseudometrics
15. *K. McKennon and J. M. Robertson,* Locally Convex Spaces
16. *M. Carmeli and S. Malin,* Representations of the Rotation and Lorentz Groups: An Introduction
17. *G. B. Seligman,* Rational Methods in Lie Algebras
18. *D. G. de Figueiredo,* Functional Analysis: Proceedings of the Brazilian Mathematical Society Symposium
19. *L. Cesari, R. Kannan, and J. D. Schuur,* Nonlinear Functional Analysis and Differential Equations: Proceedings of the Michigan State University Conference
20. *J. J. Schäffer,* Geometry of Spheres in Normed Spaces
21. *K. Yano and M. Kon,* Anti-Invariant Submanifolds
22. *W. V. Vasconcelos,* The Rings of Dimension Two
23. *R. E. Chandler,* Hausdorff Compactifications
24. *S. P. Franklin and B. V. S. Thomas,* Topology: Proceedings of the Memphis State University Conference
25. *S. K. Jain,* Ring Theory: Proceedings of the Ohio University Conference
26. *B. R. McDonald and R. A. Morris,* Ring Theory II: Proceedings of the Second Oklahoma Conference
27. *R. B. Mura and A. Rhemtulla,* Orderable Groups
28. *J. R. Graef,* Stability of Dynamical Systems: Theory and Applications
29. *H.-C. Wang,* Homogeneous Branch Algebras
30. *E. O. Roxin, P.-T. Liu, and R. L. Sternberg,* Differential Games and Control Theory II
31. *R. D. Porter,* Introduction to Fibre Bundles
32. *M. Altman,* Contractors and Contractor Directions Theory and Applications
33. *J. S. Golan,* Decomposition and Dimension in Module Categories
34. *G. Fairweather,* Finite Element Galerkin Methods for Differential Equations
35. *J. D. Sally,* Numbers of Generators of Ideals in Local Rings
36. *S. S. Miller,* Complex Analysis: Proceedings of the S.U.N.Y. Brockport Conference
37. *R. Gordon,* Representation Theory of Algebras: Proceedings of the Philadelphia Conference
38. *M. Goto and F. D. Grosshans,* Semisimple Lie Algebras
39. *A. I. Arruda, N. C. A. da Costa, and R. Chuaqui,* Mathematical Logic: Proceedings of the First Brazilian Conference
40. *F. Van Oystaeyen,* Ring Theory: Proceedings of the 1977 Antwerp Conference
41. *F. Van Oystaeyen and A. Verschoren,* Reflectors and Localization: Application to Sheaf Theory
42. *M. Satyanarayana,* Positively Ordered Semigroups
43. *D. L Russell,* Mathematics of Finite-Dimensional Control Systems
44. *P.-T. Liu and E. Roxin,* Differential Games and Control Theory III: Proceedings of the Third Kingston Conference, Part A
45. *A. Geramita and J. Seberry,* Orthogonal Designs: Quadratic Forms and Hadamard Matrices
46. *J. Cigler, V. Losert, and P. Michor,* Banach Modules and Functors on Categories of Banach Spaces
47. *P.-T. Liu and J. G. Sutinen,* Control Theory in Mathematical Economics: Proceedings of the Third Kingston Conference, Part B
48. *C. Byrnes,* Partial Differential Equations and Geometry

49. *G. Klambauer,* Problems and Propositions in Analysis
50. *J. Knopfmacher,* Analytic Arithmetic of Algebraic Function Fields
51. *F. Van Oystaeyen,* Ring Theory: Proceedings of the 1978 Antwerp Conference
52. *B. Kadem,* Binary Time Series
53. *J. Barros-Neto and R. A. Artino,* Hypoelliptic Boundary-Value Problems
54. *R. L. Sternberg, A. J. Kalinowski, and J. S. Papadakis,* Nonlinear Partial Differential Equations in Engineering and Applied Science
55. *B. R. McDonald,* Ring Theory and Algebra III: Proceedngs of the Third Oklahoma Conference
56. *J. S. Golan,* Structure Sheaves Over a Noncommutative Ring
57. *T. V. Narayana, J. G. Williams, and R. M. Mathsen,* Combinatorics, Representation Theory and Statistical Methods in Groups: YOUNG DAY Proceedings
58. *T. A. Burton,* Modeling and Differential Equations in Biology
59. *K. H. Kim and F. W. Roush,* Introduction to Mathematical Consensus Theory
60. *J. Banas and K. Goebel,* Measures of Noncompactness in Banach Spaces
61. *O. A. Nielson,* Direct Integral Theory
62. *J. E. Smith, G. O. Kenny, and R. N. Ball,* Ordered Groups: Proceedings of the Boise State Conference
63. *J. Cronin,* Mathematics of Cell Electrophysiology
64. *J. W. Brewer,* Power Series Over Commutative Rings
65. *P. K. Kamthan and M. Gupta,* Sequence Spaces and Series
66. *T. G. McLaughlin,* Regressive Sets and the Theory of Isols
67. *T. L. Herdman, S. M. Rankin III, and H. W. Stech,* Integral and Functional Differential Equations
68. *R. Draper,* Commutative Algebra: Analytic Methods
69. *W. G. McKay and J. Patera,* Tables of Dimensions, Indices, and Branching Rules for Representations of Simple Lie Algebras
70. *R. L. Devaney and Z. H. Nitecki,* Classical Mechanics and Dynamical Systems
71. *J. Van Geel,* Places and Valuations in Noncommutative Ring Theory
72. *C. Faith,* Injective Modules and Injective Quotient Rings
73. *A. Fiacco,* Mathematical Programming with Data Perturbations I
74. *P. Schultz, C. Praeger, and R. Sullivan,* Algebraic Structures and Applications: Proceedings of the First Western Australian Conference on Algebra
75. *L Bican, T. Kepka, and P. Nemec,* Rings, Modules, and Preradicals
76. *D. C. Kay and M. Breen,* Convexity and Related Combinatorial Geometry: Proceedings of the Second University of Oklahoma Conference
77. *P. Fletcher and W. F. Lindgren,* Quasi-Uniform Spaces
78. *C.-C. Yang,* Factorization Theory of Meromorphic Functions
79. *O. Taussky,* Ternary Quadratic Forms and Norms
80. *S. P. Singh and J. H. Burry,* Nonlinear Analysis and Applications
81. *K. B. Hannsgen, T. L. Herdman, H. W. Stech, and R. L. Wheeler,* Volterra and Functional Differential Equations
82. *N. L. Johnson, M. J. Kallaher, and C. T. Long,* Finite Geometries: Proceedings of a Conference in Honor of T. G. Ostrom
83. *G. I. Zapata,* Functional Analysis, Holomorphy, and Approximation Theory
84. *S. Greco and G. Valla,* Commutative Algebra: Proceedings of the Trento Conference
85. *A. V. Fiacco,* Mathematical Programming with Data Perturbations II
86. *J.-B. Hiriart-Urruty, W. Oettli, and J. Stoer,* Optimization: Theory and Algorithms
87. *A. Figa Talamanca and M. A. Picardello,* Harmonic Analysis on Free Groups
88. *M. Harada,* Factor Categories with Applications to Direct Decomposition of Modules
89. *V. I. Istrăţescu,* Strict Convexity and Complex Strict Convexity
90. *V. Lakshmikantham,* Trends in Theory and Practice of Nonlinear Differential Equations
91. *H. L. Manocha and J. B. Srivastava,* Algebra and Its Applications
92. *D. V. Chudnovsky and G. V. Chudnovsky,* Classical and Quantum Models and Arithmetic Problems
93. *J. W. Longley,* Least Squares Computations Using Orthogonalization Methods
94. *L. P. de Alcantara,* Mathematical Logic and Formal Systems
95. *C. E. Aull,* Rings of Continuous Functions
96. *R. Chuaqui,* Analysis, Geometry, and Probability
97. *L. Fuchs and L. Salce,* Modules Over Valuation Domains
98. *P. Fischer and W. R. Smith,* Chaos, Fractals, and Dynamics
99. *W. B. Powell and C. Tsinakis,* Ordered Algebraic Structures
100. *G. M. Rassias and T. M. Rassias,* Differential Geometry, Calculus of Variations, and Their Applications

101. *R.-E. Hoffmann and K. H. Hofmann,* Continuous Lattices and Their Applications
102. *J. H. Lightbourne III and S. M. Rankin III,* Physical Mathematics and Nonlinear Partial Differential Equations
103. *C. A. Baker and L. M. Batten,* Finite Geometrics
104. *J. W. Brewer, J. W. Bunce, and F. S. Van Vleck,* Linear Systems Over Commutative Rings
105. *C. McCrory and T. Shifrin,* Geometry and Topology: Manifolds, Varieties, and Knots
106. *D. W. Kueker, E. G. K. Lopez-Escobar, and C. H. Smith,* Mathematical Logic and Theoretical Computer Science
107. *B.-L. Lin and S. Simons,* Nonlinear and Convex Analysis: Proceedings in Honor of Ky Fan
108. *S. J. Lee,* Operator Methods for Optimal Control Problems
109. *V. Lakshmikantham,* Nonlinear Analysis and Applications
110. *S. F. McCormick,* Multigrid Methods: Theory, Applications, and Supercomputing
111. *M. C. Tangora,* Computers in Algebra
112. *D. V. Chudnovsky and G. V. Chudnovsky,* Search Theory: Some Recent Developments
113. *D. V. Chudnovsky and R. D. Jenks,* Computer Algebra
114. *M. C. Tangora,* Computers in Geometry and Topology
115. *P. Nelson, V. Faber, T. A. Manteuffel, D. L. Seth, and A. B. White, Jr.,* Transport Theory, Invariant Imbedding, and Integral Equations: Proceedings in Honor of G. M. Wing's 65th Birthday
116. *P. Clément, S. Invernizzi, E. Mitidieri, and I. I. Vrabie,* Semigroup Theory and Applications
117. *J. Vinuesa,* Orthogonal Polynomials and Their Applications: Proceedings of the International Congress
118. *C. M. Dafermos, G. Ladas, and G. Papanicolaou,* Differential Equations: Proceedings of the EQUADIFF Conference
119. *E. O. Roxin,* Modern Optimal Control: A Conference in Honor of Solomon Lefschetz and Joseph P. Lasalle
120. *J. C. Díaz,* Mathematics for Large Scale Computing
121. *P. S. Milojevič,* Nonlinear Functional Analysis
122. *C. Sadosky,* Analysis and Partial Differential Equations: A Collection of Papers Dedicated to Mischa Cotlar
123. *R. M. Shortt,* General Topology and Applications: Proceedings of the 1988 Northeast Conference
124. *R. Wong,* Asymptotic and Computational Analysis: Conference in Honor of Frank W. J. Olver's 65th Birthday
125. *D. V. Chudnovsky and R. D. Jenks,* Computers in Mathematics
126. *W. D. Wallis, H. Shen, W. Wei, and L. Zhu,* Combinatorial Designs and Applications
127. *S. Elaydi,* Differential Equations: Stability and Control
128. *G. Chen, E. B. Lee, W. Littman, and L. Markus,* Distributed Parameter Control Systems: New Trends and Applications
129. *W. N. Everitt,* Inequalities: Fifty Years On from Hardy, Littlewood and Pólya
130. *H. G. Kaper and M. Garbey,* Asymptotic Analysis and the Numerical Solution of Partial Differential Equations
131. *O. Arino, D. E. Axelrod, and M. Kimmel,* Mathematical Population Dynamics: Proceedings of the Second International Conference
132. *S. Coen,* Geometry and Complex Variables
133. *J. A. Goldstein, F. Kappel, and W. Schappacher,* Differential Equations with Applications in Biology, Physics, and Engineering
134. *S. J. Andima, R. Kopperman, P. R. Misra, J. Z. Reichman, and A. R. Todd,* General Topology and Applications
135. *P Clément, E. Mitidieri, B. de Pagter,* Semigroup Theory and Evolution Equations: The Second International Conference
136. *K. Jarosz,* Function Spaces
137. *J. M. Bayod, N. De Grande-De Kimpe, and J. Martínez-Maurica,* p-adic Functional Analysis
138. *G. A. Anastassiou,* Approximation Theory: Proceedings of the Sixth Southeastern Approximation Theorists Annual Conference
139. *R. S. Rees,* Graphs, Matrices, and Designs: Festschrift in Honor of Norman J. Pullman
140. *G. Abrams, J. Haefner, and K. M. Rangaswamy,* Methods in Module Theory
141. *G. L. Mullen and P. J.-S. Shiue,* Finite Fields, Coding Theory, and Advances in Communications and Computing
142. *M. C. Joshi and A. V. Balakrishnan,* Mathematical Theory of Control: Proceedings of the International Conference

143. *G. Komatsu and Y. Sakane*, Complex Geometry: Proceedings of the Osaka International Conference
144. *I. J. Bakelman*, Geometric Analysis and Nonlinear Partial Differential Equations
145. *T. Mabuchi and S. Mukai*, Einstein Metrics and Yang–Mills Connections: Proceedings of the 27th Taniguchi International Symposium
146. *L. Fuchs and R. Göbel*, Abelian Groups: Proceedings of the 1991 Curaçao Conference
147. *A. D. Pollington and W. Moran*, Number Theory with an Emphasis on the Markoff Spectrum
148. *G. Dore, A. Favini, E. Obrecht, and A. Venni*, Differential Equations in Banach Spaces
149. *T. West*, Continuum Theory and Dynamical Systems
150. *K. D. Bierstedt, A. Pietsch, W. Ruess, and D. Vogt*, Functional Analysis
151. *K. G. Fischer, P. Loustaunau, J. Shapiro, E. L. Green, and D. Farkas*, Computational Algebra
152. *K. D. Elworthy, W. N. Everitt, and E. B. Lee*, Differential Equations, Dynamical Systems, and Control Science
153. *P.-J. Cahen, D. L. Costa, M. Fontana, and S.-E. Kabbaj*, Commutative Ring Theory
154. *S. C. Cooper and W. J. Thron*, Continued Fractions and Orthogonal Functions: Theory and Applications
155. *P. Clément and G. Lumer*, Evolution Equations, Control Theory, and Biomathematics
156. *M. Gyllenberg and L. Persson*, Analysis, Algebra, and Computers in Mathematical Research: Proceedings of the Twenty-First Nordic Congress of Mathematicians
157. *W. O. Bray, P. S. Milojevič, and Č. V. Stanojević*, Fourier Analysis: Analytic and Geometric Aspects
158. *J. Bergen and S. Montgomery*, Advances in Hopf Algebras
159. *A. R. Magid*, Rings, Extensions, and Cohomology
160. *N. H. Pavel*, Optimal Control of Differential Equations
161. *M. Ikawa*, Spectral and Scattering Theory: Proceedings of the Taniguchi International Workshop
162. *X. Liu and D. Siegel*, Comparison Methods and Stability Theory
163. *J.-P. Zolésio*, Boundary Control and Variation
164. *M. Křížek, P. Neittaanmäki, and R. Stenberg*, Finite Element Methods: Fifty Years of the Courant Element
165. *G. Da Prato and L. Tubaro*, Control of Partial Differential Equations
166. *E. Ballico*, Projective Geometry with Applications
167. *M. Costabel, M. Dauge, and S. Nicaise*, Boundary Value Problems and Integral Equations in Nonsmooth Domains
168. *G. Ferreyra, G. R. Goldstein, and F. Neubrander*, Evolution Equations
169. *S. Huggett*, Twistor Theory
170. *H. Cook, W. T. Ingram, K. T. Kuperberg, A. Lelek, and P. Minc*, Continua: With the Houston Problem Book
171. *D. F. Anderson and D. E. Dobbs*, Zero-Dimensional Commutative Rings
172. *K. Jarosz*, Function Spaces: The Second Conference
173. *V. Ancona, E. Ballico, and A. Silva*, Complex Analysis and Geometry
174. *E. Casas*, Control of Partial Differential Equations and Applications
175. *N. Kalton, E. Saab, and S. Montgomery-Smith*, Interaction Between Functional Analysis, Harmonic Analysis, and Probability
176. *Z. Deng, Z. Liang, G. Lu, and S. Ruan*, Differential Equations and Control Theory
177. *P. Marcellini, G. Talenti, and E. Vesentini*, Partial Differential Equations and Applications: Collected Papers in Honor of Carlo Pucci

Additional Volumes in Preparation

partial differential equations and applications

Collected Papers in Honor of Carlo Pucci

edited by

Paolo Marcellini
Giorgio G. Talenti
Università di Firenze
Florence, Italy

Edoardo Vesentini
Scuola Normale Superiore
Pisa, Italy

CRC Press
Taylor & Francis Group
Boca Raton London New York

CRC Press is an imprint of the
Taylor & Francis Group, an **informa** business

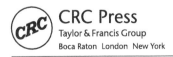

CRC Press
Taylor & Francis Group
6000 Broken Sound Parkway NW, Suite 300
Boca Raton, FL 33487-2742
First issued in hardback 2017

ISBN 13: 978-1-138-41757-1 (hbk)
ISBN 13: 978-0-8247-9698-3 (pbk)

Visit the Taylor & Francis Web site at
http://www.taylorandfrancis.com

and the CRC Press Web site at
http://www.crcpress.com

Library of Congress Cataloging-in-Publication Data

Partial differential equations and applications : collected papers in honor of Carlo Pucci / edited by Paolo Marcellini, Giorgio Talenti, Edoardo Vesentini.
p. cm. — (Lecture notes in pure and applied mathematics ; v. 177)
Includes index.
ISBN 0-8247-9698-5 (pbk. : alk. paper)
1. Differential equations, Partial. I. Pucci, Carlo. II. Marcellini, Paolo. III. Talenti, G. (Giorgio) IV. Vesentini, Edoardo. V. Series.
QA371.P36 1996
515'.353—dc20 95-50251
CIP

Preface

The present volume collects papers that have been dedicated to Carlo Pucci on the occasion of his seventieth birthday.

Carlo Pucci was a student of Giovanni Sansone and Mauro Picone, two prominent figures of the Italian school of mathematics. He took his degree in Florence, taught at the Universities of Rome, Catania and Genoa, and spent a period of time in the USA — mainly at the University of Maryland at College Park, the University of California at Berkeley, the University of Louisiana at Baton Rouge, and Rice University. Since 1968 he has been professor of mathematics at the University of Florence. He served as President of the Unione Matematica Italiana; President of the Comitato Matematico, Consiglio Nazionale delle Ricerche; Director of the Istituto di Analisi Globale e Applicazioni, Consiglio Nazionale delle Ricerche; and President of the Istituto Nazionale di Alta Matematica.

Carlo Pucci has been a scientist, a manager, an indefatigable leader. He took a great interest in various branches of mathematics, ranging from elementary topics to real analysis, convex sets, calculus of variations, and partial differential equations. He worked hard doing, teaching and advertising mathematics; he created, revitalized, and conducted academic institutions; and animated the life of the mathematical community of Italy.

His major achievements in mathematics pertain to partial differential equations and consist mainly of contributions to ill-posed problems and second-order equations of elliptic type. A sample of the contributions which earned him his scientific reputation is presented below. A complete bibliography is appended.

Problems where a partial differential equation is coupled with unfavorable boundary or initial conditions are termed ill-posed nowadays. A salient feature of these problems is the sensitivity of solutions to data, i.e., small perturbations of data may cause uncontrollable changes in the solutions. The Cauchy problem for Laplace's equation or for the heat equation, backward in time, the Cauchy problem with data on spacelike surfaces, or the Dirichlet boundary value problem for the wave equation are typical examples of ill-posed problems. Although ill-posed problems were branded by Hadamard, early in this century, as unrealistic, developments in science and technology have subsequently shown that precisely the opposite is true. For instance, problems in such areas of geophysics as locating masses and heat sources, based on the gravity and temperature field, are typically modeled by ill-posed Cauchy problems for partial differential equations of elliptic or parabolic type. A number of pioneers pointed out in the 1950s that conventional data are not enough for an ill-posed problem to have a realistic answer, and they discovered that ill-posedness can be efficaciously cured if a distinctive recipe enters the game: namely, if the data are enriched by suitable a priori information about the solutions. They then established an innovative setting which made ill-posed problems definitely tractable, even in presence of noisy data. Carlo Pucci was among these pioneers; others include F. John, M. M. Lavrentiev, L. Payne, and

A. N. Tikhonov. It is to the credit of such authors that ill-posed problems are nowadays well understood and worked on in mathematical analysis and applied mathematics.

A priori estimates are a key to boundary value problems for partial differential equations of elliptic type. The following theorem was discovered by Carlo Pucci in 1966, and by A. D. Aleksandrov and I. Bakelman, independently. Let Ω be any open bounded subset of the euclidean n-dimensional space, and

$$E = \sum_{i,j=1}^{n} a_{i,j}(x) \frac{\partial^2}{\partial x_i \partial x_j} + \sum_{i=1}^{n} b_i(x) \frac{\partial}{\partial x_i} + c(x),$$

a linear second-order partial differential operator whose coefficients obey: (i) a uniform ellipticity condition, (ii) $b_i \in L^n(\Omega)$, (iii) $c \leq 0$. Then a constant K exists such that

$$\max_{x \in \overline{\Omega}} | u(x) | \leq K \left\{ \int_{\Omega} |Eu(x)|^n dx \right\}^{1/n}$$

for every sufficiently smooth function u that vanishes on the boundary of Ω.

The main feature of this theorem is that no smoothness assumption on the coefficients of E is made. Although massive investigations about linear elliptic partial differential equations have been carried out over the years, few results are yet available in the case where the coefficients involved are merely measurable. The Aleksandrov-Bakelman-Pucci theorem is one of these and is widely recognized as crucial in several respects. The proof is simple and elegant, based upon ingredients of a geometric nature. The theorem in question is a good example of that sort of mathematics where the machinery amounts to a minimum and a sharp idea dominates the scene.

The themes mentioned above are central to the present volume — others are developed, too, that fit in well here. Contributors to the present volume include mathematicians from various countries who have kept in touch with Carlo Pucci, colleagues and former students of his.

This volume is intended to be a tribute to a mathematician who has a persistent habit of devoting all of himself to mathematics, as well as the habit of making his colleagues suffer at times but live better after all.

Paolo Marcellini
Giorgio G. Talenti
Edoardo Vesentini

Bibliography of Carlo Pucci

Un teorema di derivazione per serie. Boll. U.M.I. (3) 5 (1950), pp. 20-23.

Alcuni teoremi sulle successioni di funzioni di più variabili che possiedono derivate parziali fino all'ordine *r*. Ann. Mat. Pura Appl. (4) 31 (1950), pp. 129-141.

Derivazione per serie di funzioni a variazione limitata. Boll. U.M.I. (3) 5 (1950), pp. 281-286.

Serie di funzioni a variazione limitata. Boll. U.M.I. (3) 6 (1951), pp. 1-3.

Sulla continuità dei funzionali analitici. Rend. Mat. Pura Appl. (5) 10 (1951).

Sulla maggiorazione dell'integrale di una equazione differenziale lineare ordinaria del secondo ordine. Rend. Acc. Naz. Lincei (8) 10 (1951), pp. 300-306.

Formule di maggiorazione per un integrale di una equazione differenziale lineare del secondo ordine. Ann. Mat. Pura Appl. (4) 33 (1952), pp. 49-90.

Alcune limitazioni per gli integrali delle equazioni differenziali a derivate parziali lineari, del secondo ordine, di tipo ellittico parabolico. Rend. Acc. Naz. Lincei (8) 11 (1951), pp. 334-339.

Teoremi di esistenza e di unicità per il problema di Cauchy nella teoria delle equazioni lineari a derivate parziali. Rend. Acc. Naz. Lincei (8) 13 (1952), pp. 18-25 and 111-116.

Sul problema di Cauchy per le equazioni lineari a derivate parziali. Rend. Acc. Naz. Lincei (8) 14 (1953), pp. 198-203.

Il problema di Cauchy per le equazioni lineari a derivate parziali. Ann. Mat. Pura Appl. (4) 35 (1953), pp. 129-153.

Maggiorazione della soluzione di un problema al contorno di tipo misto relativo ad una equazione a derivate parziali lineare del secondo ordine. Rend. Acc. Naz. Lincei (8) 13 (1952), pp. 360-366.

Bounds for solutions of Laplace's equation satisfying mixed conditions. J. Rat. Mech. Anal. 2 (1953), pp. 229-302.

Sulla compattezza di successioni di funzioni reali. Rend. Acc. Naz. Lincei (8) 4 (1953), pp. 471-476.

Compattezza di successioni di funzioni e derivabilità delle funzioni limiti. Ann. Mat. Pura Appl. (4) 36 (1954), pp. 471-476.

Studio col metodo delle differenze di un problema di Cauchy relativo ad equazioni a derivate parziali del secondo ordine di tipo parabolico. Ann. Scuola Norm. Sup. Pisa (3) 7 (1953), pp. 205-215.

Nuove ricerche sul problema di Cauchy. Acc. Scienze Torino (3) 1 (1953), pp. 205-215.

A proposito di un problea isoperimetrico. Rend. Acc. Naz. Lincei (8) 17 (1954), pp. 345-346.

Un problema isoperimetrico per la determinazione della forma di una nave. Mem. Acc. Naz. Lincei (8) 4 (1955), and Rend. Acc. Naz. Lincei (8) 25 (1958).

Sui problemi di Cauchy non ben posti. Rend. Acc. Naz. Lincei (8) 18 (1955), pp. 473-477.

Sulla risoluzione di un problema isoperimetrico. Ann. Mat. Pura Appl. (4) 39 (1955), pp. 393-400.

Proprietà di massimo e minimo delle soluzioni di equazioni a derivate parziali del secondo ordine di tipo ellittico e parabolico. Rend. Acc. Naz. Lincei (8) 5 (1957) and (8) 24 (1958).

Sull'equazione del calore con dati subarmonici (with A.Weinstein). Rend. Acc. Naz. Lincei (8) 24 (1958), pp. 493-496.

Studio di un sistema di equazioni differenziali della dinamica dei gas. Rend. Acc. Naz. Lincei (8) 24 (1958), pp. 653-657.

Discussione del problema di Cauchy per le equazioni di tipo ellittico. Ann. Mat. Pura Appl. (4) 46 (1958), pp. 131-154.

The Dirichlet problem for the wave equation (with D.Fox). Ann. Mat. Pura Appl. (4) 46 (1958), pp. 156-182.

Some topics in parabolic and elliptic equations. Lecture Notes, University of Maryland (1958).

On the improperly posed Cauchy problems for parabolic equations. Symposium, Provisional International Computation Center (1959).

Alcune limitazioni per le equazioni paraboliche. Ann. Mat. Pura Appl. (4) 48 (1959), pp. 161-172.

An integrodifferential equation. Symposium, Provisional International Computation Center (1960).

Un problema variazionale per i coefficienti di equazioni differenziali di tipo ellittico. Ann. Scuola Norm. Sup. Pisa (3) 16 (1962), pp. 159-172.

Limitazioni per il gradiente di una soluzione di una equazione di tipo ellittico. Le Matematiche 16 (1961), pp. 51-54.

Sulle funzioni barriera. Le Matematiche 18 (1963), pp. 102-107.

Lectures on the Cauchy problem and on second order elliptic equations. Lecture Notes, Rice University (1963).

Regolarità alla frontiera di soluzioni di equazioni ellittiche. Ann. Mat. Pura Appl. (4) 65 (1964), pp. 311-327.

Sulla regolarità interna delle soluzioni di alcune equazioni ellittiche. Boll. U.M.I. 19 (1964), pp. 334-342.

Sulle equazioni ellittiche estremanti. Rend. Sem. Mat. Fis. Milano 35 (1965), pp. 1-11.

Operatori ellittici estremanti. Ann. Mat. Pura Appl. (4) 72 (1966), pp. 141-170.

Maximum and minimum first eigenvalue for a class of elliptic operators. Proc. Amer. Math. Soc. 17 (1966), pp. 788-795.

Su una limitazione per soluzioni di equazioni ellittiche. Boll. U.M.I. (3) 21 (1966), pp. 228-233.

Limitazioni per soluzioni di equazioni ellittiche. Ann. Mat. Pura Appl. (4) 74 (1966), pp. 15-30.

Equazioni ellittiche con soluzioni in $W^{2,p}$ $(p < 2)$. Atti Convegno Equazioni a Derivate Parziali, Bologna (1967).

Esistenza di autovalori per operatori ellittici non autoaggiunti. Atti VIII Convegno U.M.I. (1967).

Equazioni ellittiche. Lecture Notes, Università di Genova (1969).

Elliptic (second order) partial differential equations with measurable coefficients and approximating integral equations (with G.Talenti). Advances in Math. 19 (1976), pp. 48-105.

Operatori massimanti. Proc. Seminar Ist. Naz. Alta Mat. (1980), pp. 169-176.

Problemi non ben posti per l'equazione delle onde. Rend. Sem. Mat. Fis. Milano 52 (1982), pp. 473-484.

Una proprietà di massimo dell'ellisse. Archimede 4 (1982), pp. 194-207.

A not well posed problem for the wave equation (with G.Papi). Ann. Mat. Pura Appl. (4) 33 (1983), pp. 285-304.

Maximizing elliptic operators: applications and conjectures. Proc. Conference on Partial Differential Equations, Novosibirsk (1983).

Singolarità rimovibili per operatori ellittici. Boll. U.M.I. (6) 4A (1985), pp. 309-314.

Abbasso o viva Euclide? Archimede 4 (1984), pp. 168-173.

An angle's maximum principle for the gradient of solutions of elliptic equations. Boll. U.M.I. (6) 1A (1987).

Introduzione alle funzioni ed agli insiemi convessi (with M.Longinetti). Lecture Notes, Ist. Analisi Globale Applicazioni, Firenze (1986).

L'Unione Matematica Italiana del 1922 al 1944: documenti e riflessioni. Symposia Math. 28, Academic Press (1986).

A maximum principle related to level surfaces of solutions of parabolic equations. J. Austral. Math. Soc. 46 (1989), pp. 1-7.

Maximum principles related to the second derivatives of solutions of elliptic equations in the plane (with P.Manselli). Applicable Anal. 32 (1989), pp. 143-164.

Maximum length of steepest descent curves for quasi-convex functions (with P.Manselli). Geometriae Dedicata 38 (1991), pp. 211-227.

Risultati di unicità per curve evolute ed evolventi di sè stesse (with P.Manselli). Boll. U.M.I. (6) 5A (1991), pp. 373-379.

Limitazioni inferiori in L^q per le funzioni di Green relative a problemi di Dirichlet ellittici (with P.Manselli). Nonlinear Analysis, Scuola Norm. Sup. Pisa (1991).

Principio di massimo per il rapporto di Hölder di soluzioni di equazioni ellittiche (with P.Manselli). Boundary Value Problems for PDE, Masson (1993), pp. 211-222.

On the second differentiability of convex functions (with G.Bianchi & A.Colesanti). Geometriae Dedicata, to appear.

Qualitative and quantitative results for sets of singular points of convex bodies (with A.Colesanti). Forum Math., to appear.

Contents

Contributors

K. A. Ames Department of Mathematical Sciences, University of Alabama in Huntsville, Huntsville, Alabama

Orazio Arena Istituto di Matematica, Università di Firenze, Florence, Italy

Mario Bertero Dipartimento di Scienze Fisiche, Università di Genova, Genoa, Italy

Luis A. Caffarelli Institute for Advanced Study, Princeton, New Jersey

Francesco Calogero Dipartimento di Fisica, Università di Roma la Sapienza, Rome, Italy

Stefano Campi Dipartimento di Matematica Pura e Applicata, Università dell'Aquila, L'Aquila, Italy

John R. Cannon Department of Mathematics, University of Central Florida, Orlando, Florida

Andrea Colesanti Dipartimento di Matematica, Università di Firenze, Florence, Italy

Giorgio Dall'Aglio Dipartimento di Statistica, Università di Roma la Sapienza, Rome, Italy

Leonede De-Michele Dipartimento di Matematica, Università di Milano, Milan, Italy

Emmanuele DiBenedetto Mathematics Department, Northwestern University, Evanston, Illinois

David J. Diller Department of Mathematics, Northwestern University, Evanston, Illinois

Jim Douglas, Jr. Department of Mathematics, Purdue University, West Lafayette, Indiana

Paul C. Fife Department of Mathematics, University of Utah, Salt Lake City, Utah

Robert Finn Department of Mathematics, Stanford University, Stanford, California

Stefano Focardi Istituto Nazionale per la Fauna Selvatica, Ozzano dell'Emilia, Italy

Giuseppe Geymonat LMT-FNS de Cachan, Cachan, France

Francesco Gherardelli Dipartimento di Matematica, Università di Firenze, Florence, Italy

G. Gigante Dipartimento di Matematica, Università di Parma, Parma, Italy

Paolo Gronchi Dipartimento di Matematica, Università di Firenze, Florence, Italy

Richard Jordan Center for Nonlinear Analysis, Carnegie Mellon University, Pittsburgh, Pennsylvania

B. Kawohl Department of Mathematics, University of Cologne, Cologne, Germany

David Kinderlehrer Department of Mathematics, Carnegie Mellon University, Pittsburgh, Pennsylvania

Françoise Krasucki Laboratoire de Modélisation en Mécanique, Université Pierre et Marie Curie, Paris, France

M. D. Kruskal Department of Mathematics, Rutgers University, New Brunswick, New Jersey

Hung-Ju Kuo Department of Applied Mathematics, National Chung-Hsing University, Taichung, Taiwan

Jacques-Louis Lions Department of Mathematics, Collège de France, Paris, France

Fengshan Liu Department of Mathematics, Delaware State University, Dover, Delaware

E. Magenes Dipartimento di Matematica, Università di Pavia, Pavia, Italy

P. Manselli Istituto di Matematica, Università di Firenze, Florence, Italy

Paolo Marcellini Dipartimento di Matematica, Università di Firenze, Florence, Italy

Keith Miller Department of Mathematics, University of California, Berkeley, California

M. Zuhair Nashed Department of Mathematics, University of Delaware, Newark, Delaware

Vladimir Oliker Department of Mathematics and Computer Science, Emory University, Atlanta, Georgia

Lawrence E. Payne Department of Mathematics, Cornell University, Ithaca, New York

Salvadore Perez-Esteva Instituto de Matemáticas, Universidad Nacional Autónoma de México, Mexico, D.F., Mexico

Murray H. Protter Department of Mathematics, University of California, Berkeley, California

P. Pucci Dipartimento di Matematica, Università di Perugia, Perugia, Italy

Martin Schechter Department of Mathematics, University of California, Irvine, California

James Serrin Department of Mathematics, University of Minnesota, Minneapolis, Minnesota

Paolo M. Soardi Dipartimento di Matematica, Università di Milano, Milan, Italy

Giorgio G. Talenti Dipartimento di Matematicà, Università di Firenze, Florence, Italy

Giuseppe Tomassini Dipartimento di Matematicà, Scuola Normale Superiore, Pisa, Italy

F. Tonani Istituto di Mineralogia, Perrografia e Geochimica dell'Università, Palermo, Italy

Neil S. Trudinger Center for Mathematicas and Its Applications, Australian National University, Canberra, A.C.T., Australia

Edoardo Vesentini Scuola Normale Superiore, Pisa, Italy

Giovanni Alberto Viano Departimento di Fisica, Università di Genova, Genoa, Italy

Piero Villaggio Istituto di Scienza delle Costruzioni, Università di Pisa, Pisa, Italy

Aljoša Volčič Dipartimento di Matematicà, Università di Trieste, Trieste, Italy

Rudolf Výborný Mathematics Department, University of Queensland, Queensland, Victoria, Australia

Hans F. Weinberger School of Mathematics, University of Minnesota, Minneapolis, Minnesota

Enrique Zuazua Departamento de Matemática Aplicada, Universidad Complutense, Madrid, Spain

partial differential equations
and applications

1. Spatial Decay Estimates for an Evolution Equation

K. A. Ames Department of Mathematical Sciences, University of Alabama in Huntsville, Huntsville, Alabama

Lawrence E. Payne Department of Mathematics, Cornell University, Ithaca, New York

I. INTRODUCTION.

One of the many research areas to which Professor Carlo Pucci has contributed is that of ill posed problems in partial differential equations (see e.g. Pucci [8–10]). Not only has he done pioneering work in this area, but he has also continued his research interest throughout his career. The second author first became interested in such problems through a series of seminar lectures on the subject given by Professor Pucci at the University of Maryland in 1958 (Pucci [9]).

Many methods have been proposed in the literature for regularizing and/or finding approximate solutions to Cauchy problems for the backward heat equation. One such method proposes perturbing the problem into a well posed problem and using the solution of this well posed problem to generate an approximate solution of the original ill posed problem (see Ames (1982), Ewing (1975), Gajewski and Zacharias (1972), Lattes and Lions (1969), Miller (1973), Showalter (1974), (1975), Showalter and Ting (1976) and the references cited in Ames and Straughan (1995) and Payne (1975)). Most of these authors have dealt with either purely interior or purely exterior problems, but in this paper we consider spatial domains that extend to infinity in one direction. We investigate the possible behavior of solutions of one of these perturbed problems as this spatial variable tends to infinity.

Let $u(x,t)$ be a solution of the following initial–boundary value problem:

$$-\alpha\Delta u_{,t} + \Delta u + u_{,t} = 0 \quad \text{in} \quad D \times (0,\infty) \times (0,T) \tag{1.1}$$

$$u = 0 \quad \text{on} \quad \partial D \times (0,\infty) \times [0,T] \tag{1.2}$$

$$u = f \quad \text{on} \quad \overline{D} \times \{0\} \times [0,T] \tag{1.3}$$

$$u = 0 \quad \text{on} \quad D \times (0,\infty) \times \{0\} \tag{1.4}$$

where Δ is the Laplace operator, α is a positive constant (considered to be small), f is a prescribed function, and a comma indicates differentiation. The domain $D \subset \mathbb{R}^{n-1}$ is assumed to be bounded with Lipschitz boundary ∂D. We further assume for compatibility that

$$f = 0 \quad \text{on} \quad \partial D \times \{0\} \times [0,T] \tag{1.5}$$

and

$$f = 0 \quad \text{on} \quad \overline{D} \times [0,\infty) \times \{0\}. \tag{1.6}$$

Throughout this paper the coordinate axes will be chosen so that in the spatial domain

$$\begin{gathered} x_1 \in (0,\infty) \\ (x_2, x_1, \ldots, x_n) \in D. \end{gathered} \tag{1.7}$$

We also adopt the convention of summing over repeated indices with Latin indices running from 1 to n and Greek indices running from 2 to n. Thus, for instance, (1.2)–(1.4) would be written with arguments as

$$u(x,t) \equiv u(x_1, x_\beta, t) = 0, \; x_1 \in (0,\infty), \; x_\beta \in \partial D, \; t \in [0,T] \tag{1.2$'$}$$

$$u(0, x_\beta, t) = f(x_\beta, t), \; x_\beta \in \overline{D}, \; t \in [0,T] \tag{1.3$'$}$$

$$u(x,0) \equiv u(x_1, x_\beta, 0) = 0, \; x_1 \in (0,\infty), \; x_\beta \in D. \tag{1.4$'$}$$

It is easily shown that with weak assumptions on f, solutions of (1.1)–(1.4) exist, but, of course, they are not unique unless some type of behavior as $x_1 \to \infty$ is specified. The Phragmen–Lindelöf type results we derive in the next section will show that in the class of functions that grow less slowly than a certain exponential as $x_1 \to \infty$, solutions are unique. In fact, we prove that in appropriate measures, solutions of (1.1)–(1.4) either grow exponentially or decay exponentially as $x_1 \to \infty$. Furthermore, we derive in Section 3 explicit bounds in the case of decay. As one expects, these bounds fail as $\alpha \to 0$ since the problem with $\alpha = 0$ is ill posed.

Since we will need it in the next section we let λ denote the first eigenvalue of

$$\begin{aligned} u_{,\alpha\alpha} + \lambda u &= 0 \quad \text{in} \quad D \\ u &= 0 \quad \text{on} \quad \partial D. \end{aligned} \tag{1.8}$$

2. Growth and Decay Rates.

In this section we establish the following theorem:

Theorem. *Let $u(x,t)$ be a solution of (1.1)–(1.6); then either*

$$\text{a)}\ \lim_{x_1\to\infty}\left(e^{-\gamma_\alpha x_1}\left[\int_0^T\int_0^{x_1}\int_D\{\alpha^2 v_{,i\eta}\,v_{,i\eta}+\alpha v_{,\eta}^2\}d\xi d\eta + \frac{1}{2}\int_0^{x_1}\int_D v^2(\xi,T)d\xi\right]\right)\geq K, \tag{2.1}$$

or $E(0)$ is bounded and

$$\text{b)}\qquad E(x_1)\leq E(0)e^{-\gamma_\alpha x_1}. \tag{2.2}$$

Here

$$\gamma_\alpha = 2\alpha^{-1/2}(1+\lambda\alpha)^{1/2} \tag{2.3}$$

with λ defined by (1.8), and

$$E(x_1)=\int_0^T\int_{x_1}^\infty\int_D\{\alpha^2 v_{,i\eta}\,v_{,i\eta}+\alpha v_{,\eta}^2\}d\xi d\eta+\frac{1}{2}\int_{x_1}^\infty\int_D v^2(\xi,T)d\xi. \tag{2.4}$$

We have denoted the element of volume by $d\xi$.
To prove the theorem we first set

$$u(x,t)=e^{t/\alpha}v(x,t). \tag{2.5}$$

Then $v(x,t)$ satisfies

$$-\alpha\Delta v_{,t}+v_{,t}+\alpha^{-1}v=0. \tag{2.6}$$

We next introduce the function

$$\Psi(x_1)=\alpha^2\int_0^T\int_{D(x_1)}v_{,\eta}\,v_{,1\eta}\,dAd\eta, \tag{2.7}$$

where the symbol $D(x_1)$ indicates that the integral is to be taken over D for fixed x_1. Now

$$\Psi(x_1)=\Psi(0)+\int_0^T\int_0^{x_1}\int_{D(\xi)}\{\alpha^2 v_{,i\eta}\,v_{,i\eta}+\alpha v_{,\eta}^2\}d\xi d\eta+\frac{1}{2}\int_0^{x_1}\int_{D(\xi)}v^2(\xi,T)d\xi. \tag{2.8}$$

We have used integration by parts and (2.6) in deriving (2.8).
On the other hand if $|\Psi(x_1)|\to 0$ as $x_1\to\infty$ then

$$-\Psi(x_1)=E(x_1). \tag{2.9}$$

In what follows the argument of D will be dropped except when necessary for clarity.

Differentiating (2.8) it follows that

$$\Psi_{,1} = \int_0^T \int_D \{\alpha^2 v_{,i\eta}\, v_{,i\eta} + \alpha v_{,\eta}^2\} dA d\eta + \frac{1}{2}\int_D v^2(x,T) dA. \tag{2.10}$$

The arithmetic–geometric mean inequality allows us to write (2.7) as

$$|\Psi(x_1)| \leq \frac{\alpha^2}{2}\Big\{\beta \int_0^T \int_D v_{,\eta}^2 dA d\eta + \beta^{-1} \int_0^T \int_D v_{,1\eta}^2 dA d\eta\Big\} \tag{2.11}$$

for arbitrary positive constant β. Thus

$$|\Psi(x_1)| \leq \frac{\alpha^2}{2}\Big\{\beta\nu \int_0^T \int_D v_{,\eta}^2 dA d\eta + \frac{\beta(1-\nu)}{\lambda}\int_0^T \int_D v_{,\beta\eta}\, v_{,\beta\eta}\, dA d\eta + \beta^{-1}\int_0^T \int_D v_{,1\eta}^2 dA d\eta\Big\} \tag{2.12}$$

for arbitrary positive constants β and ν. Here λ is given by (1.8). Choosing

$$\beta = \frac{1}{2}\gamma_\alpha; \quad \nu = (1+\lambda\alpha)^{-1} \tag{2.13}$$

we find

$$|\Psi(x_1)| \leq \gamma_\alpha^{-1}\Psi_{,1}(x_1). \tag{2.14}$$

But (2.14) implies the two inequalities

$$\Psi(x_1) \leq \gamma_\alpha^{-1}\Psi_{,1}(x_1) \tag{2.15}$$

and

$$-\Psi(x_1) \leq \gamma_\alpha^{-1}\Psi_{,1}(x_1). \tag{2.16}$$

Integrating (2.18) with respect to x_1 for $x_1 \geq \overline{x}_1$ we obtain

$$\Psi(x_1) \geq \Psi(\overline{x}_1) e^{\gamma_\alpha(x_1-\overline{x}_1)}. \tag{2.17}$$

It is clear then that if for any $\overline{x}_1$, $\Psi(\overline{x}_1) > 0$, then $\Psi(x_1)$ must grow exponentially as $x_1 \to \infty$, and that (2.1) is satisfied. In other words, unless $\Psi(x_1) \leq 0$, $\forall x$, (2.1) must hold.

To complete the proof of the theorem we observe from an integration of (2.16) that

$$\Psi(x_1) \geq \Psi(0) e^{-\gamma_\alpha x_1} \tag{2.18}$$

or

$$-\Psi(x_1) \leq -\Psi(0) e^{-\gamma_\alpha x_1}. \tag{2.19}$$

Thus if $\Psi(x_1) \leq 0$ for all x_1, it follows immediately from (2.19) that $|\Psi(x_1)| \to 0$ as $x_1 \to \infty$, and hence, in view of (2.9), we conclude (2.2).

This completes the proof of the theorem, but to make (2.19) explicit we require a bound for $\Psi(0)$ in terms of the data. This bound is derived in the next section. We conclude this section, however, by examining some of the consequences of (2.19).

Since $v(x,t)$ vanishes for $t=0$ we have

$$\int_0^t v^2 d\eta \le \frac{t^2}{\pi^2}\int_0^t v_{,\eta}^2 d\eta \tag{2.20}$$

$$\int_0^t v_{,i}\, v_{,i}\, d\eta \le \frac{t^2}{\pi^2}\int_0^t v_{,i\eta}\, v_{,i\eta}\, d\eta. \tag{2.21}$$

Consequently, when (2.19) holds we obtain

$$\int_0^T\int_{x_1}^\infty\int_D \{\alpha^2 v_{,i}\, v_{,i} + \alpha v^2\} dx\, d\eta \le \frac{T^2}{\pi^2}E(x_1) \le \frac{T^2}{\pi^2}E(0)e^{-\gamma_\alpha x_1} \tag{2.22}$$

or from (2.5)

$$\int_0^T\int_{x_1}^\infty\int_D e^{-2\alpha^{-1}\eta}\{\alpha^2 u_{,i}\, u_{,i} + \alpha u^2\} dx\, d\eta \le \frac{T^2}{\pi^2}E(0)e^{-\gamma_\alpha x_1}. \tag{2.23}$$

This yields the cruder inequality

$$\int_0^T\int_{x_1}^\infty\int_D \{u^2 + \alpha u_{,i}\, u_{,i}\} dx\, d\eta \le \frac{T^2 e^{2\alpha^{-1}T}}{\alpha\pi^2}E(0)e^{-\gamma_\alpha x_1}. \tag{2.24}$$

Decay bounds for $\int_D u^2 dA$ can also be derived in a straightforward manner.

3. Bounds for $E(0)$.

We assume in this section that (2.19) holds, and proceed to derive a bound for $E(0)$ in terms of an arbitrary function $w(x,t)$ which satisfies the same initial and boundary conditions as $v(x,t)$, i.e.,

$$w(x,t) = v(x,t) = 0,\ x_1 \in (0,\infty),\ x_\beta \in \partial D,\ t \in [0,T] \tag{3.1}$$

$$w(0,x_\beta,t) = v(0,x_\beta,t) = e^{-\alpha^{-1}t}f(x_\beta,t),\ x_\beta \in \overline{D},\ t \in [0,T] \tag{3.2}$$

$$w(x,0) = v(x,0) = 0,\ x_1 \in (0,\infty),\ x_\beta \in D. \tag{3.3}$$

Note that $w(x,t)$ is not required to satisfy (2.6), but we do insist that $w(x,t) \to 0$ uniformly in x_β and t as $x_1 \to \infty$. Now we may write

$$\begin{aligned} E(0) = -\Psi(0) &= -\alpha^2 \int_0^T \int_{D(0)} w_{,\eta}\, v_{,1\eta}\, dA d\eta \\ &= \alpha^2 \int_0^T \int_0^\infty \int_D w_{,i\eta}\, v_{,i\eta}\, dx d\eta + \int_0^T \int_0^\infty \int_D w_{,\eta}\, \{\alpha v_{,\eta} + v\} dx d\eta \\ &= \int_0^T \int_0^\infty \int_D \{\alpha^2 w_{,i\eta}\, v_{,i\eta} + (\alpha w_{,\eta} - w) v_{,\eta}\} dx d\eta + \int_0^\infty \int_D vw dx \Big|_{t=T} \end{aligned} \quad (3.4)$$

Using the arithmetic–geometric mean inequality we find after some rearrangement

$$\begin{aligned} E(0) &\leq \int_0^T \int_0^\infty \int_D \{\alpha^2 w_{,i\eta}\, w_{,i\eta} + \alpha(w_{,\eta} - \alpha^{-1} w)^2\} dx d\eta + 2 \int_0^\infty \int_D w^2(x,T) dx \\ &= \int_0^T \int_0^\infty \int_D \{\alpha^2 w_{,i\eta}\, w_{,i\eta} + \alpha(w_{,\eta} + \alpha^{-1} w)^2\} dx d\eta \\ &= \Phi(w). \end{aligned} \quad (3.5)$$

An appropriate choice for $w(x,t)$ is

$$w(x,t) = e^{-\sigma x_1} e^{-t/\alpha} f(x_\beta, t) \quad (3.6)$$

which from assumptions (1.5)–(1.6) satisfies (3.1)–(3.3). Here σ is a positive constant to be determined. Using (3.6), $\Phi(w)$ becomes

$$\begin{aligned} \Phi(w) &= \alpha^2 \int_0^T \int_0^\infty \int_D e^{-2(\sigma x_1 + \alpha^{-1}\eta)} \{(-\alpha^{-1} f_{,\beta} + f_{,\beta\eta})(-\alpha^{-1} f_{,\beta} + f_{,\beta\eta}) \\ &\quad + \sigma^2(-\alpha^{-1} f + f_{,\eta})^2\} dx d\eta + \alpha \int_0^T \int_0^\infty \int_D e^{-2(\sigma x_1 + \alpha^{-1}\eta)} f_{,\eta}^2 dx d\eta \\ &= \frac{1}{2\sigma}[B_1 + \sigma^2 B_2] \end{aligned} \quad (3.7)$$

where

$$B_1 = \int_0^T \int_D e^{-2\eta/\alpha} \{(\alpha f_{,\beta\eta} - f_{,\beta})(\alpha f_{,\beta\eta} - f_{,\beta}) + \alpha f_{,\eta}^2\} dA d\eta \quad (3.8)$$

and

$$B_2 = \int_0^T \int_D e^{-2\eta/\alpha} (\alpha f_{,\eta} - f)^2 dA d\eta. \quad (3.9)$$

Choosing

$$\sigma = [B_1/B_2]^{1/2} \quad (3.10)$$

we arrive at the bound

$$E(0) \leq [B_1 B_2]^{1/2}. \quad (3.11)$$

Inequality (3.11) gives the desired bound for $E(0)$ in terms of data.

References.

1. Ames, K. A. (1982). On the comparison of related properly and improperly posed Cauchy problems for first order operator equations, SIAM J. Math. Anal., 13: 594–606.
2. Ames, K. A. and Straughan, B., Contemporary Non-standard and Improperly Posed Problems for Partial Differential Equations, Academic Press (to appear).
3. Ewing, R. E. (1975). The approximation of certain parabolic equations backward in time by Sobolev equations, SIAM J. Math. Anal., 6: 283–294.
4. Gajewski, H. and Zacharias, K. (1972). Zur Regularisierung einer Klasse nichtkorrecter Probleme bei Evolutionsgleichung, J. Math. Anal. Appl., 38: 784–789.
5. Lattes, R. and Lions, J. L. (1969). The Method of Quasireversibility, Applications to Partial Differential Equations, American Elsevier, New York.
6. Miller, K. (1973). Stabilized quasi-reversibility and other nearly-best possible methods for non-well-posed problems, Symp. on Non-well-posed Problems and Logarithmic Convexity, Springer Lecture Notes #316: 161–176.
7. Payne, L. E. (1975). Improperly posed problems in partial differential equations, Reg. Conf. Series in Appl. Math. #22, SIAM.
8. Pucci, C. (1955). Sui problemi di Cauchy non "ben posti", Rend. Acad. Naz. Lincei, 18: 473–477.
9. Pucci, C. (1958). Some topics in parabolic and elliptic equations, Lecture Series #36, Institute for Fluid Dynamics and Applied Mathematics, U. of Maryland, College Park, MD.
10. Pucci, C. (1958). Discussione del problema di Cauchy per le equazioni di tipo ellitico, Ann. Mat. Pure Appl. 46: 131–153; 391–412.
11. Showalter, R. E. (1974). The final value problem for evolution equations, J. Math. Anal. App. 47: 563–572.
12. Showalter, R. E. (1975). Quasireversiblity of first and second order parabolic evolution equations, Pitman Research Notes in Math. #1: 76–84.
13. Showalter, R. E. and Ting, T. W. (1976). Pseudo-parabolic partial differential equations, SIAM J. Math. Anal. 1: 1–26.

2. On the Range of Ural'tseva's Axially Symmetric Operator in Sobolev Spaces

Orazio Arena Istituto di Matematica, Università di Firenze, Florence, Italy

P. Manselli Istituto di Matematica, Università di Firenze, Florence, Italy

Introduction

In [9], N.Ural'tseva considered the second-order axially symmetric partial differential operator, which in our notations reads

(1) $$L_\alpha := \alpha\Delta + (1-3\alpha)\sum_{1}^{2}{}_{ij}\, x_i x_j \left(x_1^2+x_2^2\right)^{-1} \partial^2/\partial x_i \partial x_j$$

and is uniformly elliptic for $\alpha \in (0,1/3)$. She proved the impossibility of $W^{2,3}$ bounds for solutions in a cylinder $C \subset \mathbb{R}^3$ to the Dirichlet problem

(2) $$L_\alpha u = f \ \text{ in } C,\ u|_{\partial C} = 0\,,$$

for a suitable choice of $\alpha \in (0,1/3)$ and $f \in L^3(C)$.

In [8], problem (2) was studied for $f \in L^p(C)$ and the following statements were proved: (i) problem (2) does not have a unique solution in $W^{2,p}(C)$ if $\alpha \in (0,1/3)$ and $p \in (1, p_1(\alpha))$; (ii) problem (2) has a unique solution in $W^{2,p}(C)$ if

$$(\alpha,p) \in \mathcal{A}_1 := (0,1/3) \times (p_1(\alpha), p_2(\alpha))\,,$$

where $p_1(\alpha) < 2 < p_2(\alpha)$, $p_1(0^+) = p_2(0^+) = 2$, $p_1(1/3^-) = 1$, $p_2(1/3^-) = +\infty$.

In the present paper,

$$\mathcal{V}_{\alpha,p} := L_\alpha\left(\left\{u \in W^{2,p}(C)\colon u|_{\partial C} = 0\right\}\right),$$

the range of L_α, is studied for $C =$ a cylinder in $\mathbb{R}^3$, and

$$(\alpha,p) \in \mathcal{A} := (0,1/3) \times (p_1(\alpha), +\infty)\,,$$

an open set in the α, p-plane. Clearly, $\mathcal{V}_{\alpha,p} \supset \mathcal{V}_{\alpha,\overline{p}}$ if $p < \overline{p}$. Moreover, it turns out that:

(i) $\mathcal{A}$ is the union of disjoint sets

$$\mathcal{A} = \mathcal{A}_1 \cup \mathcal{L}_2 \cup \mathcal{A}_2 \cup \mathcal{L}_3 \cup \ldots \cup \mathcal{A}_k \cup \mathcal{L}_{k+1} \cup \ldots\,,$$

where $\mathcal{A}_k$ are open sets of the form

$$\mathcal{A}_k := \left\{(\alpha,p)\colon \alpha \in (0,\alpha_{k+1}), p_k(\alpha) < p < p_{k+1}(\alpha)\right\} \cup \left\{(\alpha,p)\colon \alpha \in [\alpha_{k+1},\alpha_k), p_k(\alpha) < p\right\}$$

and

$$\mathcal{L}_k := \left\{(\alpha,p)\colon \alpha \in (0,\alpha_k), p = p_k(\alpha)\right\}$$

(the nested intervals $(0,\alpha_k)$ and the functions $p_k(\alpha)$ will be defined below);

(ii) $\mathscr{V}_{\alpha,p} = L^p(C)$ if $(\alpha,p) \in \mathcal{A}_1$, by the quoted result; (iii) if $k > 1$ and $(\alpha,p) \in \mathcal{A}_k$, then $\mathscr{V}_{\alpha,p}$ is a closed subset of $L^p(C)$, and $\mathscr{V}_{\alpha,p}^{\perp}$ is an infinite-dimensional subset of $L^{p'}(C)$ $(p'=p/(p-1))$ spanned by all (generalized) solutions (in $L^{p'}(C)$) to $L_\alpha^* v=0$ in C, $v|_{\partial C} = 0$; (iv) if $k > 1$ and $(\alpha,p) \in \mathcal{L}_k$, then $\mathscr{V}_{\alpha,p}$ is not a closed subset of $L^p(C)$. These statements can be easily read in terms of semi-Fredholm operators.

It will also be shown that, if $1 < p < +\infty$ and $0 < \alpha < 1/3$, then $\mathscr{Z}_{\alpha,p} := L_\alpha(W^{2,p}(C))$ is a dense subset of $L^p(C)$.

Let us recall that Cerutti-Fabes-Escauriaza [3], [4] proved uniqueness of good solutions to the Dirichlet problem for operators whose coefficients are discontinuous in one point. So, by using a result of Bass and Pardoux [2] and probabilistic methods, one can prove uniqueness of good solutions to the Dirichlet problem, with right hand side in L^3 (this remark is due to Escauriaza: here 3 is the dimension of the space); see also Escauriaza [5]. In the present paper it is proved that, for suitable α and p and for $f \in L^3(C)$, the generalized solutions to problem (4) below are classical solutions, but not so regular (they can be in $W^{2,2+\epsilon}(C)$ for some $0 < \epsilon \ll 1$, if $0 < \alpha \ll 1$). Moreover, solutions to $L^* v = 0$ in C, vanishing on the boundary and belonging to $L^q(C)$, are constructed for suitable q.

Let us introduce the notations

$$x=(x_1,x_2,x_3) \in \mathbb{R}^3 \ , \ <x> := \sqrt{x_1^2 + x_2^2}$$

and $Z := \{x \in \mathbb{R}^3 \colon\ x_1 = x_2 = 0\}$. C will be the cylinder $\{x \in \mathbb{R}^3 \colon <x> < \rho,\ 0 < x_3 < \pi\}$ and S its curved boundary $\{x \in \mathbb{R}^3 \colon <x>=\rho,\ 0<x_3<\pi\}$, T will denote $\partial C \backslash \overline{S}$; we write C_ρ, S_ρ, T_ρ when we need to emphasize the dependence on $\rho > 0$. Cylindrical coordinates in $\mathbb{R}^3$ will also be used: $x_1 = r\cos\theta$, $x_2 = r\sin\theta$, $x_3 = z$; so $x = (r;\theta;z) = (r;\xi)$, where $\xi = (\cos\theta, \sin\theta, z)$. $W^{2,p}(C)$ and $W^{2,p}_{loc}(C \cup T)$ are standard Sobolev spaces; $W^{2,p}_{\gamma_0}(C)=\{u \in W^{2,p}(C)\colon u|_{\partial C} = 0\}$.

The operator L_α can be written in cylindrical coordinates as

$$L_\alpha u = (1-2\alpha)\left[u_{rr} + (\mu/r)u_r + (\mu/r^2)u_{\theta\theta} + \mu u_{zz}\right],$$

where $\mu = \mu(\alpha) := \alpha/(1-2\alpha)$ is an increasing function of $\alpha \in (0,1/3)$, valued on $(0,1)$.

For later use, let us define a set of parameters. If $\alpha \in (0,1/3)$,

$$k_0 = k_0(\alpha) := (\mu - 1)/2,$$

$$k_\nu = k_\nu(\alpha) := (1\text{-}\mu)^2/4 + \mu\nu^2]^{1/2} \qquad \nu \in \mathbb{Z}\backslash\{0\},$$

$$\lambda_\nu = \lambda_\nu(\alpha) := (1-\mu)/2 + k_\nu(\alpha) \qquad \nu \in \mathbb{Z}.$$

k_ν, λ_ν are increasing functions of α in $(0,1/3)$; $k_\nu = k_{-\nu}$, $\lambda_\nu = \lambda_{-\nu}$, $\lambda_0 \equiv 0$, $\lambda_1 = \lambda_{-1} \equiv 1$, $\lambda_\nu(0^+) = 1$, $\lambda_\nu(1/3^-) = |\nu|$, and $\lambda_\nu(\alpha) < \lambda_{\nu+1}(\alpha)$ if $\nu > 0$. The parameter $p_1(\alpha)$, quoted above, is defined as

$$p_1(\alpha) := [2(1-2\alpha)]/(1-\alpha) = 2/(\mu+1);$$

p_1 is decreasing in $(0,1/3)$ and valued on $(1,2)$. Let $\alpha_\nu := 2/(2+\nu^2)$ $(|\nu|>1)$; then $\lambda_\nu(\alpha_\nu) = 2$, the intervals $(0,\alpha_{|\nu|})$ are nested as $|\nu|$ increases, $(0,\alpha_2) = (0,1/3)$ and $\alpha_\nu \to 0$ as $|\nu| \to \infty$. For $\alpha \in (0,\alpha_\nu)$, we define

$$p_\nu(\alpha) := 2/(2-\lambda_\nu(\alpha));$$

then $p_\nu(0^+) = 2$, $p_\nu(\alpha_\nu{}^-) = +\infty$ and $p_{|\nu|}(\alpha) < p_{|\nu|+1}(\alpha)$ in $(0, \alpha_{|\nu|+1})$.

For the sake of brevity, the results of our paper are presented here in a short form and we refer the readers to [1] for more detailed proofs.

In a paper in progress the Ural'tseva operator will be studied in $W^{2,p}$, $1<p<p_1(\alpha)$, with data on the boundary and on the singular axis.

The present paper depends on results from [7] and [8], here quoted as propositions.

1 The homogeneous Dirichlet problem

Definition. Let $1 \le s < +\infty$, $u \in C^0(C \cup T)$, $g \in L^s(S)$. The notation $u|_S = g$ means

$$\lim_{r \to \rho-} \int_{S_\rho} |\, u(r;\theta;z) - g(\rho;\theta;z)\,|^s \, d\sigma = 0 .$$

Proposition 1 (see [7], [8]). Let $\alpha \in (0,1/3)$, $1 \le p < +\infty$. The following problem

(3) $$u \in W^{2,p}_{loc}(C \cup T), \quad L_\alpha u = 0 \text{ in } C, \quad u|_T = 0, \quad u|_S = 0,$$

has the unique solution $u \equiv 0$ if $p > p_1(\alpha)$. If $p < p_1(\alpha)$, problem (3) has infinitely many nonzero solutions.

Proposition 2 (see [7],[8]). Let $\alpha \in (0,1/3)$, $1 \le s < +\infty$, $g \in L^s(S)$, and $p_1(\alpha) < p < p_2(\alpha)$. The following problem

(4) $$L_\alpha u = 0 \ \text{ in } C, \quad u|_T = 0, \quad u|_S = g,$$

has only one solution $u \in C^2((C \cup T)\backslash Z) \cap C^1(C \cup T) \cap W^{2,p}_{loc}(C \cup T)$ satisfying

(5) $$\| [\mathfrak{G}_{r,\rho}\, g](r;\cdot\,) \|_{L^s(S_1)} \le \| g(\rho;\cdot\,) \|_{L^s(S_1)} ,$$

where $\mathfrak{G}_{r,\rho}\, g(\rho;\cdot\,) := u(r;\cdot\,)$.

For later use, explicit expressions of the solution to (4) are needed. To do this we introduce classes of special functions.

Let $R_{\nu n}(r,\rho)$ $(0 \le r \le \rho,\ \nu \in \mathbb{Z},\ n \in \mathbb{N})$ be the solutions to the differential equation

(6) $$\mathfrak{L}_{\nu n} y := y'' + (\mu/r)y' - \mu(n^2 + \nu^2/r^2)y = 0 \quad \text{in } (0,\rho)$$

having the form

(7) $$R_{\nu n}(r,\rho) = \left(\frac{r}{\rho}\right)^{(\mu-1)/2} \frac{I_{k_\nu}(\sqrt{\mu}nr)}{I_{k_\nu}(\sqrt{\mu}n\rho)} = \left(\frac{r}{\rho}\right)^{\lambda_\nu} \frac{\sigma_\nu(\sqrt{\mu}nr)}{\sigma_\nu(\sqrt{\mu}n\rho)} ,$$

where I_{k_ν} are the modified Bessel functions of the first kind, and

(8) $$\sigma_\nu(x) = \sum_{m=0}^{\infty} \frac{\Gamma(1+k_\nu)}{m!\Gamma(1+m+k_\nu)} \left(\frac{x}{2}\right)^{2m}.$$

The functions above have the following properties (see [7]):

(α) $$0 \le (\partial/\partial r) R_{\nu n}(r,\rho) \le r(n^2+\nu^2/r^2) R_{\nu n}(r,\rho) \quad (0 < r < \rho,\ \nu \in \mathbb{Z},\ n \in \mathbb{N});$$

$$0 \le \partial^2/\partial r^2 R_{\nu n}(r,\rho) \le \mu(n^2+\nu^2/r^2) R_{\nu n}(r,\rho) \quad (0 < r < \rho,\ \nu \in \mathbb{Z},\ n \in \mathbb{N});$$

(β) $$0 \le R_{\nu n}(r,\rho) \le \left(\tfrac{r}{\rho}\right)^{\mu|\nu|/2} e^{-\mu n(\rho-r)/2} \quad (0<r<\rho,\ \nu\in\mathbb{Z}\setminus\{0\},\ n\in\mathbb{N});$$

(γ) $$0 \le \sigma_0(\sqrt{\mu}nr)/\sigma_0(\sqrt{\mu}n\rho) \le \sqrt{\mu}n\rho[\cosh(\sqrt{\mu}nr)/\sinh(\sqrt{\mu}n\rho)] \quad (0<r<\rho,\ n\in\mathbb{N});$$

(δ) $$0 \le \sigma_\nu(\sqrt{\mu}nr)/\sigma_\nu(\sqrt{\mu}n\rho) \le (\sqrt{\mu}n\rho)^{2|\nu|}[\cosh(\sqrt{\mu}nr)/\cosh(\sqrt{\mu}n\rho)]$$
$(0<r<\rho,\ \nu\in\mathbb{Z}\setminus\{0\},\ n\in\mathbb{N},\ n>n(\rho)$, depending on $\rho>0)$.

Proposition 3 (see [8]). The solution to problem (4) can be written in the following ways.

(i) Let $g(\rho;\xi) \sim \sum_{-\infty}^{+\infty}{}_\nu \sum_1^{\infty}{}_n\, g_{\nu n}\, e^{i\nu\theta}\sin nz$, where $\xi=(\cos\theta,\sin\theta,z)\in S_1$; then

(9) $$u(r;\xi)=\sum_{-\infty}^{+\infty}{}_\nu \sum_1^{\infty}{}_n\, g_{\nu n}\, R_{\nu n}(r,\rho)\, e^{i\nu\theta}\sin nz\ .$$

By (α)-(δ), the series (9) converges uniformly in C_t for $0<t<\rho$.

(ii) Let $r^{\lambda_\nu}\omega_\nu(r,z) := \sum_1^{\infty}{}_n\, g_{\nu n}\, R_{\nu n}(r,\rho)\sin nz$; then

(10) $$\rho^{\lambda_\nu}\omega_\nu(\rho,z)=(2\pi)^{-1}\int_0^{2\pi} g(\rho;\theta;z)e^{-i\nu\theta}d\theta$$

and

(11) $$u(r;\xi)=\sum_{-\infty}^{+\infty}{}_\nu\, r^{\lambda_\nu}\,\omega_\nu(r,z)e^{i\nu\theta}=u_H(r;\xi)+\sum_{|\nu|<4/\mu} r^{\lambda_\nu}\,\omega_\nu(r,z)e^{i\nu\theta}\ .$$

Moreover, if $0<t<\rho$, there exists $K_{(t)}$ such that

$$\|u_H\|_{C^2(C_t\cup T_t)} \le K_{(t)}\,\|g\|_{L^1(S)},$$
$$\|\omega_\nu\|_{C^2(C_t\cup T_t)} \le K_{(t)}\,\|g\|_{L^1(S)},$$

and

(12) $$\left(\int_0^\pi |\omega_\nu(r,z)|^s dz\right)^{1/s} \le \left(\int_0^\pi |\omega_\nu(\rho,s)|^s dz\right)^{1/s} \quad (0<r<\rho,\ \nu\in\mathbb{Z}).$$

As in theorem 2 from [8], one can show that

Remark 1. Let $|\nu_0|\ge 2$, and $1\le s<+\infty$. There exists K_1 such that

(13) $$\left(\tfrac{r}{\rho}\right)^{-\lambda_{\nu_0}}\|[\mathfrak{G}_{r,\rho}g](r;\cdot)\|_{L^1(S_1)} \le K_1\,\|g(\rho;\cdot)\|_{L^s(S_1)} \quad (0<r\le\rho),$$

if $g\in L^s(S)$ and satisfies $g_{\nu n}=0$ ($|\nu|<\nu_0$, $n\in\mathbb{N}$).

2 The nonhomogeneous Dirichlet problem

By using properties of Bessel functions, one can prove the following facts.

Remark 2. (i) The span of the set of functions having the form

(14) $$r^{|\nu|+2m}e^{i\nu\theta}\sin nz \qquad (\nu\in\mathbb{Z},\ m\in\mathbb{N}\cup\{0\},\ n\in\mathbb{N})$$

is dense in $L^p(C)$, if $1<p<+\infty$.

(ii) Let

$$\zeta_{\nu mn}(r) := \sum_0^{\infty}{}_j \frac{(n^2/4)^j\,\Gamma(m+1)\Gamma(|\nu|+m+1)}{4\Gamma(j+m+2)\Gamma(|\nu|+j+m+2)}\, r^{2j+|\nu|+2m+2}\ ;$$

the span of the set of functions having the form

(15) $$[\zeta_{\nu mn}(r)-(\zeta_{\nu mn}(\rho)\, I_{|\nu|}(nr)/I_{|\nu|}(n\rho))]e^{i\nu\theta}\sin nz \quad (\nu\in\mathbb{Z},\ m\in\mathbb{N}\cup\{0\},\ n\in\mathbb{N})$$

is dense in $W^{2,p}_{\gamma_0}(C)$, if $1 < p < +\infty$.

Remark 3. Let $\nu \in \mathbb{Z}$, $m \in \mathbb{N} \cup \{0\}$, $n \in \mathbb{N}$, and

$$\eta_{\nu mn}(r) := \sum_0^\infty {}_j \frac{(\mu n^2/4)^j\, \Gamma\left(\frac{|\nu|-\lambda_\nu}{2}+m+1\right)\Gamma\left(\frac{|\nu|+\lambda_\nu+\mu-1}{2}+m+1\right)}{4\,\Gamma\left(\frac{|\nu|-\lambda_\nu}{2}+j+m+2\right)\Gamma\left(\frac{|\nu|+\lambda_\nu+\mu-1}{2}+j+m+2\right)}\, r^{2j+|\nu|+2m+2},$$

$$w_{\nu mn}(r;\theta;z) := \eta_{\nu mn}(r)\, e^{i\nu\theta} \sin nz.$$

Then $\eta_{\nu mn}(r)$ is an entire function of $r \in \mathbb{C}$, $w_{\nu mn} \in C^\infty(\mathbb{R}^3)$ and

$$L_\alpha\, w_{\nu mn} = r^{|\nu|+2m} e^{i\nu\theta} \sin nz \quad (\nu \in \mathbb{Z},\ m \in \mathbb{N} \cup \{0\},\ n \in \mathbb{N}). \tag{16}$$

The next lemma is an easy consequence of remark 3.

Lemma 1. Let

$$f(r;\theta;z) := \sum_{-\infty}^{+\infty}{}_\nu \sum_1^\infty{}_n \sum_0^\infty{}_m\, a_{\nu mn}\, r^{|\nu|+2m}\, e^{i\nu\theta} \sin nz, \tag{17}$$

where $a_{\nu mn} \neq 0$ for finitely many indices $\nu \in \mathbb{Z}$, $m \in \mathbb{N} \cup \{0\}$, $n \in \mathbb{N}$. Then f is in $C^\infty(\mathbb{R}^3)$. Moreover: (i) the function

$$u_1(r;\theta;z) := \sum_{-\infty}^{+\infty}{}_\nu \sum_1^\infty{}_n \sum_0^\infty{}_m\, a_{\nu mn}\, \eta_{\nu mn}(r)\, e^{i\nu\theta} \sin nz$$

is in $C^\infty(\mathbb{R}^3)$ and satisfies $L_\alpha u_1 = f/(1-2\alpha)$ in $\mathbb{R}^3$; (ii) the function

$$u_2(r;\theta;z) := \sum_{-\infty}^{+\infty}{}_\nu \sum_1^\infty{}_n \left\{ \sum_0^\infty{}_m\, a_{\nu mn}\, [\eta_{\nu mn}(r) - \eta_{\nu mn}(\rho) R_{\nu n}(r,\rho)] \right\} e^{i\nu\theta} \sin nz \tag{ii}$$

satisfies $L_\alpha u_2 = f/(1-2\alpha)$ in $C\backslash Z$, $u_2|_{\partial C} = 0$.

Theorem 1. For every $p \in (1,+\infty)$, let $\mathfrak{Z}_{\alpha,p} := L_\alpha(W^{2,p}(C))$. Then

$$\overline{\mathfrak{Z}}_{\alpha,p} = L^p(C).$$

Proof. By remark 2, for every $f \in L^p(C)$ there exist f_k such that f_k has the form (17) and $f_k \to f$. By lemma 1, $f_k \in \mathfrak{Z}_{\alpha,p}$. The thesis follows. □

Theorem 2. Let $\nu \in \mathbb{Z}$, $|\nu| > 1$, $0 < \alpha < \alpha_\nu$, $p_\nu(\alpha) < p < \infty$, $n \in \mathbb{N}$. Then the problem

$$L^*_\alpha v = 0 \text{ in } C, \quad v|_{\partial C} = 0$$

has infinitely many weak solutions $v_{\nu n} L^{p'}(C)$ of the form

$$v_{\nu n}(r;\theta;z) := y_{\nu n}(r) e^{i\nu\theta} \sin nz, \tag{18}$$

where

$$y_{\nu n}(r) = r^{\mu+\lambda_\nu-1}\, \sigma_\nu(\sqrt{\mu}nr) \int_r^\rho t^{-\mu-2\lambda_\nu} \frac{dt}{[\sigma_\nu(\sqrt{\mu}nt)]^2}. \tag{19}$$

Proof. Notice that

$$y_{\nu n}(r) = (2k_\nu)^{-1}\, r^{-\lambda_\nu}\, (1+o(1)) \quad\text{and}\quad y'_{\nu n}(r) = \lambda_\nu (2k_\nu)^{-1}\, r^{-1-\lambda_\nu}(-1+o(1))$$

as $r \to 0$; therefore $v_{\nu n} \in L^{p'}(C)$.

Let us show that $L^*_\alpha v_{\nu n} = 0$ in C. Let $u \in W^{2,p}_{\gamma_0}(C)$, $0 < \epsilon < \rho$ and

$$\mathfrak{I}_\epsilon(v_{\nu n}; u) := \int_{C\backslash C_\epsilon} v_{\nu n}\, [(L_\alpha u)/(1-2\alpha)]\, dx.$$

Using cylindrical coordinates and assuming u has the form

$$u = \zeta_{\nu_1 m n_1}\, e^{i\nu_1\theta} \sin(n_1 z) \quad (\nu_1 \in \mathbb{Z},\ m \in \mathbb{N} \cup \{0\},\ n_1 \in \mathbb{N}),$$

where

$$\tilde{\zeta}_{\nu mn}(r) = [\zeta_{\nu mn}(r) - (\zeta_{\nu mn}(\rho) I_{|\nu|}(nr))/I_{|\nu|}(n\rho)],$$

then integrating by parts, we get $\mathcal{I}_\epsilon(v_{\nu n}; u) = O(\epsilon^{-\lambda_\nu + |\nu|}) \to 0$ as $\epsilon \to 0$.

As $\mathcal{I}_\epsilon(v_{\nu n}; u) \to \int_{C'} v_{\nu n}[(L_\alpha u)/(1-2\alpha)]\, dx$ when $\epsilon \to 0$, the latter quantity vanishes for all u having the form (15); by remark 2(ii), the thesis follows. □

To study the behaviour of L_α in Sobolev spaces $W^{2,p}_{\gamma_0}(C)$, $p \geq p_2(\alpha)$, we need to recall representation formulas from [8].

Proposition 4. Let $\alpha \in (0, 1/3)$, $u \in C_0^\infty(\mathbb{R}^3)$, supp $u \subset C$, and

$$u(r; \theta; z) = \sum_{-\infty}^{+\infty}{}_\nu \sum_1^\infty{}_n \, \omega_{\nu n}(r)\, e^{i\nu\theta} \sin nz \,. \tag{20}$$

Then: (i) $f := L_\alpha u$ can be written as

$$f(r; \theta; z) = \sum_{-\infty}^{+\infty}{}_\nu \sum_1^\infty{}_n \, f_{\nu n}(r)\, e^{i\nu\theta} \sin nz \,,$$

where $\mathcal{L}_{\nu n}\omega_{\nu n} = f_{\nu n}/(1-2\alpha)$ in $(0,\rho)$, $\omega_{\nu n}(\rho) = 0$;

(ii) $\omega_{\nu n}(0) = 0$ $(\nu \neq 0)$, $\omega'_{0n}(0) = 0$ and $\omega_{\nu n}$ can be written also as

$$\omega_{\nu n}(r) = \int_{+\infty}^{r} R_{\nu n}(r,t)\, dt \int_0^t R_{\nu n}(t',t) \left(\frac{t'}{t}\right)^\mu f_{\nu n}(t') dt' \,; \tag{21}$$

(iii) equation (21) gives

$$u(r; \cdot) = \int_{+\infty}^{r} \mathcal{G}_{r,t}\, dt \int_0^t \left(\tfrac{t'}{t}\right)^\mu \mathcal{G}_{t',t}\, f(t'; \cdot)\, dt' \,. \tag{22}$$

Theorem 3. Let $\alpha \in (0, 1/3)$, $p > p_2(\alpha)$ and $p \neq p_\nu(\alpha)$ for all $\nu \in \mathbb{Z}$ such that $\alpha_\nu > \alpha$. Then, there exists K such that the inequality

$$\left\| D^2 u \right\|_{L^p(C)} \leq K \left\| L_\alpha u \right\|_{L^p(C)} \tag{23}$$

holds for every $u \in W^{2,p}_{\gamma_0}(C)$.

Proof. As $p \in (p_2(\alpha), +\infty)$, then $2 - 2/p > \lambda_2(\alpha)$. Thus, there exists an integer $\overline{\nu} \geq 3$ such that $\lambda_{\overline{\nu}-1}(\alpha) < 2 - 2/p < \lambda_{\overline{\nu}}(\alpha)$; $\overline{\nu}$ is the largest integer such that $p_{\overline{\nu}-1}(\alpha) < p$.

Because of standard techniques (see, e.g., [8], lemma 3), it is enough to show (23) for functions $u \in C_0^\infty(\mathbb{R}^3)$ such that supp $u \subset C$.

As in proposition 4, let us write

$$u(r; \theta; z) = \sum_{-\infty}^{+\infty}{}_\nu \Big(\sum_1^\infty{}_n \, \omega_{\nu n}(r) \sin nz\Big) e^{i\nu\theta} = u_{\mathcal{H}} + \sum_{|\nu| < \overline{\nu}} u_\nu \,,$$

where

$$u_\nu(r; \theta; z) = \Big(\sum_1^\infty{}_n \, \omega_{\nu n}(r) \sin nz\Big) e^{i\nu\theta} \,.$$

The function $f := L_\alpha u$ can be written as

$$f(r;\theta;z) = \sum_{-\infty}^{+\infty}{}_\nu \Big(\sum_1^\infty{}_n \, f_{\nu n}(r) \sin nz\Big) e^{i\nu\theta} = f_{\mathcal{H}} + \sum_{|\nu| < \overline{\nu}} f_\nu \,,$$

where

$$f_\nu(r; \theta; z) = \Big(\sum_1^\infty{}_n \, f_{\nu n}(r) \sin nz\Big) e^{i\nu\theta} \,.$$

By theorems 5 and 4 of [8],

$$\left\| D^2 \sum_{|\nu|<2} u_\nu \right\|_{L^p(C)} \le K_2 \left\| \sum_{|\nu|<2} f_\nu \right\|_{L^p(C)},$$

$$\left\| D^2\Big(u - \sum_{|\nu|<2} u_\nu\Big) \right\|_{L^p(C)} \le$$

$$K_3 \left\{ \left\| f - \sum_{|\nu|<2} f_\nu \right\|_{L^p(C)} + \left\| \Big(u - \sum_{|\nu|<2} u_\nu\Big)/<\cdot>^2 \right\|_{L^p(C)} \right\}.$$

To prove the thesis we have to show the following inequalities

$$\left\| u_{\mathcal{H}}/<\cdot>^2 \right\|_{L^p(C)} \le K_4 \left\| f_{\mathcal{H}} \right\|_{L^p(C)}, \tag{24}$$

$$\left\| u_\nu/<\cdot>^2 \right\|_{L^p(C)} \le K_5 \left\| f_\nu \right\|_{L^p(C)} \qquad (2 \le |\nu| \le \overline{\nu}-1). \tag{25}$$

By (22),

$$u_{\mathcal{H}}(r;\cdot) = \int_{+\infty}^{r} \mathcal{G}_{r,t}\, dt \int_0^t \left(\frac{t'}{t}\right)^\mu \mathcal{G}_{t',t}\, f_{\mathcal{H}}(t';\cdot)\, dt'\,.$$

Remark 1 (with $\nu_0 = \overline{\nu}$) and (5) give

$$\left\| u_{\mathcal{H}}(r;\cdot) \right\|_{L^p(S_1)} \le K_1 \int_r^\infty \left(\frac{r}{t}\right)^{\lambda_{\overline{\nu}}} dt \int_0^t \left(\frac{t'}{t}\right)^\mu \left\| f_{\mathcal{H}}(t';\cdot) \right\|_{L^p(S_1)} dt'.$$

As $2+p(\lambda_{\overline{\nu}}-2) = p(2/p-2+\lambda_{\overline{\nu}}) > 0$ and $2-p(\mu+1) = 2(1-p/p_1(\alpha)) < 0$, by Hardy inequalities we get

$$\left\{ \int_0^\infty r \left[r^{-2} \left\| u_{\mathcal{H}}(r;\cdot) \right\|_{L^p(S_1)} \right]^p dr \right\}^{1/p} \le$$

$$\le k \frac{p^2}{[2+p(\lambda_{\overline{\nu}}-2)][p(1+\mu)-2]} \left\{ \int_0^\infty r \left[\left\| f_{\mathcal{H}}(r;\cdot) \right\|_{L^p(S_1)} \right]^p dr \right\}^{1/p},$$

i.e. (24). Let $2 \le |\nu| \le \overline{\nu}-1$; (21) and (7) yield $[\omega_{\nu n}(r) e^{i\nu\theta}]/\sigma_\nu(\sqrt{\mu} n r) = O(r^{|\nu|})$ $(r \to 0)$. We have:

$$\omega_{\nu n}(r) = \int_0^r R_{\nu n}(r,t)\, dt \int_0^t R_{\nu n}(t',t) \left(\frac{t'}{t}\right)^\mu f_{\nu n}(t') dt',$$

i.e.,

$$u_\nu(r;\ \cdot\ ) = \int_0^r \mathcal{G}_{r,t}\, dt \int_0^t \left(\frac{t'}{t}\right)^\mu \mathcal{G}_{t',t}\, f_\nu(t';\ \cdot\ )\, dt'\,.$$

By using remark 1 (with $\nu_0 = \overline{\nu}$) and (5), we get:

$$\left\| u_\nu(r;\cdot) \right\|_{L^p(S_1)} \le K_1 \int_0^r \left(\frac{r}{t}\right)^{\lambda_{\overline{\nu}}} dt \int_0^t \left(\frac{t'}{t}\right)^\mu \left\| f_\nu(t';\ \cdot\ ) \right\|_{L^p(S_1)} dt'$$

for $2 \le |\nu| \le \overline{\nu}-1$; as $2+p(\lambda_\nu-2) = p(2/p-2+\lambda_\nu) < 0$ and $2-p(\mu+1) = 2(1-p/p_1(\alpha)) < 0$, using twice Hardy's inequality gives

$$\left\{ \int_0^\infty r \left[r^{-2} \left\| u_\nu(r;\cdot) \right\|_{L^p(S_1)} \right]^p dr \right\}^{1/p} \le K_1 \left\{ \int_0^\infty r \left[\left\| f_\nu(r;\cdot) \right\|_{L^p(S_1)} \right]^p dr \right\}^{1/p},$$

for $2 \le |\nu| \le \overline{\nu}-1$. These inequalities imply (25). □

To prove the next theorem, let us state an approximation lemma.

Lemma 2. Let $\alpha \in (0,1/3)$, $p > p_2(\alpha)$ and $p \ne p_\nu(\alpha)$ for all $\nu \in \mathbb{Z}$ satisfying $\alpha_\nu > \alpha$. Let $\overline{\nu}$ be the largest integer satisfying $\overline{\nu} \ge 3$ and $p_{\overline{\nu}-1}(\alpha) < p$. Denote by $\mathcal{I}_{\overline{\nu}}$ the closure (in $L^{p'}(C)$, $p' = p/(p-1)$) of Span $\{v_{\nu n}$: $n \in \mathbb{N}$, $\nu \in \mathbb{Z}$, $2 \le |\nu| \le \overline{\nu}-1\}$, and assume

$f \in L^p(C)$, $f \in \mathcal{S}_{\overline{\nu}}^{\perp}$. Then, for every $\epsilon > 0$, there exists g of the form (17), such that $g \in \mathcal{S}_{\overline{\nu}}^{\perp}$ and $\|f - g\|_{L^p(C)} < \epsilon$.

Theorem 4. Let $\alpha \in (0, 1/3)$, $p > p_2(\alpha)$ and $p \neq p_\nu(\alpha)$ for all $\nu \in \mathbb{Z}$ satisfying $\alpha_\nu > \alpha$. Let $\overline{\nu}$ be the largest integer satisfying $\overline{\nu} \geq 3$ and $p_{\overline{\nu}-1}(\alpha) < p$. Then $\mathcal{V}_{\alpha,p}$ is a closed subset of $L^p(C)$. Its orthogonal complement is given by $\mathcal{V}_{\alpha,p}^{\perp} = \mathcal{S}_{\overline{\nu}} :=$ Closure (in $L^{p'}(C)$, $p' = p/(p-1)$) of Span $\{v_{\nu n}\colon\ n \in \mathbb{N},\ \nu \in \mathbb{Z},\ 2 \leq |\nu| \leq \overline{\nu} - 1\}$ (functions $v_{\nu n}$, solutions to the problem $L_\alpha^* v = 0$ in C, $v|_{\partial C} = 0$, are defined in theorem 2).

Proof. Theorem 3 implies that $\mathcal{V}_{\alpha,p}$ is a closed subset of $L^p(C)$.

By theorem 2, if $u \in W^{2,p}_{\gamma_0}(C)$ satisfies: $\int_C v_{\nu n}\, L_\alpha u\, dx = 0$ $(n \in \mathbb{N},\ \nu \in \mathbb{Z},\ 2 \leq |\nu| \leq \overline{\nu}-1)$, then $\mathcal{V}_{\alpha,p} \subset \mathcal{S}_{\overline{\nu}}^{\perp}$.

Now, let f have the form (17) and assume $\int_C v_{\nu n}\, f\, dx = 0$ $(n \in \mathbb{N},\ \nu \in \mathbb{Z},\ 2 \leq |\nu| \leq \overline{\nu}-1)$, i.e., $f \in \mathcal{S}_{\overline{\nu}}^{\perp}$. By proposition 4, there is a unique $u_2 \in W^{2,2}_{\gamma_0}(C)$ such that $L_\alpha u = f$ in C; by proposition 3, u_2 must be the function given by lemma 1(ii). Since $\eta_{\nu mn}(\epsilon) = o(R_{\nu n}(\epsilon, \rho))$ and $\eta'_{\nu mn}(\epsilon) = o(R'_{\nu n}(\epsilon, \rho))$ $(\epsilon \to 0)$, and

$$R_{\nu n}(r,\rho) = c_2 r^{\lambda_\nu}(1+o(1)), \quad R'_{\nu n}(r,\rho) = \lambda_\nu\, c_2\, r^{\lambda_\nu - 1}(1+o(1)) \quad (c_2 \neq 0)$$

by (7), integrating by parts as in theorem 2 gives

$$(26) \qquad 0 = \sum_{0}^{\infty}{}_m\, a_{\nu mn}\, \eta_{\nu mn}(\rho) \qquad (2 \leq |\nu| \leq \overline{\nu} - 1).$$

Let

$$\overline{u}_2(r;\theta;z) =$$

$$\left\{\Sigma_{|\nu| \leq 1} + \Sigma_{|\nu| \geq \overline{\nu}}\right\} \sum_{1}^{\infty}{}_n \left\{\sum_{0}^{\infty}{}_m\, a_{\nu mn}\, [\eta_{\nu mn}(r) - \eta_{\nu mn}(\rho) R_{\nu n}(r,\rho)]\right\} e^{i\nu\theta} \sin nz\,.$$

As $\lambda_0 = 0$, $\lambda_1 = \lambda_{-1} = 1$ and $(\lambda_\nu - 2)p + 1 < -1$ if $|\nu| \geq \overline{\nu}$, we have $\overline{u}_2 \in W^{2,p}_{\gamma_0}(C)$; moreover, (26) tells us that

$$(u_2 - \overline{u}_2)(r;\theta;z) = \Sigma_{1<|\nu|<\overline{\nu}} \sum_{1}^{\infty}{}_n \sum_{0}^{\infty}{}_m\, a_{\nu mn}\, \eta_{\nu mn}(r)\, e^{i\nu\theta} \sin nz$$

is a smooth function. Therefore u_2 is in $W^{2,p}_{\gamma_0}(C)$, consequently $f = L_\alpha u_2 \in \mathcal{V}_{\alpha,p}$. Now, by the previous lemma, every f in $L^p(C)$, orthogonal to $\mathcal{S}_{\overline{\nu}}$, can be approximated by functions of the form (17), orthogonal to $\mathcal{S}_{\overline{\nu}}$. We have just proved that the latter ones are in $\mathcal{V}_{\alpha,p}$, then: $\mathcal{S}_{\overline{\nu}}^{\perp} \subset \mathcal{V}_{\alpha,p}$. □

Theorem 5. Let $|\nu| > 1$, $0 < \alpha < \alpha_\nu$; then:

$$(27) \qquad \inf\left\{ \|L_\alpha u\|_{L^{p_\nu}(C)} / \|D^2 u\|_{L^{p_\nu}(C)}\colon\ u \in W^{2,p_\nu}(C)\backslash\{0\},\ u=0 \text{ on } \partial C\right\} = 0\,;$$

moreover, $\mathcal{V}_{\alpha,p_\nu}$ is <u>not</u> a closed subset of $L^{p_\nu}(C)$.

Proof. Let us assume $\rho = 1$; in the proof c will be a positive constant, not necessarily the same at every occurrence.

Let $u_0(r;\theta;z) := R_{\nu 1}(r,1) e^{i\nu\theta} \sin z$; then $u_0 \in C^\infty(\overline{C}\backslash Z)$ and $L_\alpha u_0 \equiv 0$. By the definition of $R_{\nu 1}(r,1)$,

$$u_0(r;\theta;z) = c(1+o(1)) r^{\lambda_\nu}, \quad |Du_0(r;\theta;z)| = c(1+o(1)) r^{\lambda_\nu - 1}.$$

$$|D^2u_0(r;\theta;z)| = c(1+o(1))\,r^{\lambda_\nu-2}$$

as $r\to 0$.

Let $\phi\in C^\infty[0,1]$ be a cutoff function: $\phi\equiv 1$ in $[0,3/4]$, $\phi\equiv 0$ in a neigbourhood of 1; $L_\alpha(\phi u_0)$ vanishes in $(0,3/4)$ and is continuous in $\bar{C}\backslash Z$. Let $\epsilon\in(0,3/4)$, $u_\epsilon\in C^{1,1}(\bar{C})$, $u_\epsilon=0$ on ∂C, and let u_ϵ be defined by

$$u_\epsilon=\phi u_0 \quad \text{for } \epsilon\le r\le 1,\ u_\epsilon=(\gamma_1 r^2+\gamma_2 r^4)\,e^{i\nu\theta}\sin z \quad \text{for } 0\le r\le\epsilon,$$

where γ_1,γ_2 are suitably chosen. We have $|L_\alpha u_\epsilon|\le c\epsilon^{\lambda_\nu-2}$ as $0\le r\le\epsilon$, and

(28) $$\|L_\alpha u_\epsilon\|^{p_\nu}_{L^{p_\nu}(C)}\le c\Big(1+\epsilon^{(\lambda_\nu-2)p_\nu}\int_0^\epsilon r\,dr\Big)=\tfrac{3}{2}c\ ;$$

therefore, $\|L_\alpha u_\epsilon\|_{L^{p_\nu}(C)}$ is bounded as $\epsilon\to 0$.

On the other hand, (α) implies $|R''_{\nu 1}(r,1)|\ge cr^{\lambda_\nu-2}$ and

(29) $$\|D^2u_\epsilon\|^{p_\nu}_{L^{p_\nu}(C)}\ge\|(u_\epsilon)_{rr}\|^{p_\nu}_{L^{p_\nu}(C)}\ge c\int_\epsilon^{3/4} r^{(\lambda_\nu-2)p_\nu}\,r\,dr=c\int_\epsilon^{3/4}\frac{dr}{r}\to+\infty$$

as $\epsilon\to 0$. Thus, (28) and (29) imply (27).

Furthermore, notice that $u_\epsilon\in W^{2,p_\nu}_{\gamma_0}(C)$, $L_\alpha u_\epsilon\in\mathcal{V}_{\alpha,p_\nu}$, $L_\alpha u_\epsilon\to L_\alpha(\phi u_0)$ in $L^{p_\nu}(C)$, but there exists no function $v\in W^{2,p_\nu}_{\gamma_0}(C)$ such that $L_\alpha v=L_\alpha(\phi u_0)$. If this was the case, $w:=v-\phi u_0$ would solve $w\in W^{2,p}_{loc}(C\cup T)$, $L_\alpha w=0$ in C, $w|_T=0$, $w|_S=0$, where $p>p_1(\alpha)$; by proposition 1 this implies $w\equiv 0$. Consequently, $\mathcal{V}_{\alpha,p_\nu}$ is not a closed subset of $L^{p_\nu}(C)$. □

References

[1] **O.Arena-P.Manselli** (1995). On the Ural'tseva axially symmetric operator, to appear.

[2] **R.F.Bass-E.Pardoux** (1987). Uniqueness for diffusions with piecewise constant coefficients, Prob. Th. Rel. Fields 76, 557-572.

[3] **C.Cerutti-L.Escauriaza-E.B.Fabes** (1993). Uniqueness for the Dirichlet problem for elliptic operators with discontinuous coefficients, Annali di Matematica Pura e Appl. (IV), 163, 161-180.

[4] **C.Cerutti-L.Escauriaza-E.B.Fabes** (1991). Uniqueness for some diffusions with discontinuous coefficients, Annals of Probability 19, n.2, 525-537 .

[5] **L.Escauriaza** (1992). Uniqueness in the Dirichlet problem for time independent elliptic operators, IMA Math. Appl. vol. 45, Springer .

[6] **G.Hardy-J.Littlewood-G.Polya** (1934). Inequalities, Cambridge Univ. Press.

[7] **L.Lamberti-P.Manselli** (1983). Existence, uniqueness and counterexamples for an axially symmetrric elliptic operator, B.U.M.I. (6) 2-B, 431-443.

[8] **P.Manselli-G.Papi** (1984). Nonhomogeneous Dirichlet problem for elliptic operators with axially discontinuous coefficients, J. Diff. Eqns, v.55, n.3, 368-384 .

[9] **N.Ural'tseva** (1967). Impossibility of $W^{2,p}$ bounds for multidimensional elliptic operators with discontinuous coefficients, L.O.M.I. 5, 250-254.

[10] **H.F.Weinberger** (1965). Partial differential equations, Blaisdell.

3. The Use of "A Priori" Information in the Solution of Ill-Posed Problems

Mario Bertero Dipartimento di Scienze Fisiche, Università di Genova, Genoa, Italy

1 INTRODUCTION

A peculiar feature of ill-posed problems is the great variety of methods which have been proposed and are used for their solution. The definition of a regularization algorithm introduced by A.N.Tikhonov [1,2] provides a rather general setting for describing most of these methods, at least the deterministic ones, i.e. the methods where statistical properties of the data and of the solution are not used. This definition, however, even if it shows important relationships between methods for solving ill-posed problems and approximation theory, does not explain why so many different methods can be used for solving the same problem and why this situation is necessary and desirable.

The essential point which provides the answer to this question can already be found in the pioneering papers of C.Pucci [3,4] and F.John [5] on the solution of some ill-posed problems for partial differential equations. In fact in these papers it is shown that it is possible to construct, from approximate data, sequences of approximate solutions converging to the solution of the problem, if the solution satisfies some additional conditions. The need of additional information on the solution is clearly stated and investigated in subsequent papers of the same authors [6,7]. In these papers the additional information consists, in general, of an upper or of a lower bound (positivity) on the solution. Nowadays many other problems have been considered and many other kinds of additional information have been introduced. Therefore the great variety of methods is, in some sense, related to the great variety of additional information.

The need of additional information, which now is usually called a *priori* information, can be easily understood by the following remarks.

As it is known ill-posedness is a typical feature of inverse problems. Now a linear inverse problem which is also ill-posed can be reduced, in general, to the solution of a first kind functional equation

$$Lf = g \tag{1}$$

where $L : X \to Y$ is a linear continuous operator (X and Y are Hilbert spaces) whose range is not closed in Y. The operator L is assumed to be known, as well as the function $g \in Y$, the data of the problem. The problem consists in determining the function $f \in X$. Very well-known examples are provided by the inversion of the Radon transform, which is basic in tomography and medical imaging, by the inversion of the Laplace transform or by the problem of image restoration, where L is a convolution operator.

When g is affected by errors, as in the case of measured data, g is not in the range of the operator L and therefore no solution of equation (1) exists. For this reason one is obliged to look for approximate solutions, i.e. for functions $f \in X$ such that the residual

$$r = g - Lf \tag{2}$$

is not zero. Since the range of L is not closed, the set of the approximate solutions such that the residual has a norm which does not exceed some prescribed quantity $\varepsilon, \| r \|_Y \leq \varepsilon$, is not bounded. This means, of course, that there are too many approximate solutions of our equations: most of them are completely unphysical. At this point the need of a *priori* information clearly appears: this must provide the definition of a set of physically acceptable solutions such that the intersection of this set with the set of approximate solutions is sufficiently "small".

In this short note we discuss a few results about the method of constrained least squares solutions which has a rather wide applicability and which is rather flexible for taking into account many kinds of a *priori* information. This is the content of Sect.2.

In Sect.3 we discuss the method of successive approximations which provides a sequence of functions, in principle computable, converging to the constrained least squares solution, provided that some suitable conditions are satisfied.

2 CONSTRAINED LEAST SQUARES SOLUTIONS

We first give examples of constraints which are frequently used in solving a linear inverse problem. Each constraint is defined in terms of a set of functions where one has to look for the approximate solutions. The following examples are mainly drawn from $L^2(\mathbb{R}^n)$, the space of square integrable functions of n variables:

i) the set of all the functions satisfying the "energy constraint"

$$\int |f(x)|^2 dx \leq E^2 \tag{3}$$

i.e. the ball of radius E in $L^2(\mathbb{R}^n)$;

ii) the set of all the functions which belong to a closed ball of some suitable Sobolev space; for instance, if D is some linear differential operator, the set of the functions such that

$$\int |(Df)(x)|^2 dx \leq E^2 \ ; \tag{4}$$

iii) the set of all the functions which are zero a.e. outside a bounded region $\mathcal{D}$ of $\mathbb{R}^n$; such a constraint is used, for instance, in the problem of restoring diffraction limited images [8] and in the problem of limited angle tomography [9];

iv) the set of all the functions whose Fourier transform is zero a.e. outside a bounded domain $\mathcal{B}$ of $\mathbb{R}^n$, i.e. the set of the so-called band-limited functions with band $\mathcal{B}$; this constraint is used, for instance in the problem of extrapolating band-limited signals or images [10];

v) the set of all the functions which are non-negative a.e.; this constraint is important in many problems of imaging, and, in particular, in image restoration [11];

vi) the set of all the functions which coincide a.e. on a bounded region $\mathcal{D} \subset \mathbb{R}^n$, with a prescribed function g;

vii) the set of all the functions whose Fourier transforms coincide a.e. on a bounded region $\mathcal{D} \subset \mathbb{R}^4$ with a prescribed function $\hat{g}$;

viii) any intersection of the sets i)-vii);

ix) in the case of $L^2(a, b)$ the set of all the non-decreasing functions with values in $[0, 1]$, i.e. the set containing all the distribution functions of random variables with values in $[a, b]$; this constraint is also important in the solution of Abel equation.

A common feature of these sets is that they are closed and convex. Moreover the sets i), ii) and ix) are bounded (the sets ii) and ix) are also compact in L^2), the sets iii), iv) are closed subspaces, the set v) is a convex cone, while the sets vi), vii) are affine subspaces (translations of closed subspaces).

Anyway the previous examples suggest that, in many ill-posed problems, the additional information consists in requiring that the *approximate solution belongs to a closed and convex set C*. Then it is quite natural to look for those functions in C which minimize the norm of the residual (2). These are the so-called *constrained least squares solutions*, i.e. the solutions $\tilde{f}_C$ of the problem

$$\| L\tilde{f}_C - g \|_Y = \inf_{f \in C} \| Lf - g \|_Y \quad . \tag{5}$$

The well-posedness of this problem depends on additional properties both of L and of C. We have the following result [12]:

Theorem 1 - If the closed and convex set C is also bounded and if the linear and continuous operator L is injective (i.e. the equation $Lf = 0$ has the unique solution $f = 0$), then there exists a unique solution of problem (5).

The proof is rather simple and is contained in the following remarks: since the linear operator L is continuous, the set LC is also bounded, closed and convex; then these exists a unique $\tilde{g} \in LC$ having minimal distance from g; finally the equation $Lf = \tilde{g}$ has a unique solution $\tilde{f}$ which is also the unique solution of problem (5).

It is also possible to prove that this solution depends continuously on the data when the set C is compact (otherwise, under the conditions of Theorem 1, the continuity is only weak), but we do not discuss this point in this short note. Results can be found in [12].

The condition of boundedness of the set C can be removed if the operator L has a bounded inverse. In fact, in such a case we have:

Theorem 2 - If the set C is closed and convex and if the linear and continuous operator L has a continuous inverse L^{-1}, then there exists a unique solution of problem (5).

The proof is similar to that of Theorem 1 if one remarks that the set LC is closed thank to the assumption of continuity of the operator L^{-1}.

At first glance this theorem seems not interesting for ill-posed problems because we assume the continuity of the inverse operator. There are however two cases where it can be applied. The first is that of a discretized problem. In such a case the inverse operator is always continuous, even if it can be extremely ill-conditioned (and also the constrained least squares problem can be ill-conditioned). The second is the case of regularized solutions. In fact it is well-known that they can be defined as least squares solutions: if $Z = Y \oplus X$ and if the linear operator $L_\alpha : X \to Z$ is defined by ($\alpha > 0$)

$$L_\alpha f = \left\{Lf, \sqrt{\alpha} f\right\} \tag{6}$$

then the (unconstrained) least squares solution f_α defined by

$$\| L_\alpha f_\alpha - h \|_Z = \inf_{f \in X} \| L_\alpha f - h \|_Z \tag{7}$$

where $h = \{g, 0\}$, coincides with the regularized solutions f_α which minimizes the functional [13]

$$\Phi(f) = \| Af - g \|_Y^2 + \alpha \| f \|_X^2 \quad . \tag{8}$$

Then Theorem 2 assures the existence and uniqueness of a *constrained regularized solution* $f_{\alpha,C}$, solution of the problem

$$\| L_\alpha f_{\alpha,C} - h \|_Z = \inf_{f \in C} \| L_\alpha f - h \|_Z \quad . \tag{9}$$

In fact, the operator L_α^{-1} is given by

$$L_\alpha^{-1} h = L_\alpha^{-1}\{g, 0\} = (L^* L + \alpha I)^{-1} L^* g \tag{10}$$

(where L^* is the adjoint operator) and therefore is continuous.

3 THE METHOD OF SUCCESSIVE APPROXIMATIONS

The previous Section contains a few results about existence and uniqueness of constrained least squares solutions but does not provide a method for computing or approximating these solutions. Such a method can be obtained by a modification of the well-known method of successive approximations which is used in the case of unconstrained least squares solutions, i.e. of the functions $\tilde{f}$ such that

$$\| L\tilde{f} - g \|_Y = \inf_{f \in X} \| Lf - g \|_Y \quad . \tag{11}$$

If the operator L is not injective, uniqueness does not hold and the set of least squares solutions coincides with the set of the solutions of the Euler equation

$$L^* L \tilde{f} = L^* g \quad . \tag{12}$$

In the case of an ill-posed problem this equation has solutions if and only if $g \in R(L) \oplus R(L)^\perp$ [14], where $R(L)$ denotes the range of L and $R(L)^\perp$ its orthogonal complement. If the solution is not unique, there exists a unique solution of minimal norm, the so-called *generalized solution*, denoted by f^+.

The *method of successive approximations* is the iterative method defined by

$$f_{n+1} = f_n + \tau(L^* g - L^* L f_n) \tag{13}$$

where τ, the relaxation parameter, must satisfy the conditions

$$0 < \tau < \frac{2}{\| L \|^2} \quad . \tag{14}$$

If we introduce the nonlinear continuous operator

$$T(f) = \tau L^* g + (I - \tau L^* L) f \tag{15}$$

then the recursive equation (13) can be written as follows

$$f_{n+1} = T(f_n) \tag{16}$$

and therefore, if the sequence f_n has a limit, this limit is a fixed point of T, hence a solution of eq.(12).

When the operator L has a continuous inverse L^{-1}, it is easy to show that the operator T is a contraction

$$\| T(f_2) - T(f_1) \|_X \leq \rho \| f_2 - f_1 \|_X \tag{17}$$

where ($\sigma(L^* L)$ denotes the spectrum of $L^* L$)

$$\rho = \sup_{\lambda \in \sigma(L^* L)} | 1 - \tau\lambda | < 1 \quad . \tag{18}$$

Then, from the contraction mapping Theorem, it follows that f_n converges to the unique fixed point of T, independently of the initial estimate f_0.

When $\lambda = 0$ is a point in the spectrum of $L^* L$, i.e. the problem is ill-posed, then the operator T is nonexpansive

$$\| T(f_2) - T(f_1) \|_X \leq \| f_2 - f_1 \|_X \quad . \tag{19}$$

In such a case it is not easy to obtain results on the convergence of f_n from general results on nonexpansive mappings. However, from a careful investigation of the particular operator (15), it is possible to obtain the following result which is proved in [15-17]:

Theorem 3 - If $g \in R(L) \oplus R(L)^\perp$, then, for any given initial estimate f_0, f_n strongly converges to a least squares solution of the problem. Moreover, if $f_0 = 0$, then f_n strongly converges to the generalized solution f^+.

If g does not satisfy the conditions stated in the Theorem, the sequence f_n does not converge but has interesting approximation properties. Assume, indead, that g is an approximation of the exact data $\bar{g}$ and that there exists an exact solution $\bar{f}$ corresponding to $\bar{g}$. Then, if f_n is the sequence computed using g, consider the error $\delta_n = \| f_n - \bar{f} \|_X$. It is possible to prove that δ_n has a minimum for some value of n, since it is decreasing for small values of n and increasing for large values of n. Therefore in the case of a practical application with real data affected by errors, the problem is to find criteria for stopping the iterations at some optimum value of n. This is a rather difficult task but, fortunately, f_n does not change too much for values of n around the optimal one and therefore it is possible to reasonably stop the iterations by simply looking at f_n.

In the case of the constrained least squares problem (5), the method of successive approximations can be modified as follows: let us denote by P_C the projection operator onto the convex and closed set C; then we replace eq.(16) by the following one

$$f_{n+1} = P_C T(f_n) \quad . \tag{20}$$

When the operator L has a continuous inverse L^{-1}, the operator T is a contraction - see eq.(17) - and therefore the operator $P_C T$ is also a contraction, since P_C is nonexpansive. It follows that f_n converges to the unique fixed point of $P_C T$.

In such a case it is easy to prove the following Theorem:

Theorem 4 - If the operator L has a continuous inverse L^{-1}, then the unique fixed point of the operator $P_C T$ is also the unique constrained least squares solution $\tilde{f}_C$, solution of the problem (5).

The proof is given in [18] in the case of a finite dimensional problem but it holds true for a general linear and continuous operator. It consists essentially in proving that, if $\tilde{f}_C$ is not a solution of eq.(12) (i.e. if the unconstrained least squares solution does not belong to C), then the gradient of the functional $\| Lf - g \|_Y$ computed at $f = \tilde{f}_C$ is also orthogonal to the support plane of C at the support point $\tilde{f}_C$.

The previous Theorem has some interesting implications. It implies, in fact, that the constrained regularized solutions, considered at the end of Sect.2, can be computed by means of the recursive equation (20) if in eq.(15) the operator L is replaced by the operator L_α, eq.(6). In the case of a discrete problem this method can be easily implemented if the operator P_C has a simple expression; in fact, the computation of Tf requires only matrix-vector multiplication.

However, in the case of a general linear operator L, we only know that the mapping $T_C = P_C T$ is nonexpansive - see eq.(19). It should be important to investigate, in such a case, the set of the fixed points of this operator and the relationship between this set and the set of the constrained least squares solutions. In particular, as we know from Theorem 1, when the set C is bounded and the operator L is invertible, the constrained least squares solution $\tilde{f}_C$ exists and is unique for any g. It seems natural to conjecture that $\tilde{f}_C$ is also the unique fixed point of the operator $P_C T$.

Anyway, from our numerical experience, in the case of simulated data, it seems to be reasonable to state that the sequence f_n obtained from eq.(20) has approximation properties quite similar to those of the sequence f_n derived from eq.(16) and discussed after Theorem 3.

REFERENCES

1. A.N.Tikhonov (1963). Solution of incorrectly formulated problems and the regularization method, *Soviet Math. Dokl., 4*: 1035-1038.

2. A.N.Tikhonov and V.Y.Arsenin (1977). *Solutions of Ill-Posed Problems*, Winston/Wiley, Washington.

3. C.Pucci (1953). Studio col metodo delle differenze di un problema di Cauchy relativo ad equazioni a derivate parziali del secondo ordine di tipo parabolico, *Ann. Scuola Normale Sup. Pisa, Serie III, 7*: 205-215.

4. C.Pucci (1955). Sui problemi di Cauchy non "ben posti", *Atti Acc. Naz. Lincei, 18*: 473-477.

5. F.John (1955). Numerical solution of the equation of heat conduction for preceding times, *Ann. Mat. Pura Appl., 40*: 129-142.

6. D.Fox and C.Pucci (1958). The Dirichlet problem for the wave equation, *Ann. Mat. Pura Appl., 46*: 155-182.

7. F.John (1960). Continuous dependence on data for solutions of partial differential equations with a prescribed bound, *Comm. Pure Appl. Math., 13*: 551-585.

8. M.Bertero and E.R.Pike (1982). Resolution in diffraction limited imaging, *Optica Acta, 29*: 727-746.

9. R.W.Gerchberg (1974). Super-resolution through error energy reduction, *Optica Acta, 21*: 709-720.

10. A.Lent and H.Tuy (1981). An iterative method for the extrapolation of band-limited functions, *J. Math. Anal. Appl., 83*: 554-565.

11. J.Biemond, R.L.Lagendijk and R.M.Mersereau (1990). Iterative methods for image deblurring, *Proc. IEEE, 78*: 856-883.

12. V.K.Ivanov (1962). On linear problems which are not well-posed, *Soviet Math. Dokl., 3*: 981-983.

13. K.Miller (1970). Least squares methods for ill-posed problems with a prescribed bound, *SIAM J. Math. Anal., 1*: 52-74.

14. C.W.Groetsch (1984). *Generalized Inverses of Linear Operators*, Dekker, New York.

15. L.Landweber (1951). An iteration formula for Fredholm integral equations of the first kind, *Amer. J. Math., 73*: 615-624.

16. V.Fridman (1956). A method of successive approximation for Fredholm integral equations of the first kind, *Usp. Math. Nauk.*, *11*: 232-234.

17. H.Bialy (1959). Iterative behandlung linearer funktionalgleichungen, *Arc. Rat. Mech. Anal.*, *4*: 166-176.

18. R.L.Lagendijk, J.Biemond and D.E.Boekee (1988). Regularized iterative image restoration with ringing reduction, *IEEE Trans. Acoust., Speech, Signal Processing*, *ASSP-36*: 1874-1888.

4. Allocation Maps with General Cost Functions

Luis A. Caffarelli Institute for Advanced Study, Princeton, New Jersey

Introduction: In the theory of allocations one is interested to transfer a domain Ω_1 into a domain Ω_2 (one can think on sending k points $X_1, \ldots, X_k$ in a one to one fashion to k points Y_1, Y_k), minimizing certain cost function $C(X, Y)$. (See for instance Rachev ([R]). For $C(X, Y) = \frac{1}{2}|X - Y|^2$, this problem has interest in several fields, and a complete existence theory was given by Brenier ([B]) and a regularity theory was developed by the author ([C1], [C2]).

Here we redo Brenier's theory for general strictly convex cost function $C(X - Y)$.

In a conversation R. McCann told us that he had obtained similar results in collaboration with W. Gangbo.

In all of this, note f and g will be two bounded functions with compact support Ω_1 and Ω_2 respectively, and the compatibility condition

$$\int_{\Omega_1} f(X)dX = \int_{\Omega_2} g(Y)dY.$$

We will consider a cost function $C(X - Y)$, smooth (say $C^{1,\alpha}$) and strictly convex.

In particular the map (for fixed Y_0)

$$X \longrightarrow \nabla_X C(X - Y)$$

is continuous with a continuous inverse.

We are interested to find, among all transformations $Y(X)$ that carry the density f onto the density g (i.e., for every continuous function h,

$$\int h(Y)g(Y)dY = \int h(Y(X))f(X)dX$$

the one that minimizes the allocation cost

$$\int C(X - Y(X))f(X)dX.$$

We will do that through a heuristically standard dual problem, following the ideas of Brenier ([B]) in the case of $C(X - Y) = |X - Y|^2$ providing the necessary details to complete this program. The dual problem is the following:

Problem 1: Among all pairs of continuous functions $\varphi(X), \psi(Y)$ defined in Ω_1, Ω_2, and satisfying

$$\varphi(X) + \psi(Y) \geq -C(X - Y)$$

minimize

$$\int \varphi(X)f(X)dX + \int \psi(Y)g(Y)dY.$$

We will show that this problem has a solution, that the map

$$Y = X + (\nabla C)^{-1}(\nabla\varphi(X))$$

is well defined and "nice" and that it solves the allocation problem. Of particular interest is the case

$$C(X - Y) = |X - Y|^p \quad (1 < p < \infty)$$

and $p = 1 + \epsilon$, since, for ϵ going to zero, this provides a "smooth" approximation to the solution of the Monge mass transfer problem.

We start with the following simple

Theorem 1.

(a) *Problem 1 has a minimizing pair* φ_0, ψ_0.

(b) φ_0, ψ_0 *are* C*-convex, i.e.,*

$$\varphi_0(X) = \sup_X -C(X-Y) - \psi_0(Y)$$
$$\psi_0(Y) = \sup_X -C(X-Y) - \varphi_0(Y).$$

In particular φ_0, ψ_0 *are Lipschitz and "*C^1 *by below" (i.e., have a uniformly* C^1 *function supporting them by below).*

Proof. Given a pair φ, ψ, of admissible functions, its energy can be improved by replacing φ by

$$\varphi^* = \sup_Y [-C(X-Y) - \psi(Y)]$$

and vice versa.

Therefore we may restrict ourselves to C-convex pairs φ, ψ.

Such pairs must be normalized because if φ, ψ are admissible, then for any constant $\lambda, \varphi + \lambda, \psi - \lambda$ are admissible and with the same energy.

Thus, we may impose that $\varphi(X_0) = 0$.

This bounds ψ by below by

$$\psi(Y) \geq -C(X_0 - Y)$$

and φ by above and below by diam $(\Omega_1) \cdot [\sup_{Y \in \Omega_2} \|C(X,Y)\|_{Lip}]$. Further, if M is large φ and $\overline{\psi} = M$ form an admissible pair, as long as

$$M \geq 2 \sup_{X,Y \in \Omega_1, \Omega_2} |C(X-Y)|$$

Thus $\varphi, \min(\psi, M)$ are a new admissible pair.

We can restrict therefore our minimization to a family of uniformly bounded uniformly Lipschitz pairs φ, ψ, and thus we can extract a minimizing sequence φ_k, ψ_k that converges uniformly to a minimizer φ_0, ψ_0.

We now study the uniqueness and differentiability properties of φ_0, ψ_0.

For that purpose we will assign to every X_0 in Ω_1 the set of "images"

$$K(X_0) = \{Y \in \overline{\Omega}_2 : \varphi(Y_0) + \psi(Y) = -C(X_0, Y)\}.$$

If $Y \in K(X_0), -\psi(Y) - C(X,Y)$ supports (is tangent by below) $\varphi(X)$ at X_0. Thus heuristically

$$\nabla_X \varphi(X_0) = -\nabla C(X_0 - Y).$$

Since ∇C is a bicontinuous invertible map with inverse $\nabla C^*(P) \quad (P \in R^n)$ we should recuperate Y as $Y = (\nabla C^*)(\nabla\varphi(X_0) + X_0$. That is, always heuristically,

$$K(X_0) = \nabla C^*(\nabla\varphi(X_0)) + X_0.$$

Before the next theorem, a few elementary remarks about $\varphi(X)$ and $\psi(Y)$.

We say that a function w is $C^{1,\alpha}$ convex if

$$w(X) = \sup_{\lambda \in \Lambda} g_\lambda(X),$$

with $g_\lambda(X)$ in $C^{1,\alpha}$, uniformly in λ.

Lemma 1. *Let w be $C^{1,\alpha}$ convex, then*

(a) at every point $X_0, w(X)$ is asymptotic to the convex cone

$$\Gamma_{X_0} = \sup_{\substack{g_\lambda(X): \\ g_\lambda(X) = w(Y_0)}} \langle \nabla g_\lambda(X_0, X - Y_0\rangle + w(X_0).$$

More precisely

$$\Gamma_{Y_0}(X) - C|X - X_0|^{1+\alpha} \leq w(X) \leq \Gamma_{X_0}(X) + o(|X - X_0|).$$

(b) If, say, 0 is a supporting gradient to Γ_{X_0} at X_0, (i.e., $0 = \nabla g_{\lambda_1}(X_0)$, and $Z = \nabla g_{\lambda_2}(X_1)$ is a supporting gradient at X_1, then

$$\langle Z, X_1 - X_0 \rangle \geq -|X_1 - X_0|^{1+\alpha}.$$

(c) A point of Lebesge differentiability for ∇w is a point of continuity.

(d) The sets of points that have more than one supporting g are a set of measure zero.

Proof. (a) is clear.

(b) follows from the fact that:

$$\langle Z, X_1 - X_0 \rangle \geq w(X_1) - w(X_0) \geq$$
$$\Gamma_{X_0}(X_1) - w(X_0) \geq -C|X_1 - X_0|^{1+\alpha},$$

since 0 is a supporting gradient.

(c) If X_0 is not a point of continuity it has two supporting g'_s, with gradients (say) 0 and te_1.

Let us split $X - X_0 = \lambda e_1 + \mu e_2$ ($e_2 \perp e_1$). If X stays in the cone

$$\lambda \leq -M|\mu|,$$

from (b) we get

$$-\lambda(z_1 \pm \frac{\lambda}{M} z_2 \geq -C\lambda^{1+\alpha}$$

or

$$z_1 \leq \frac{1}{M}|z_2| + C\lambda^{\alpha}.$$

On the other hdna, if X stays in the cone $\lambda \geq M|\mu|$ we get

$$z_1 - t \geq \frac{1}{M}|z_2| - C\lambda^{\alpha}.$$

For M large, λ small, depending on t and $\sup_{\lambda} \nabla g_{\lambda}$, these sets stay away from each other and X_0 cannot be a point of Lebesgue differentiability for ∇w. (d) follows from (c).

Corollary. *$K(X_0)$ is single valued and continuous almost everywhere in X.*

We are now ready to prove the change of variable formula.

Theorem. *Let h be a continuous function in the support of g. Then*

$$\int h(Y)g(Y)dY = \int h(K(X))f(X)dY.$$

Proof. We consider, in the optimization problem the perturbation

$$\psi_{\epsilon}(Y) = \psi(Y) + \epsilon h(Y)$$

$$\varphi_{\epsilon}(X) = \sup_{Y} -C(X-Y) - \psi(Y) - \epsilon h(Y).$$

The energy variation should thus be positive

$$0 \leq \delta E_{\epsilon} = \epsilon \int h(Y)g(Y)dY + \\ + \int (\varphi_{\epsilon}(X) - \varphi(X))f(X)dX.$$

Notice that by definition $|\varphi_{\epsilon}(X) - \varphi(X)| \leq \epsilon \sup_{Y} |h(Y)|$.

Therefore we divide by ϵ and we get

$$0 \leq \int h(Y)g(Y)dY + \int \frac{[\varphi_{\epsilon}(X) - \varphi(X)]}{\epsilon} f(X)dY.$$

The second term consists of uniformly bounded functions of which will be enough to compute the a.e. limit to obtain the limiting integrand.

For that we just consider the points X of continuity for $K(X)$.

We notice that

$$-\epsilon h(K(X)) \leq \varphi_\epsilon(X) - \varphi(X) \leq -\epsilon h(Y_\epsilon)$$

for Y_ϵ the point that realizes the value of φ_ϵ. But if a subsequence Y_ϵ converges to a Y_0 different to $(K(X))$ in the limit we will have a second supporting function $-C(X - Y_0) - \psi(Y_0)$ for φ at X, a contradiction. Thus the a.e. limit of $\frac{\varphi_\epsilon(X)-\varphi(X)}{\epsilon}$ is $h(K(X))$.

To complete this presentation, we show that the map $Y(X)$ minimizes the allocation integral.

Theorem. *Among all measurable maps $Y(X)$ that satisfy the change of variable formula, $K(X)$ is the unique minimizer of the cost function*

$$\int C(Y(X) - X)\ f(X)dX.$$

In particular, the pair φ, ψ is unique.

Proof. We write $I = \int \psi(Y)g(Y)dY$ using both changes of coordinates:

$$0 = \int [\psi(Y(X)) - \psi(K(X))]f(X)dX$$
$$\geq \int [-C(Y(X) - X) - \varphi(X)]$$
$$-[-C(K(X) - X) - \varphi(X)]f(X)dX.$$

(since for one we have inequality and the other equality) with equality in the integrals if and only if we have that $Y(X) \in K(X)$ a.e., that is $Y(X) = K(X)$, a.e.

References

[B] Brenier, Y., Polar Factorization and Monotone Rearrangement of Vector-Valued Functions, C.P.A.M., Vol. 44, 1991, pp. 375-417.

[C1] Caffarelli, L.A., The Regularity of Mappings with Convex Potentials, J. AMS 5, 1992, pp. 99-104.

[C2] Caffarelli, L.A., Boundary Regularity of Maps with Convex Potentials, C.P.A.M., Vol. 45, 1992, pp. 1141-1151.

[R] Rachev, S.T., Probability Metrics and the Stability of Stochastic Models, Wiley, 1991.

5. An Elementary Theorem in Plane Geometry and Its Multidimensional Extension

Francesco Calogero Dipartimento di Fisica, Università di Roma la Sapienza, Rome, Italy

M. D. Kruskal Department of Mathematics, Rutgers University, New Brunswick, New Jersey

This paper is dedicated to Carlo Pucci, in recognition of his continued interest in the popularization of elementary mathematics.

SUMMARY

Take any 4 points in the plane, and select 3 of them (or, equivalently, 1). Let T be the area of the triangle formed by these 3 points, and A the area of the annulus comprised between the circle on which these 3 points lie and the concentric circle on which the fourth point lies. The value of the product $T \cdot A$ is then the same, for the 4 different selections of the 3 points. We prove this theorem and we formulate and prove a multidimensional extension, as well as a generalization (circles replaced by quadrics).

Mathematics Subject Classification: 51M05, 51M25, 51N20

1. INTRODUCTION AND STATEMENT OF RESULTS

Theorem 1. Let $x^{(1)}$, $x^{(2)}$ and $x^{(3)}$ indicate 3 points on a straight line. Choose any one of them, say $x^{(3)}$, and let $X^{(3)} = \frac{1}{2}(x^{(1)} + x^{(2)})$ be the center of the segment $[x^{(1)}, x^{(2)}]$, of length $2R_3 = |x^{(1)} - x^{(2)}|$; also let $r_3 = |x^{(3)} - X^{(3)}|$ be the distance of the point $x^{(3)}$ from $X^{(3)}$. Let C_3 and c_3 be the two coplanar concentric circles centered at $X^{(3)}$, having respectively radii R_3 (so that $x^{(1)}$ and $x^{(2)}$ lie on C_3) and r_3 (so that $x^{(3)}$ lies on c_3). Let $A_3 = \pi|R_3^2 - r_3^2|$ be the area of the plane annulus comprised between C_3 and c_3. Let $P_3 = R_3 \cdot A_3$ be the volume of the annular cylinder characterized by the height R_3 and the radii R_3 and r_3. Let P_1 resp. P_2 be the analogous quantities, corresponding to the choice of $x^{(2)}$ resp. $x^{(1)}$ in place of $x^{(3)}$. Then $P_1 = P_2 = P_3$. □

Proof: by elementary computation (or see below). □

This theorem could be viewed as an (elementary if amusing) result in 3-dimensional geometry (since the annular cylinders of volume P_j are 3-dimensional objects), or in 2-dimensional geometry (since the annuli of area A_j are 2-dimensional objects), or in 1-dimensional geometry (since the 3 points $x^{(j)}$ are restricted to lie on a straight line). We choose the latter interpretation, and give a generalization of this result to d-dimensional space. In this Section the theorem is stated for $d = 2$ and for $d = 3$; the formulation for any d is then obvious. In the following Section the theorem is proven for any d. The last Section contains an extension, and includes a formulation of the extended theorem for any d.

Theorem 2 (plane geometry). Let $\mathbf{x}^{(j)}, j = 1,..,4$ denote 4 points in the plane. Select any one of them, say $\mathbf{x}^{(4)}$. Let T_4 be the area of the triangle with vertices $\mathbf{x}^{(1)}, \mathbf{x}^{(2)}$ and $\mathbf{x}^{(3)}, C_4$ the circle on which these 3 points lie, and c_4 the concentric circle on which $\mathbf{x}^{(4)}$ lies. Let A_4 be the area of the annulus comprised between these two circles, and $P_4 = T_4 \cdot A_4$. Let $P_j, j = 1, 2, 3$ be analogously defined, by replacing the role of $\mathbf{x}^{(4)}$ with $\mathbf{x}^{(j)}, j = 1, 2, 3$ respectively. Then $P_1 = P_2 = P_3 = P_4$. □

Theorem 3 (3-dimensional geometry). Let $\mathbf{x}^{(j)}, j = 1,..,5$, indicate 5 points in 3-dimensional space. Select any one of them, say $\mathbf{x}^{(5)}$. Let T_5 be the volume of the tetrahedron with vertices $\mathbf{x}^{(1)}, \mathbf{x}^{(2)}, \mathbf{x}^{(3)}$, and $\mathbf{x}^{(4)}, C_5$ the sphere on which these 4 points lie, and c_5 the concentric sphere on which $\mathbf{x}^{(5)}$ lies. Let A_5 be the area of the plane annulus comprised between two circles having the same radii as the two spheres C_5 and c_5, and $P_5 = T_5 \cdot A_5$. Let $P_j, j = 1,..,4$ be analogously defined, by replacing the role of $\mathbf{x}^{(5)}$ with $\mathbf{x}^{(j)}, j = 1,..,4$ respectively. Then $P_1 = P_2 = P_3 = P_4 = P_5$. □

Remark 1. The quantity A_j always indicates the *area* of a *plane* annulus, irrespective of the dimensionality d for which the theorem is formulated. Hence this theorem is particularly neat for $d = 2$, when it is most naturally formulated as a result in plane geometry (see Summary). □

Remark 2. The theorem is only valid in a limit sense in special cases (for instance, for $d = 2$, if 3 of the 4 points lie on a straight line, then the area of the corresponding triangle vanishes, while the area of the corresponding annulus is infinite); and it is trivial in some cases (for instance, for $d = 2$, if the 4 points all lie on a circle, then $P_j = 0, j = 1, .., 4$). □

2. PROOF

Let $\mathbf{x}^{(j)}, j = 1, .., N$ indicate $N \equiv d+2$ points in d-dimensional space; $x_k^{(j)}, k = 1, .., d$, their cartesian coordinates; and Δ the $(N \times N)$-determinant

$$\Delta = \begin{vmatrix} 1 & x_1^{(1)} & x_2^{(2)} & \cdots & x_d^{(1)} & r^2(\mathbf{x}^{(1)}) \\ 1 & x_1^{(2)} & x_2^{(2)} & \cdots & x_d^{(2)} & r^2(\mathbf{x}^{(2)}) \\ \vdots & \vdots & \vdots & \vdots & \vdots & \vdots \\ 1 & x_1^{(N)} & x_2^{(N)} & \cdots & x_d^{(N)} & r^2(\mathbf{x}^{(N)}) \end{vmatrix} ,$$

where

$$r^2(\mathbf{x}) \equiv \sum_{k=1}^{d} x_k^2 \qquad . \tag{1}$$

The value of Δ is clearly invariant under translations of the reference frame (in fact, also under rotations; but this is not needed for the proof). Indeed, the effect of translations is to add, to each column (except the first and last), a column proportional to the first one, and moreover to add, to the last column, $N-1 = d+1$ columns each of which is proportional to one of the first $N-1$ columns. To evaluate the determinant Δ we may therefore choose any conveniently translated frame of reference. Let us then translate the origin of the cartesian reference frame to the center of the hypersphere in d-dimensional space, characterized by the requirement that all but one of the N points $\mathbf{x}^{(j)}$ lie on it (say, all of them except $\mathbf{x}^{(m)}$). Then, in this coordinate system, Δ reads as follows:

$$\Delta = \begin{vmatrix} 1 & x_1^{(1)} & x_2^{(1)} & \cdots & x_d^{(1)} & R_m^2 \\ 1 & x_1^{(2)} & x_2^{(2)} & \cdots & x_d^{(2)} & R_m^2 \\ \vdots & \vdots & \vdots & \vdots & \vdots & \vdots \\ 1 & x_1^{(m)} & x_2^{(m)} & \cdots & x_d^{(m)} & r_m^2 \\ \vdots & \vdots & \vdots & \vdots & \vdots & \vdots \\ 1 & x_1^{(N)} & x_2^{(N)} & \cdots & x_d^{(N)} & R_m^2 \end{vmatrix} ,$$

where R_m is the radius of the hypersphere mentioned above ($R_m = r(\mathbf{x}^{(j)}), j = 1, .., m-1, m+1, .., N$), and r_m is the radius of the concentric hypersphere on which $\mathbf{x}^{(m)}$ lies ($r_m = r(\mathbf{x}^{(m)})$). Now, by subtracting the first column multiplied by R_m^2 from the last, it is immediately seen that, up to a sign s,

$$\Delta = s \cdot (R_m^2 - r_m^2) \cdot T_m$$

with

$$T_m = \begin{vmatrix} 1 & x_1^{(1)} & x_2^{(1)} & \cdots & x_d^{(1)} \\ 1 & x_1^{(2)} & x_2^{(2)} & \cdots & x_d^{(2)} \\ \vdots & \vdots & \vdots & \vdots & \vdots \\ 1 & x_1^{(m-1)} & x_2^{(m-1)} & \cdots & x_d^{(m-1)} \\ 1 & x_1^{(m+1)} & x_2^{(m+1)} & \cdots & x_d^{(m+1)} \\ \vdots & \vdots & \vdots & \vdots & \vdots \\ 1 & x-1^{(N)} & x_2^{(N)} & \cdots & x_d^{(N)} \end{vmatrix} .$$

Clearly, up to the factor $(d!)^{-1}$ and to a sign, T_m is the volume of the "hyper-tetrahedron" having, in d-dimensional space, the $d+1=N-1$ points $\mathbf{x}^{(j)}, j = 1, 2, .., m-1, m+1, .., N$ as its $N-1$ vertices; while the difference $R_m^2 - r_m^2$ is, up to the factor π, the area of the plane annulus comprised between the circles of radii R_m and r_m. But the value of Δ is independent of our selection of the particular point $\mathbf{x}^{(m)}$ to evaluate it. This proves the theorem. □

3. EXTENSIONS AND OUTLOOK

It is easy to extend the proof given above, by replacing the function $r^2(\mathbf{x})$ (which defines via the equation $r^2(\mathbf{x}) = R^2$ the hypersphere of radius R centered at the origin, on which the point $\mathbf{x}$ lies; see (1)), with the function

$$\rho^2(\mathbf{x}) \equiv \sum_{k_1,k_2=1}^{d} \alpha_{k_1k_2} x_{k_1} x_{k_2} , \tag{2}$$

where the coefficients $\alpha_{k_1k_2}$ are hereafter supposed to be fixed and to define, via the equation $\rho^2(\mathbf{x}) = R^2$, the hyperellipsoid (associated with the real symmetrical matrix $\alpha_{k_1k_2}$) of "mean radius" R centered at the origin, on which the point $\mathbf{x}$ lies (a necessary and sufficient condition to guarantee tha this be indeed a hyperellipsoid is that all the eigenvalues of the real symmetric matric $\alpha_{k_1k_2}$ be positive; but this requirement is not essential for the validity of the following theorem, see below). This justifies the extension of the theorem proven above, corresponding to the replacement of the two hyperspheres (ordinary spheres for $d=3$, coplanar plane circles for $d=2$ and $d=1$) by the two hyperellipsoids associated with the equations $\rho^2(\mathbf{x}^{(j)}) = R_m^2$, $j = 1, .., m-1, m+1, .., N$ and $\rho^2(\mathbf{x}^{(m)}) = r_m^2$, the first being characterized by the requirement that the $N-1$ points $\mathbf{x}^{(j)}, j = 1, .., m-1, m+1, ..N$, lie on it (this determines the "mean radius" R_m and the appropriately translated cartesian frame of reference), the second being characterized by the requirement that $\mathbf{x}^{(m)}$ lie on it (this identifies the "mean radius" r_m, in the same reference frame); while the planar annulus is characterized, as before, by the two radii R_m and r_m (for $d=2$, one might as well consider the elliptical annulus comprised between the two ellipses, which again makes for a neater geometrical formulation in this case).

As implied by this argument, the requirement that the equation $\rho^2(x) = R^2$ define a hyperellipsoid (rather than a generic quadric in d-dimensional space) is in fact unessential for the validity of the theorem, whose extended formulation therefore reads as follows:

Theorem (d-dimensional geometry). Consider $N \equiv d+2$ points $\mathbf{x}^{(j)}$, $j = 1, ..., N$, and a quadric characterized, in the cartesian coordinates x_k, by the equation $\rho^2(\mathbf{x}) = R^2$, see (2). In the following the real symmetrical matrix $\alpha_{k_1 k_2}$ is held fixed, and the quadric associated to it via this equation is referred to as "The Quadric", with R its "Mean Radius" (if The Quadric is not a hyperellipsoid, R^2 might be negative). Now select any one point (say, $\mathbf{x}^{(m)}$) out of the set of N points $\mathbf{x}^{(j)}$, and determine R_m via the requirement that The Quadric, with Mean Radius R_m, and with an appropriate translation (without rotations) of the cartesian frame of reference, pass through the $N-1 = d+1$ points $\mathbf{x}^{(j)}, j = 1, .., m-1, m+1, .., N$ (i.e., all of them except $\mathbf{x}^{(m)}$). Then determine a second quantity r_m via the requirement that, in the same (translated) reference frame, The Quadric with Mean Radius r_m go through the point $\mathbf{x}^{(m)}$. Compute finally the quantity $P_m = |R_m^2 - r_m^2| \cdot T_m$, where T_m is, as before, the volume of the hypertetrahedron in d-dimensional space having as vertices the $d+1 = N-1$ points $\mathbf{x}^{(j)}, j = 1, .., m-1, m+1, .., N$. Of course N quantities P_j can be computed in this manner, corresponding to the N possible different selections of the point $\mathbf{x}^{(m)}$ from the set $\{\mathbf{x}^{(j)}, j = 1, .., N\}$. The theorem states that these N quantities P_j are all equal. □

Let us finally note that the above method of proof suggests an approach to obtain other results of this kind; for instance, for $d = 2$, one might start by considering a (5×5)-determinant whose 5 lines have the entries $1, x_1^{(j)}, x_2^{(j)}, [x_1^{(j)}]^2 - [x_2^{(j)}]^2, x_1^{(j)} x_2^{(j)}$, with $j = 1, .., 5$ and $x_1^{(j)}, x_2^{(j)}$ the 2 cartesian components of the 5 points $\mathbf{x}^{(j)}$ in the plane; and so on. Perhaps such results will be reported in subsequent papers.

ACKNOWLEDGEMENTS

These results were found and written during the meeting in Oberwolfach on "Nonlinear Evolution Equations, Solitons and the Inverse Scattering Transform", 11-17 July 1993.

6. Minimum Problems for Volumes of Convex Bodies

Stefano Campi Dipartimento di Matematica Pura e Applicata, Università dell'Aquila, L'Aquila, Italy

Andrea Colesanti Dipartimento di Matematica, Università di Firenze, Florence, Italy

Paolo Gronchi Dipartimento di Matematica, Università di Firenze, Florence, Italy

0. INTRODUCTION.

A nice result in plane geometry states that in the class of all convex sets of constant width d, the Reuleaux triangle has least area. The first proofs of this theorem are contained in the papers by Lebesgue (1914) and by Blaschke (1915). Subsequently alternative proofs were given by several authors: Fujiwara (1927, 1931), Mayer (1934-35), Eggleston (1952), Besicovitch (1963) and Chakerian (1966).

Surprisingly enough there are convex sets having all widths exceeding d, and area smaller than the one of the Reuleaux triangle of constant width d. Precisely Pál (1921) showed that the regular triangle has least area among all convex sets of minimal width, the *thickness*, equal to d.

What are the possible extensions of these results to higher dimension? Every width of a plane convex set Σ can be seen as the measure of the orthogonal projection of Σ onto a line. Hence the notion of width for an n-dimensional convex body Ω admits $n-1$ natural extensions as the measures of orthogonal projections of Ω onto subspaces of fixed dimension j, $j = 1, 2, \ldots, n-1$.

Here we restrict our attention to the three-dimensional case. Given a convex body Ω in $\mathbb{R}^3$, for every direction u we consider the following quantities: the *width* along u, i.e., the distance between two support planes orthogonal to u, and the *brightness* along u, i.e., the area of the orthogonal projection of Ω onto a plane orthogonal to u.

Then four problems arise naturally: find a convex body of minimum volume in each of the following classes:

A) The class of convex bodies with constant width d;

B) The class of convex bodies with thickness d;

C) The class of convex bodies with constant brightness b;

D) The class of convex bodies with minimal brightness b.

Throughout we will refer to the above problems as to problem A, B, C and D respectively.

Notice that the existence of solutions for each of these problems is guaranteed by standard compactness arguments.

Problems A,B,C and D are still unsolved. On the other hand, several authors turned their attention to the classes of convex bodies involved in those problems; for instance we refer to Blaschke (1916), Firey (1965), Chakerian and Groemer (1983), Schramm (1988), Heil and Martini (1993), Martini (1994) and Gardner (1995).

In particular, as a solution of Problem A, Bonnesen and Fenchel (1934) conjectured a sort of *Reuleaux tetrahedron* constructed by Meissner (1912) (see also Hilbert and Cohn–Vossen (1932)). Such a body is obtained by slightly reducing the intersection of four spheres centered in the vertices of a regular tetrahedron of edge's length d.

A candidate to solve Problem B was proposed by Heil (1978) (see also Gruber and Schneider (1979)); in this case the construction is based upon a tetrahedron, too. Namely the Heil body is the convex hull of six circular arcs of radius d, centered at the mid–points of the edges of a regular tetrahedron of edge's length $d\sqrt{2}$, and the four vertices of a rescaled tetrahedron of edge's length $d(2\sqrt{6}-\sqrt{2})/3$.

Problem C involves the class of bodies of constant brightness. In the literature, the first example of non–spherical three–dimensional convex body of constant brightness is a body of revolution described by Blaschke (1916). Firey (1965) gave a method for constructing families of bodies of revolution with constant brightness in arbitrary dimension.

As an intermediate step in attacking Problems A–D one may tend to restrict oneself to the case of bodies of revolution. Sharp lower bounds for the volume of bodies of this kind, involving averages of widths and brightnesses, were found by Hadwiger (1949).

In the first part of the present paper we prove that in this class the solutions to Problems A and B are provided by the bodies of revolution of the Reuleaux triangle and the regular triangle respectively (see Sections 1 and 2). As far as Problems C and D are concerned, we conjecture that the solutions are the Blaschke body and a suitable cone respectively.

Section 3 contains some remarks on the monotonicity of the volume of a convex body with respect to its widths or brightnesses. From this point of view we prove a characterizing property of quadrangles.

The main result of Section 4 is a necessary condition for a convex body Ω in order to be a solution to Problem C. Such a condition concerns the distribution of the area measure of Ω. This suggests the construction of a body that we conjecture to be the unique solution to Problem B (Section 5). We conjecture also that the solution to Problem D is given by the regular tetrahedron.

1. ROTATING THE REULEAUX TRIANGLE.

In this section we prove that among all convex bodies of revolution having constant width d, there is a unique body of minimum volume. Such a body is obtained by rotating a Reuleaux triangle around one of its axes of symmetry.

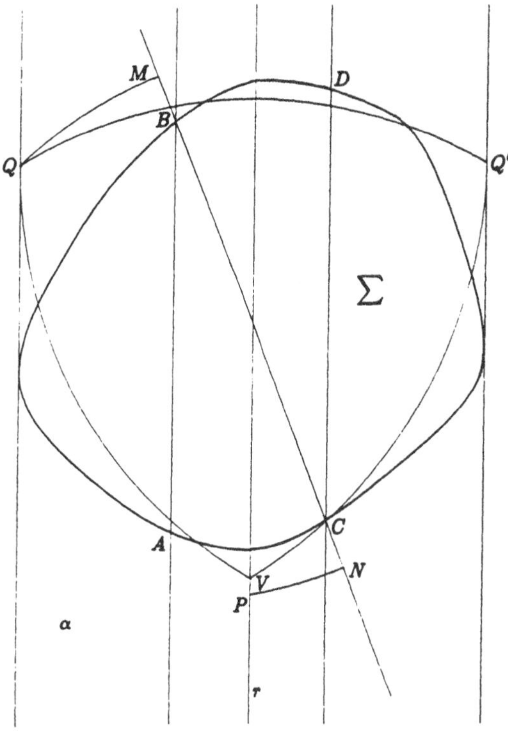

FIGURE 1

We derive this result from a property of the Reuleaux triangle which supplies also a new proof of the Blaschke–Lebesgue theorem.

We denote with $\mathfrak{G}(d)$ the class of all plane convex sets (up to rigid motions) with constant width d. Let Σ be in $\mathfrak{G}(d)$, for every vector u of $\mathbb{R}^2$ there exists a chord of Σ of length d, parallel to u, i.e. a diameter of Σ. By a continuity argument this fact implies that there exists a diameter of Σ which has the distance $\frac{d}{2}$ from each support line parallel to it. For every $\Sigma \in \mathfrak{G}(d)$ we fix a diameter $\delta(\Sigma)$ having this property. Let $x \in (0, \frac{d}{2})$ and consider the pair of chords of Σ at a distance x from $\delta(\Sigma)$ parallel to it. With $f_\Sigma(x)$ we denote the sum of the lengths of these chords.

PROPOSITION 1.1. *Let $\Sigma \in \mathfrak{G}(d)$, then*

$$f_\Sigma(x) \geq f_\mathcal{R}(x), \quad \forall x \in \left(0, \frac{d}{2}\right),$$

where $\mathcal{R}$ is the Reuleaux triangle from $\mathfrak{G}(d)$. Furthermore if $f_\Sigma(x_0) = f_\mathcal{R}(x_0)$ for some $x_0 \in (0, \frac{d}{2})$ then $\Sigma = \mathcal{R}$.

Proof. Assume that

$$f_\Sigma(\bar{x}) < f_\mathcal{R}(\bar{x}) \tag{1.1}$$

for some $\bar{x} \in \left(0, \frac{d}{2}\right)$. Let r be the line containing $\delta(\Sigma)$ and AB, CD the chords of Σ parallel to r at a distance $\bar{x}$ from r (see Figure 1). We consider the quadrangle $ABCD$

and assume that BC is its shortest diagonal. Since B and C belong to $\partial\Sigma$, there are two diameters, say b and c, through B and C respectively; b and c must lie in the same half-plane, say α, bounded by the line containing BC. Construct the Reuleaux triangle $\mathcal{R}$ of vertices V, Q and Q', having r as an axis of symmetry, C as a boundary point and the vertex V on $\alpha \cap r$. Inequality (1.1) implies that B is an interior point of $\mathcal{R}$. Let M and N be two points on the line containing BC at a distance d from C and B respectively and P a point on r at the same distance from B. Now the circular arc $\widehat{QM}$ of radius d contains a point of $\partial\Sigma$, namely one of the endpoints of c; analogously another point of $\partial\Sigma$ must be on the circular arc $\widehat{PN}$ of the same radius. On the other hand

$$\operatorname{dist}(\widehat{QM}, \widehat{PN}) = \operatorname{dist}(Q, P) > \operatorname{dist}(Q, V) = d,$$

which gives a contradiction.

If $f_\Sigma(x_0) = f_{\mathcal{R}}(x_0)$ for some $x_0 \in \left(0, \frac{d}{2}\right)$, then $P = V$ and the same argument used above shows that $V, Q \in \partial\Sigma$. Furthermore the characterizing property of $\delta(\Sigma)$ implies that $Q' \in \partial\Sigma$, too. Therefore $\Sigma \subset \mathcal{R}$ which is impossible unless $\Sigma = \mathcal{R}$. □

The area of a body $\Sigma \in \mathfrak{G}(d)$ is the integral of f_Σ over $\left(-\frac{d}{2}, \frac{d}{2}\right)$, hence the Blaschke–Lebesgue theorem follows from Proposition 1.1.

Consider now a three-dimensional convex body Ω of revolution and constant width d; Ω can be obtained by rotating a body from $\mathfrak{G}(d)$ having an axis of symmetry around such an axis. Then a second consequence of Proposition 1.1 is the following

THEOREM 1.2. *Among all three-dimensional convex bodies of revolution of constant width d, the unique body of minimum volume is the one generated by rotating the Reuleaux triangle around one of its axes.*

2. SHAKING CONVEX BODIES.

Let Ω be a convex body in $\mathbb{R}^3$ and v a direction from the unit sphere S^2. Fix a plane τ orthogonal to v and a closed half-space α bounded by τ. The *shaken body* $S\Omega_v$ of Ω with respect to v, is the set contained in α such that for every line r parallel to v, $S\Omega_v \cap r$ is either a segment having an endpoint on τ and the same length of $\Omega \cap r$, or the empty set, whether r intersects Ω or not (see Bonnesen and Fenchel (1934)). Obviously $S\Omega_v$ is defined up to reflections. We call *base* of $S\Omega_v$ the set $S\Omega_v \cap \tau$.

It is easy to see that $S\Omega_v$ is a convex body with the same volume as Ω. Is it possible to compare widths and brightnesses of Ω and $S\Omega_v$? In the next proposition we give an answer to this question in the case of bodies of revolution. For a given $u \in S^2$ we denote with $w(\Omega; u)$ the width of Ω along u and by $b(\Omega; u)$ the brightness of Ω along u.

PROPOSITION 2.1. *Let Ω be a three-dimensional convex body of revolution and let r be its axis of revolution. If v is a direction parallel to r, then*

$$w(\Omega; u) \le w(S\Omega_v; u) \quad \forall u \in S^2, \tag{2.1}$$

$$b(\Omega; u) \le b(S\Omega_v; u) \quad \forall u \in S^2. \tag{2.2}$$

Proof. Clearly $S\Omega_v$ is a body of revolution with the same axis as Ω. Thus in order to prove (2.1), we may restrict ourselves to the sections $\Omega \cap \pi$ and $S\Omega_v \cap \pi$ where π is a plane containing r. Let z be a direction of π, we define $c(\Omega; z)$ to be the maximum length of a chord of Ω parallel to z. Let AB be a maximal chord of Ω parallel to u and A' and B' be the symmetric points of A and B with respect to r. Assume that dist$(A, A') \geq$dist(B, B'). Call H the projection of B onto the segment AA'; clearly $S\Omega_v \cap \pi$ contains the triangle of vertices A, B and H (up to reflections). Then

$$c(S\Omega_v; z) \geq c(\Omega; z)$$

for every direction z of π.

For every fixed direction u of π, call l_1 and l_2 the support lines to $\Omega \cap \pi$, orthogonal to u and let P_i be a point from $\Omega \cap l_i$, $i = 1, 2$. If t is a direction parallel to P_1P_2, then

$$\begin{aligned} w(\Omega; u) &= |\langle u, t\rangle| c(\Omega; t) \\ &\leq |\langle u, t\rangle| c(S\Omega_v; t) \\ &\leq w(S\Omega_v; u), \end{aligned}$$

where $\langle \cdot, \cdot \rangle$ denotes the scalar product. Inequality (2.1) is then proved.

Let Σ be an arbitrary plane convex body with axis of symmetry r. Given a point $P \in \Sigma \cap r$ denote with Δ_P the disk centered at P and having as a diameter the chord of Σ orthogonal to r at P. We set

$$D(\Sigma) := \bigcup_{P \in \Sigma \cap r} \Delta_P .$$

It is easy to see that $D(\Sigma)$ is a convex set with r as axis of symmetry. We shall prove that

$$\text{area}(D(S\Sigma_v)) \geq \text{area}(D(\Sigma)) \tag{2.3}$$

where v is parallel to r. Firstly let us see that (2.3) implies (2.2). In $\mathbb{R}^3$ we choose a coordinate system so that the z–axis coincides with r, and $\pi \equiv \{y = 0\}$. Let $\Sigma = \Omega \cap \pi$ and $\Sigma' = S\Omega_v \cap \pi$. Fix a direction u from the plane $\{x = 0\}$ and denote with $\Sigma|_u$ the projection of Σ parallel to u onto $\{z = 0\}$. Observe that the projection of Ω parallel to u onto $\{z = 0\}$ coincides with $D(\Sigma|_u)$ (where the y–axis is chosen as symmetry axis of $\Sigma|_u$). Consequently

$$b(\Omega; u) = \text{area}(D(\Sigma|_u))|\langle u, v\rangle| . \tag{2.4}$$

By the same argument

$$b(S\Omega_v; u) = \text{area}(D(\Sigma'|_u))|\langle u, v\rangle| . \tag{2.5}$$

On the other hand $\Sigma'|_u$ is the shaken body of $\Sigma|_u$ along the y–axis. Hence inserting (2.4) and (2.5) into (2.3) gives (2.2).

Finally we prove (2.3). Let q be a line parallel to r intersecting $D(\Sigma)$ and A_1 and A_2 be the endpoints of $D(\Sigma) \cap q$. There are two points P_1 and P_2 from $\Sigma \cap r$ such that $A_i \in \Delta_{P_i}$, $i = 1, 2$. Assume that the radius of Δ_{P_1} does not exceed the one of Δ_{P_2}. Let Q be a point of $S\Sigma_v$ at a distance equal to dist(P_1, P_2) from the base of $S\Sigma_v$. Evidently the chord of $S\Sigma_v$ orthogonal to r at Q has length greater than or equal to the diameter of Δ_{P_1}; hence the length of $D(S\Sigma_v) \cap q$ is greater than or equal to dist(A_1, A_2). Due to the arbitraryness of q, (2.3) follows. □

Proposition 2.1 ensures that solutions to problems B and D for bodies of revolution can be found in the class of shaken bodies.

THEOREM 2.2. *Among all three-dimensional convex bodies of revolution with given thickness d, the unique body of minimum volume is the cone generated by the revolution of a regular triangle of side $\frac{2d}{\sqrt{3}}$, around one of its axes of symmetry.*

Proof. Let Ω be a shaken body from the class under consideration. Let r be the axis of revolution of Ω, τ the plane containing the base of Ω and ρ the radius of the base. Obviously $\rho \geq \frac{d}{2}$; moreover we may assume $\rho \leq \frac{d}{\sqrt{3}}$. Indeed, if $\rho > \frac{d}{\sqrt{3}}$, then

$$\mathrm{vol}(\Omega) \geq \frac{\pi\rho^2 d}{3} > \mathrm{vol}(\tilde{\Omega})$$

where $\tilde{\Omega}$ is the cone claimed as the minimizer.

Let Σ be the intersection of Ω with a plane through the axis of revolution and denote the endpoints of the segment $\Sigma \cap \tau$ with A_1 and A_2. Call α the half plane bounded by the line through A_1 and A_2 and containing Σ. On $r \cap \alpha$ take a point V at a distance d from the segment A_1A_2. Let $Q \in \alpha$ be the point such that $\mathrm{dist}(A_1, Q) = d$, and the angle $\widehat{QA_1A_2}$ equals $\frac{\pi}{6}$. Finally Q' is the symmetric point of Q with respect to r. Since the thickness of Ω is d, it is easy to check that V, Q and Q' belong to Σ. Hence the volume of Ω is greater than or equal to

$$\frac{\pi d}{12}\left(4\rho^2 - 3\sqrt{3}\rho d + 3d^2\right), \tag{2.6}$$

which is the volume of the body Γ_ρ obtained by rotating around r the convex hull of A_1, A_2, Q, V and Q'.

The function in (2.6) attains its minimum at $\frac{d}{\sqrt{3}}$ in $[\frac{d}{2}, \frac{d}{\sqrt{3}}]$. On the other hand, for $\rho = \frac{d}{\sqrt{3}}$, $\Gamma_\rho = \tilde{\Omega}$. This proves that $\tilde{\Omega}$ is a minimizer.

Now let Λ be a minimizer; according to Proposition 2.1, $S\Lambda_v$ is still a minimizer, where v is parallel to r. Then by the first part of the present proof

$$S\Lambda_v = \tilde{\Omega}.$$

Hence Λ must be the convex hull of a disk of radius $\frac{d}{\sqrt{3}}$ and a segment of length d orthogonal to the disk and passing through its center. The unique body of this form and thickness d is $\tilde{\Omega}$. □

3. A CHARACTERIZATION OF QUADRANGLES.

The Pál theorem shows that in general the area does not depend monotonically on the width function. Analogously in the three-dimensional case the volume does not depend monotonically either on the width function or on the brightness function.

If we restrict ourselves to the class of centrally symmetric bodies, we notice at once that the monotonicity with respect to the width function is recovered. This is not true for brightnesses, unless the involved bodies are *zonoids*. Recall that a zonoid is the limit of finite Minkowski sums of segments, in the sense of the Hausdorff metric (see for instance Goodey and Weil (1993)).

More precisely:

i) Let Ω and Λ be three-dimensional convex bodies. If Λ is a zonoid and

$$b(\Lambda; u) \geq b(\Omega; u), \quad \forall u \in S^2,$$

then $\text{vol}(\Lambda) \geq \text{vol}(\Omega)$ (Petty (1967), Schneider (1967)).

ii) Let Λ be a three-dimensional convex body with the boundary of class C^6 and positive Gaussian curvature. If Λ is not a zonoid, a centered convex body Ω exists, such that

$$b(\Omega; u) \geq b(\Lambda; u), \quad \forall u \in S^2,$$

but $\text{vol}(\Lambda) \geq \text{vol}(\Omega)$ (Schneider (1967)).

Notice that statements *i*) and *ii*) are particular cases of more general results proven in $\mathbb{R}^n$, with $n \geq 3$.

For solving problems B and D, it would be helpful to characterize the classes $\mathfrak{C}_1$ and $\mathfrak{C}_2$ defined as follows:

$\Lambda \in \mathfrak{C}_1$, if and only if $w(\Lambda; u) \leq w(\Omega; u)$ for every direction u implies $\text{vol}(\Lambda) \leq \text{vol}(\Omega)$, for every convex body Ω.

$\Lambda \in \mathfrak{C}_2$, if and only if $b(\Lambda; u) \leq b(\Omega; u)$ for every direction u implies $\text{vol}(\Lambda) \leq \text{vol}(\Omega)$, for every convex body Ω.

From the inequalities found by Rogers and Shephard (1957) and by Zhang (1991) it follows that the family of simplices of $\mathbb{R}^3$ is contained in both $\mathfrak{C}_1$ and $\mathfrak{C}_2$.

Notice that $\mathfrak{C}_1$ and $\mathfrak{C}_2$ are closed in the Hausdorff metric as well as affinely invariant.

Here we consider the two-dimensional case where the functions w and b coincide and consequently $\mathfrak{C}_1$ and $\mathfrak{C}_2$ are the same class, denoted with $\mathfrak{C}$.

THEOREM 3.1. *A plane convex body Σ belongs to the class $\mathfrak{C}$ if and only if Σ is a quadrangle, possibly degenerate.*

Proof. Let $\Sigma \in \mathfrak{C}$, then Σ has least area among all convex bodies with the same width function as Σ. Thus according to the results of Sholander (1947) and Chakerian (1966), Σ must be a *triarc*; i.e. $\partial\Sigma$ contains three points P_1, P_2 and P_3, called *base points*, verifying the following properties:

i) the segment P_iP_j, $i \neq j$, is a maximal chord of Σ;

ii) for every direction u there exists a maximal chord of Σ parallel to u, containing one of the P_i's;

iii) at least one of the P_i's lies on each pair of parallel support lines of Σ.

Let Σ' be the intersection of Σ with the closed half-plane containing P_3 and bounded by the line through P_1 and P_2; if Σ is not a quadrangle (possibly degenerate) we may assume $\Sigma \neq \Sigma'$. Up to affinity, we may assume that the lines orthogonal to the segment P_1P_2 at P_1 and P_2, are support lines of Σ. Let $h > 0$ be the maximum distance of a point of $\Sigma \setminus \Sigma'$ from the segment P_1P_2. Let Q be a point outside Σ such that the segment P_3Q has length h, and is orthogonal to P_1P_2 (see fig. 2). With $\bar{\Sigma}$ we denote the convex hull of Σ' and Q. We shall prove that

$$w(\bar{\Sigma}; u) \geq w(\Sigma; u), \tag{3.1}$$

for every direction u, and

$$\text{area}(\bar{\Sigma}) \leq \text{area}(\Sigma). \tag{3.2}$$

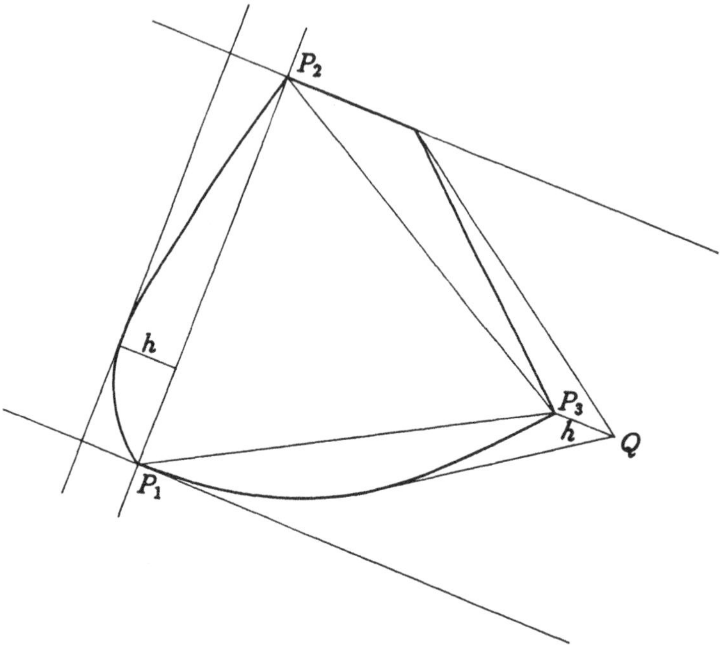

FIGURE 2

Let l, m be the support lines of Σ orthogonal to an arbitrary direction u; by virtue of property *iii*) their union contains at least one of the P_i's. If they contain P_1 or P_2 then l and m intersect $\bar{\Sigma}$, hence (3.1) is verified in this case. If the union of l and m contains P_3, then we can write

$$w(\bar{\Sigma};u) = w(\Sigma';u) + w(P_3Q;u) = w(\Sigma' + P_3Q;u)$$

where $\Sigma' + P_3Q$ is the Minkowski sum of the set Σ' and the segment P_3Q. Since Σ is contained (up to translation) in $\Sigma' + P_3Q$, (3.1) follows.

To prove (3.2), notice that

$$\text{area}(\Sigma \setminus \Sigma') \geq \frac{h}{2}\,\text{dist}(P_1, P_2) \geq \text{area}(\bar{\Sigma} \setminus \Sigma'). \tag{3.3}$$

Since $\Sigma \in \mathfrak{C}$, (3.2) and a fortiori the two inequalities in (3.3), must be an equality. Then by the first relation in (3.3) we deduce that $\Sigma \setminus \Sigma'$ is a triangle. By repeating the same argument starting from the chords P_2P_3 and P_3P_1, we deduce that Σ must be a poligon with six vertices at the most. Assume that Σ is a hexagon. The above construction produces a pentagon $\bar{\Sigma}$ which by virtue of (3.1) and (3.2) is still in $\mathfrak{C}$. Proving that $\mathfrak{C}$ does not contain pentagons concludes the first part of the proof.

Let Σ be a pentagon from $\mathfrak{C}$ and P_1, P_2 and P_3 its base points. Let P_2P_3 be an edge of Σ. We may assume that the opposite edge to P_2 is not parallel to P_2P_3. Then there exist two parallel support lines of Σ touching $\partial\Sigma$ only at P_1 and P_2. Starting from such lines and the chord P_1P_2, the above construction leads to a quadrangle $\bar{\Sigma}$ such that

$$\frac{h}{2}\,\mathrm{dist}(P_1, P_2) > \mathrm{area}(\bar{\Sigma} \setminus \Sigma).$$

This implies, via (3.3), that the area of $\bar{\Sigma}$ is strictly less than that of Σ. This contradicts the fact that Σ is in $\mathfrak{C}$.

Now let Σ be a quadrangle, possibly degenerate. Up to affinity, Σ is the convex hull of a pair of orthogonal segments s_1 and s_2 having a common point.

We recall that for a general convex set Γ

$$c(\Gamma; z) = \min_{u\in S_z} \frac{w(\Gamma; u)}{|\langle z, u\rangle|}, \tag{3.4}$$

where $c(\Gamma; z)$ denotes the length of a maximal chord of Γ parallel to the direction z and $S_z = \{u \in I\!R^2 \mid \|u\| = 1, \langle u, z\rangle \neq 0\}$.

If Λ is a convex set such that

$$w(\Lambda; u) \geq w(\Sigma; u)$$

for every direction u, then by virtue of (3.4)

$$c(\Lambda; z) \geq c(\Sigma; z)$$

for every direction z. Therefore

$$\begin{aligned} 2 \cdot \mathrm{area}(\Lambda) &\geq c(\Lambda; u_1)c(\Lambda; u_2) \\ &\geq c(\Sigma; u_1)c(\Sigma; u_2) = 2 \cdot \mathrm{area}(\Sigma), \end{aligned}$$

where u_1 and u_2 are directions parallel to the segments s_1 and s_2 respectively. Hence Σ is in $\mathfrak{C}$. □

In the same way as Theorem 3.1, one can prove a result analogous to the Pál theorem, namely:

THEOREM 3.2. *For every plane convex body Σ there exists a triangle $\mathcal{T}$ such that*

$$w(\Sigma; u) \leq w(\mathcal{T}; u) \tag{3.6}$$

for every direction u, and

$$\mathrm{area}(\Sigma) \geq \mathrm{area}(\mathcal{T}). \tag{3.7}$$

Notice that unless Σ is a quadrangle, possibly degenerate, a triangle $\mathcal{T}$ exists for which (3.6) and (3.7) are strict.

4. CONSTANT BRIGHTNESS AND MINIMUM VOLUME.

In this section we prove a necessary condition concerning the area measure of convex bodies of minimum volume and prescribed constant brightness. The two-dimensional counterpart of this result is contained in a paper by Kallay (1974). The proof we give here in the three-dimensional case can easily be extended to an arbitrary dimension.

Let us recall some basic notions from the Brunn-Minkowski theory; for an exhaustive treatise on this topic, see Schneider (1993).

Let $\mathfrak{B}$ be the family of Borel subsets of the unit sphere S^2. For a general convex body Ω of $\mathbb{R}^3$, a finite non-negative Borel measure σ_Ω is defined, called the *area measure* of Ω: for every $\beta \in \mathfrak{B}$, $\sigma_\Omega(\beta)$ is the two-dimensional Hausdorff measure of the subset of $\partial\Omega$ which is the reverse image of β through the Gauss map.

Let Ω be a convex body and u a direction, the area of the orthogonal projection of Ω onto a plane orthogonal to u is given by the formula

$$b(\Omega; u) = \frac{1}{2} \int_{S^2} |\langle u, z\rangle| d\sigma_\Omega(z) . \tag{4.1}$$

Henceforth the brightness function of Ω is uniquely determined by the even part of σ_Ω. A theorem by A. D. Aleksandrov ensures that the converse of this result is also true.

A central result regarding area measures of convex bodies is the following:

THEOREM *(Aleksandrov-Minkowski). A finite non-negative Borel measure σ is the area measure of a convex body if and only if σ satisfies the following conditions:*
i) The support of σ is not entirely contained in a great circle of S^2;
ii)

$$\int_{S^2} z d\sigma(z) = 0 .$$

Moreover, if σ satisfies i) and ii), then there is a unique (up to translation) convex body having σ as area measure.

Based on this result is the definition of the *Blaschke sum* $r \cdot \Omega \# s \cdot \Gamma$ as the unique convex body whose area measure is $r\sigma_\Omega + s\sigma_\Gamma$, where Ω and Γ are non-degenerate convex bodies, and r and s are non-negative real numbers.

The Blaschke symmetrized body $B\Omega$ of Ω is defined by

$$B\Omega := \frac{1}{2} \cdot [\Omega \#(-\Omega)] ,$$

where $(-\Omega)$ denotes the symmetric body of Ω with respect to the origin. Clearly $B\Omega$ is centrally symmetric; moreover, according to (4.1), the brightness function of Ω and $B\Omega$ coincide. Consequently, if Ω has constant brightness, then $B\Omega$, being a centrally symmetric body with constant brightness, is a ball.

The behavior of the volume of the Blaschke sum can be studied by means of the Kneser-Süss inequality:

$$[\mathrm{vol}(t \cdot \Gamma \#(1-t) \cdot \Lambda)]^{\frac{2}{3}} \geq t\, [\mathrm{vol}(\Gamma)]^{\frac{2}{3}} + (1-t)\, [\mathrm{vol}(\Lambda)]^{\frac{2}{3}} , \tag{4.2}$$

where equality holds if and only if Γ and Λ are homothetic.

A straightforward consequence of this inequality is that the ball is the unique body of maximum volume and given constant brightness.

In the following we deal with bodies of minimum volume.

THEOREM 4.1. *Let Ω be a body of minimum volume in the class of all convex bodies having constant brightness b. Then σ_Ω has the following form*

$$\sigma_\Omega(\omega) = \int_\omega s_\Omega(z)dz\,, \quad \forall \omega \in \mathfrak{B}\,, \tag{4.3}$$

where s_Ω is a non-negative function in $L_1(S^2)$ such that

$$s_\Omega(z) + s_\Omega(-z) = \frac{2b}{\pi} \quad \text{a.e.} \quad \textit{in} \quad S^2\,, \tag{4.4}$$

$$s_\Omega(z) s_\Omega(-z) = 0 \quad \text{a.e.} \quad \textit{in} \quad S^2\,. \tag{4.5}$$

Proof. Let $\mathcal{L}$ denote the Lebesgue measure on S^2. Since $B\Omega$ is the ball of radius $\sqrt{\frac{b}{\pi}}$, we have

$$\sigma_\Omega(\omega) + \sigma_\Omega(-\omega) = \frac{2b}{\pi}\mathcal{L}(\omega)\,, \quad \forall \omega \in \mathfrak{B}\,, \tag{4.6}$$

where $-\omega$ is the antipodal set of ω. Due to the non-negativity of σ_Ω, (4.6) implies that σ_Ω is absolutely continuous with respect to $\mathcal{L}$. The existence of $s_\Omega \in L_1(S^2)$ is then a consequence of the Radon–Nikodym theorem. Besides, (4.4) follows from the Lebesgue–Besicovitch theorem on differentiability of absolutely continuous measures (see Evans and Gariepy (1992)).

Let us turn to (4.5). Let

$$\beta_n := \left\{ z \in S^2 \,|\, \frac{1}{n} \le s_\Omega(z) \le \frac{2b}{\pi} - \frac{1}{n} \right\}, \quad n = 1, 2, \ldots\,;$$

β_n is a symmetric measurable subset of S^2. Assume that $\mathcal{L}(\beta_{\bar n}) > 0$ for some $\bar n$. Call η the intersection of $\beta_{\bar n}$ with a fixed open hemisphere of S^2. Clearly $\mathcal{L}(\eta) > 0$. A function $g \in L_1(\eta)$ exists such that

$$\mathcal{L}(\{z \in \eta \,|\, |g(z)| > 0\}) > 0\,, \tag{4.7}$$

$$|g(z)| \le \frac{1}{\bar n} \quad \text{a.e.} \quad \text{in}\, \eta\,, \tag{4.8}$$

$$\int_\eta z g(z) dz = 0\,. \tag{4.9}$$

Next we introduce the functions s_i, $i = 1, 2$, defined on S^2 as follows

$$s_i(z) = \begin{cases} s_\Omega(z) + (-1)^i g(z) & \text{if} \quad z \in \eta \\ s_\Omega(z) - (-1)^i g(-z) & \text{if} \quad z \in -\eta \\ s_\Omega(z) & \text{elsewhere} \end{cases}\,.$$

Set

$$\sigma_i(\omega) = \int_\omega s_i(z)\mathrm{d}z$$

for every $\omega \in \mathfrak{B}$. Since $\eta \subset \beta_{\bar{n}}$, (4.8) implies that $\sigma_i(\omega) \geq 0$, $i = 1,2$, for every $\omega \in \mathfrak{B}$. Moreover, from (4.9)

$$\int_{S^2} z\mathrm{d}\sigma_i(z) = 0\,, \quad i = 1,2\,.$$

Therefore σ_1 and σ_2 are finite non–negative Borel measures satisfying the conditions of the Aleksandrov–Minkowski theorem.

Let Ω_i be the convex body such that $\sigma_{\Omega_i} = \sigma_i$, $i = 1,2$. The measures σ_1 and σ_2 have the same even part as σ_Ω, thus Ω_1 and Ω_2 have constant brightness b. Moreover, from the definition of s_1 and s_2

$$\Omega = \frac{1}{2}\cdot(\Omega_1 \# \Omega_2)\,.$$

Notice that by virtue of (4.7), Ω_1 and Ω_2 are not homothetic. Applying inequality (4.2) gives

$$\mathrm{vol}(\Omega) > \min\{\mathrm{vol}(\Omega_1), \mathrm{vol}(\Omega_2)\}\,,$$

which contradicts the fact that the volume of Ω is a minimum. This implies that $\mathcal{L}(\beta_n) = 0$ for every n, and then (4.5) follows. □

5. CONJECTURES.

The result stated in Theorem 4.1 is a first step to determine solutions to Problem C. It can be viewed as a condition on the Gaussian curvature of minimizers. Roughly speaking, if z and z' are antipodal points (i.e. where the outwards normal vectors are opposite) on the boundary of a convex body Ω of constant brightness b and minimum volume, then the Gaussian curvature of $\partial\Omega$ at z and z' equals $\frac{\pi}{2b}$ and infinity respectively. What is needed in order to characterize a minimizer Ω is the knowledge of the support of the distribution of σ_Ω, i.e. the image on S^2, under the Gauss map, of all the points of $\partial\Omega$ with finite curvature.

In the plane case, for the body of constant width and minimum area, i.e. for the Reuleaux triangle $\mathcal{R}$, the support of $\sigma_{\mathcal{R}}$ has the same automorphism group as the regular triangle. This suggests the following conjecture in the three–dimensional case. Consider three pairwise orthogonal great circles on S^2, they determine eight open regions. Define s as a piecewise constant function whose values are 0 and $\frac{2b}{\pi}$ on those regions alternately. Let $\bar{\Omega}$ be the unique convex body such that s is the distribution of its area measure. Clearly $\bar{\Omega}$ has constant brightness b, and the Gaussian curvature is $\frac{\pi}{2b}$ at each regular point of $\partial\bar{\Omega}$. We conjecture that $\bar{\Omega}$ has minimum volume among all convex bodies of constant brightness b.

The same idea leads to a similar conjecture in the case of bodies of revolution. Fix on S^2 a great circle as an equator and take the pair of parallel circles at a spherical distance $\frac{\pi}{4}$ from it. In such a way S^2 is divided into four open regions. Define s to be a piecewise constant function whose values are 0 and $\frac{2b}{\pi}$ on those regions alternately; s is the distribution of the area measure of the convex body of revolution with constant brightness constructed by Blaschke (1916). We conjecture that this body has minimum volume among all convex bodies of revolution with constant brightness b.

Analogously, starting from the Pál Theorem, the natural candidate for solving problem D is the simplex. This conjecture is supported also by the Zhang inequality mentioned in Section 3.

Finally, by taking Proposition 2.1 into account, we conjecture that the solution to problem D for bodies of revolution is a suitable cone.

REFERENCES

Besicovitch, A. S. (1963), *Minimum area of a set of constant width*, Proc. of Symposia in Pure Math. Vol. 7, Convexity, Amer. Math. Soc, 13–4.

Blaschke, W. (1915), *Konvexe Bereiche gegebener konstanter Breite und kleinsten Inhalts*, Math. Annalen **76**, 504-13.

Blaschke, W. (1916), *Kreis und Kugel*, (Veit, Leipzig), 2nd Ed. W. de Gruyter, Berlin, 1956.

Bonnesen, T., Fenchel, W. (1934), *Theorie der konvexen Körper*, Springer, Berlin.

Chakerian, G. D. (1966), *Sets of constant width*, Pacific J. Math. **19,1**, 13–21.

Chakerian, G. D., Groemer, H. (1983), *Convex bodies of constant width*, Convexity and its applications (P. Gruber, J. M. Wills, eds.), Birkhäuser, Basel, pp. 49–96.

Eggleston, H. G. (1952), *A proof of Blaschke's theorem on the Reuleaux triangle*, Quart. J. Math. Oxford Ser. **2**, **3**, 296–7.

Evans, L. C., Gariepy, R. F. (1992), *Measure theory and fine properties of functions*, CRC press, Boca Raton.

Firey, W. J. (1965), *The brightness of convex bodies*, Tech. Report 19, Oregon State Univ., Corvallis, Oregon.

Fujiwara, M. (1927, 1931), *Analytical proof of Blaschke's theorem on the curve of constant breadth with minimum area, I and II*, Proc. Imp. Acad. Japan **3 (1927)**, **7 (1931)**, 307–9 (1927), 300–2 (1931).

Gardner R. J. (1995), *Geometric tomography*, Cambridge University Press, New York.

Goodey, P., Weil, W. (1993), *Zonoids and generalizations*, Handbook of convex geometry (P. Gruber, J. M. Wills, eds.), North–Holland, Amsterdam, pp. 1297–326.

Gruber, P., Schneider, R. (1979), *Problems in geometric convexity*, Contributions to geometry (J. Tölke, J. M. Wills, eds.), Birkhäuser, Berlin, pp. 255–78.

Hadwiger, H. (1949), *Elementare Studie über konvexe Körper*, Math. Nachr. **2**, 114–23.

Heil, E. (1978), *Kleinste konvexe Körper gegebener Dicke*, Preprint no. 453, Technische Hochschule Darmstadt.

Heil, E., Martini, H. (1993), *Special convex bodies*, Handbook of convex geometry (P. Gruber, J. M. Wills, eds.), North–Holland, Amsterdam, pp. 347–85.

Hilbert, D., Cohn–Vossen, S. (1932), *Anschauliche Geometrie*, Springer, Berlin.

Kallay, M. (1974), *Reconstruction of a plane convex body from the curvature of its boundary*, Israel J. Math. **17**, 149–61.

Lebesgue, H. (1914), *Sur le problème des isopérimètres et sur le domaines de largeur constante*, Bull. Soc. Math. Franc., C.R., 72–76.

Mayer, A. E. (1934–35), *Der Inhalt der Gleichdicke*, Math. Annalen **110**, 97–127.

Martini, H. (1994), *Cross–sectional measures*, Colloquia Mathematica Societatis János Bolyai, vol. 63, Intuitive Geometry, North–Holland, Amsterdam, pp. 269–310.

Meissner, E. (1912), *Drei Gipsmodelle von Flächen konstanter Breite*, Z. Math. Phys. **60**, 92–4.

Pál, J. (1921), *Ein Minimumproblem für Ovale*, Math. Annalen **83**, 311-9.

Petty, C. M. (1967), *Projection bodies*, Proc. Colloquium Convexity, Copenhagen, 1965, Københavns Univ. Math. Inst., pp. 234–41.

Rogers, C. A., Shephard G. C. (1957), *The difference body of a convex body*, Arch. Math. **8**, 220–33.

Schneider, R. (1967), *Zu einem Problem von Shephard über die Projektionen konvexer Körper*, Math. Zeitschr. **101**, 71–82.

Schneider, R. (1993), *Convex bodies: the Brunn–Minkowski theory*, Cambridge University Press, Cambridge.

Schramm, O. (1988), *On the volume of sets having constant width*, Israel J. Math. **63**, **2**, 178–82.

Sholander, M. (1952), *On certain minimum problems in the theory of convex curves*, Trans. Amer. Math. Soc. **73**, 139–73.

Zhang, G. (1991), *Restricted chord projection and affine inequalities*, Geom. Dedicata **39**, 213–22.

7. On the Continuous Dependence of the Solution of a Linear Parabolic Partial Differential Equation on the Boundary Data and the Solution at an Interior Spatial Point

John R. Cannon Department of Mathematics, University of Central Florida, Orlando, Florida

Salvadore Perez-Esteva Instituto de Matemáticas, Universidad Nacional Autónoma de México, Mexico, D.F., Mexico

To Carlø, thank you for all the encouragement and help, Best Wishes.

John and Salvador.

ABSTRACT

We consider the equation $u_t = (1 + a(x,t))\, u_{xx} + b(x,t)u_x + c(x,t)u + f(x,t)$, $0 < x < 1$, $0 < t \leq T$, subject to the condition $u(0,t) = \varphi(t)$, $u(1,t) = \psi(t)$, $u(\xi,t) = g(t)$, $0 < t < T_m$, $T_m \leq T$, where ξ is an irrational number in $0 < x < 1$. Under the additional conditions that the $C^{2+\alpha,1+\alpha/2}$ norm of u is bounded by M, $0 < x < 1$, where M is a specified positive constant, we demostrate that u depends continouously upon the data f, φ, ψ, g and M provided that the coefficients a, b, and c tend to zero sufficiently fast as t tends to zero. An interesting subset of the analysis is an estimate of the L^p norm of the theta function for $1 \leq p < 3$.

[1]The research was supported in part by the National Science Foundation Grant # DMS - 9401847

1 INTRODUCTION

In [1], the problem

$$\begin{aligned} u_t &= u_{xx}, \quad 0 < x < 1, \qquad 0 < t \leq T, \\ u(0,t) &= \phi(t), \quad 0 < t \leq T, \\ u(1,t) &= \psi(t), \quad 0 < t \leq T, \\ u(\xi,t) &= g(t), \quad 0 < t \leq T_m \leq T, \\ |u(x,0)| &< M, \end{aligned}$$

where ϕ, ψ and g are known functions, M is a known positive constant, and $0 < \xi < 1$, was discussed. Here and in the following, $u_x = \frac{\partial u}{\partial x}$, $u_t = \frac{\partial u}{\partial t}$, $u_{xx} = \frac{\partial^2}{\partial x^2}$, etc. It was shown [3] that the solution is unique provided that ξ is an irrational number in $0 < x < 1$ and it was shown [1] that the solution could be estimated in terms of the data with continuous dependence following if ξ were irrational.

In this paper we consider the uniqueness and the continuous dependence of the solution $u = u(x,t)$ of the problem.

$$u_t = (1 + a(x,t))\, u_{xx} + b(x,t)u_x + c(x,t)u + f(x,t), \quad 0 < x < 1, \quad 0 < t \leq T, \tag{1.1}$$

$$u(0,t) = \varphi(t), \qquad 0 \leq t \leq T, \tag{1.2}$$

$$u(1,t) = \psi(t), \qquad 0 \leq t \leq T, \tag{1.3}$$

$$u(\xi,t) = g(t), \qquad 0 \leq t \leq T_m \leq T, \tag{1.4}$$

where ξ, $0 < \xi < 1$, is irrational $0 < T_m$, and $-1 < -\delta \leq a(x,t)$.

Under the additional assumption that the $C^{2+\alpha,1+\alpha/2}$ norm of u is bounded by a known positive constant M, we shall show that the solution of (1.1)–(1.4) depends continuously upon φ, ψ, g and f provided that the coefficients a, b, and c tend to zero sufficiently fast as t tends to zero. From this follows the unicity of any "bounded" solution of (1.1)–(1.4).

The paper is organized as follows. In Section 2 we define the necessary notations, list our assumptions and state our theorem, which is an a priori estimate of a "bounded" solution u of (1.1)–(1.4), and a corollary of the theorem which is the continuous dependence result.

An estimate of the L^p norm of the theta function, which is needed in our analysis, is presented in Section 3. Section 4 deals with the derivation of the a priori estimate and Section 5 deals with its application for continuous dependence.

2 NOTATION AND STATEMENTS OF RESULTS

Let $Q_T = (0,1) \times (0,T)$ and $\overline{Q}_T = [0,1] \times [0,T]$. For $0 < \alpha < 1$ and $R > 0$, we shall denote by $H^\alpha([0,R])$ the space of real valued functions h defined on $[0,R]$ such that

$$\|h\|_{H^\alpha([0,R])} = \sup_{0 \leq s \leq R} |h(s)| + \sup_{\substack{s,s' \in [0,R] \\ s \neq s'}} \frac{|h(s) - h(s')|}{|s - s'|} < \infty\,. \tag{2.1}$$

Likewise, $H^{1+\alpha}([0,R])$ is the space of continuously differentiable real valued functions defined on $[0,R]$ such that

$$\|h\|_{H^{1+\alpha}([0,R])} = \sup_{0 \leq s \leq 1} |h(s)| + \|h'\|_{H^\alpha([0,R])} < \infty\,. \tag{2.2}$$

For functions $v: \overline{Q}_T \to \mathbf{R}$, the real numbers, let

$$H^T_{x,\alpha}(v) = \sup_{\substack{x,x' \in [0,1] \\ x \neq x' \\ t \in [0,T]}} \frac{|v(x,t) - v(x',t)|}{|x - x'|^\alpha} \tag{2.3}$$

$$H^T_{t,\alpha}(v) = \sup_{\substack{x \in [0,1] \\ t \neq t' \\ t,t' \in [0,T]}} \frac{|v(x,t) - v(x,t')|}{|t - t'|^\alpha} \tag{2.4}$$

and

$$\|v\|_t = \sup_{(x,s) \in Q_t} |v(x,s)|, \qquad 0 < t \leq T. \tag{2.5}$$

We denote by $H^{\alpha,\alpha/2}(\overline{Q}_T)$ the space of functions $v: \overline{Q}_T \to \mathbf{R}$ such that

$$\|v\|_T^{\alpha,\alpha/2} = \|v\|_T + H^T_{x,\alpha}(v) + H^T_{t,\alpha/2}(v) < \infty. \tag{2.6}$$

Likewise, we denote by $H_{2+\alpha,1+\alpha/2}\left(\overline{Q}_T\right)$ the space of functions $v: \overline{Q}_T \to \mathbf{R}$ such that

$$\begin{aligned} \|v\|_T^{2+\alpha,1+\alpha/2} &= \|v\|_T + \|v_x\|_T + H^T_{t,\frac{1+\alpha}{2}}(v_x) + \|v_t\|_T^{\alpha,\alpha/2} + \\ &\quad + \|v_{xx}\|_T^{\alpha,\alpha/2} < \infty. \end{aligned} \tag{2.7}$$

It will be convenient to use

$$\|h\|_t = \sup_{s \in [0,t]} |h(s)| \tag{2.8}$$

for functions $h: [0,t] \to \mathbf{R}$ such as φ, ψ and g.

We state now our assumptions on the data and the solution u:

A1 These is a positive number $\delta < 1$ so that $-1 < -\delta \leq a(x,t)$ for $(x,t) \in \overline{Q}_T$;

A2 The functions a, b, c and f belong to $H^{\alpha,\alpha/2}\left(\overline{Q}_T\right)$;

A3 The functions φ and ψ belong to $H^{1+\alpha}([0,T])$;

A4 The solution $u \in H^{2+\alpha,1+\alpha/2}\left(\overline{Q}_T\right)$ and there exists a known positive constant M such that

$$\|u\|_T^{2+\alpha,1+\alpha/2} \leq M; \tag{2.9}$$

A5 As t tends to zero the functions $a, b,$ and c tend to zero in the following manner: Let

$$\mathcal{E}(t) = \|a\|_{L^q(Q_t)} + \|b\|_{L^q(Q_t)} + \|c\|_{L^q(Q_t)}, \tag{2.10}$$

where

$$\|h\|_{L^q(Q_t)} = \left[\int_0^t \int_0^1 |h(x,\tau)|^q \, dx \, d\tau\right]^{\frac{1}{q}} \tag{2.11}$$

for any function $h: Q_t \to \mathbf{R}$ and where $\frac{3}{2} < q \leq \infty$.

Then, we assume that

$$\lim_{t\to 0} \{t^{\alpha_p}\mathcal{E}(t)\}^{\exp\{-\frac{1}{t}\}} = 0\,, \tag{2.12}$$

where

$$\alpha_p = \frac{3}{2p} - \frac{1}{2} \tag{2.13}$$

with $1 \leq p < 3$.

We are now able to state our result.

THEOREM 1 *Assume that assumptions* A1, A2, A3 *and* A4 *hold. For each positive integer* N *and each positive* t *in* $0 < t < T_m$ *every solution* u *of* (1.1)–(1.4) *with* ξ *irrational,* $0 < \xi < 1$, *satisfies*

$$\begin{aligned} \|u\|_T \leq\ & \exp\{\lambda T\}\Big[C(t,N)\mu_1(t)^{\exp\{-\frac{1}{t}\}} + \\ & + C(t,N)\{M_5 t^{\alpha_p}\mathcal{E}(t)\}^{\exp\{-\frac{1}{t}\}} + \frac{2m_3}{\pi^2 N}\Big] + \mu_2(T)\,, \end{aligned} \tag{2.14}$$

where

$$\lambda = \|c\|_T + 1\,, \tag{2.15}$$

$$C(t,N) = M_6\left[\sum_{n=1}^{N} |\sin n\pi\xi|^{-1}\exp\left\{\frac{n^2\pi^2 t}{2}\right\}\right], \tag{2.16}$$

$$\begin{aligned} \mu_1(t) = \ & \|g\|_t + \exp\{\lambda t\}\,[\|\varphi\|_t + \|\psi\|_t + \|f\|_t + \\ & + M_2(|\varphi(0)| + |\varphi'(0)| + |\psi(0)| + |\psi'(0)|)]\,, \end{aligned} \tag{2.17}$$

$$\begin{aligned} \mu_2(T) = \ & \exp\{\lambda T\}\,[\|\varphi\|_T + \|\psi\|_T + \|f\|_T + \\ & + M_2(|\varphi(0)| + |\varphi'(0)| + |\psi(0)| + |\psi'(0)|)]\,, \end{aligned} \tag{2.18}$$

and the constants M_2, M_3 *and* M_6 *depend only on* M, T, p *and the* $H^{\alpha,\alpha/2}$ *norm of* a, b *and* c *and where* $1 \leq p < 3$, $\alpha_p = \frac{3}{2p} - \frac{1}{2}$, $\mathcal{E}(t)$ *is defined by 2.10 in assumption A5 and* q *is the Hölder conjugate of* p; *ie,*

$$\frac{1}{p} + \frac{1}{q} = 1\,. \tag{2.19}$$

COROLLARY 1 *If in addition to the assumption given in the theorem we add assumption A5, then any solution* u *of (1.1)–(1.4) is unique and depends continuously upon the data* φ, ψ, f *and* g.

Proof: The proof of the Theorem is given in section 4 and that of the Corollary is given in Section 5.

As the analysis utilizes the representation of the inhomogeneous heat equation, it is convenient to develop an estimate of the $L^p(Q_t)$ norm of the theta function in Section 3 for use in Section 4.

3 AN ESTIMATE OF THE $L^p(Q_t)$ NORM OF THE THETA FUNCTION

As stated above it is convenient to consider an estimate of

$$\Theta(x,t) = \sum_{m=-\infty}^{\infty} K(x+2m,t)\,, \tag{3.1}$$

where for $-\infty < x < \infty$,

$$K(x,t) = (4\pi t)^{-1/2} \exp\left\{-\frac{x^2}{4t}\right\}, \quad t > 0\,. \tag{3.2}$$

Now, for $-1 < x < 1$,

$$\begin{aligned}
\Theta(x,t) &= \sum_{m=-\infty}^{\infty} K(x+2m,t) \\
&= K(x,t) + \sum_{m=1}^{\infty} K(x-2m,t) + \sum_{m=1}^{\infty} K(x+2m,t) \\
&= K(x,t)\left\{1 + \sum_{m=1}^{\infty}\left(\exp\left\{\frac{xm-m^2}{t}\right\}\right.\right. \\
&\quad \left.\left. +\exp\left\{\frac{-xm-m^2}{t}\right\}\right)\right\} \\
&\le K(x,t)\left\{1 + \exp\left\{\frac{x-1}{t}\right\} + \exp\left\{\frac{-x-1}{t}\right\}\right. \\
&\quad +\exp\left\{\frac{2x-4}{t}\right\} + \exp\left\{\frac{-2x-4}{t}\right\} + \\
&\quad \left. +\sum_{m=3}^{\infty}\exp\left\{\frac{xm-m^2}{t}\right\} + \sum_{m=3}^{\infty}\exp\left\{\frac{-xm-m^2}{t}\right\}\right\} \\
&\le K(x,t)\left\{5 + 2\sum_{m=3}^{\infty}\exp\left\{\frac{-(m-1)^2}{t}\right\}\right\}
\end{aligned} \tag{3.3}$$

Since, for $-1 < x < 1$,

$$m^2 - xm \ge (m-1)\,m \ge (m-1)^2\,,$$

$$m^2 + xm \ge (m-1)\,m \ge (m-1)^2\,,$$

and $\exp\{-x\}$ is decreasing.

Using the inequality $\exp\{-x\} \le \frac{n!}{x^n}$, we see that

$$\Theta(x,t) \le K(x,t)\left\{5 + 2t\sum_{m=2}^{\infty}\frac{1}{m^2}\right\}. \tag{3.4}$$

Consequently,

$$\Theta(x,t) \le M_4 K(x,t)\,, \tag{3.5}$$

where M_4 depends only on T. Thus we obtain

$$\Theta(\xi - y, t - \tau) \leq M_4 K(\xi - y, t - \tau) \tag{3.6}$$

and

$$\Theta(\xi + y, t - \tau) \leq M_4 K(\xi + y, t - \tau) \tag{3.7}$$

As $\xi + y \geq \xi - y$, for $y > 0$ it follows that

$$K(\xi + y, t - \tau) \leq K(\xi - y, t - \tau)$$

Hence, we obtain the estimate

$$|\Theta(\xi - y, t - \tau) \pm \Theta(\xi + y, t - \tau)| \leq 2M_4 K(\xi - y, t - \tau). \tag{3.8}$$

We are now in a position to estimate the L^p norm of

$$\{\Theta(\xi - y, t - \tau) \pm \Theta(\xi + y, t - \tau)\}.$$

LEMMA 1 *Let $0 < x < 1$ and $1 \leq p < 3$. There exists a constant C which depends only upon, p and t as t tends to infinity such that*

$$\|\Theta(x - \cdot, t - \cdot) \pm \Theta(x + \cdot, t - \cdot)\|_{L^p(Q_t)} \leq Ct^{\alpha_p}. \tag{3.9}$$

where $\alpha_p = \frac{3}{2p} - \frac{1}{2}$.

Proof: From (3.6), we see that

$$\|\Theta(x - \cdot, t - \cdot) \pm \Theta(x + \cdot, t - \cdot)\|_{L^p(Q_t)} \leq 2M_4 \|K(x - \cdot, t - \cdot)\|_{L^p(Q_t)}$$

$$\begin{aligned} &\leq 2M_4 \left(\int_0^t \int_0^1 K(x - y, t - \tau)^p dy d\tau \right)^{\frac{1}{p}} \\ &\leq \frac{M_4}{\sqrt{p}\,(4\pi)^{\frac{p-1}{2}}} \left(\int_0^t \frac{1}{(t - \tau)^{\frac{p-1}{2}}} \int_{-\infty}^{\infty} K(x - y, \frac{t - \tau}{p}) dy d\tau \right)^{\frac{1}{p}} \\ &= Ct^{\alpha_p}, \end{aligned} \tag{3.10}$$

where

$$C = 2M_4 p^{-1/2} (4\pi)^{1/2 - p/2} \left(\frac{3}{2} - \frac{p}{2} \right)^{-\frac{1}{p}}. \tag{3.11}$$

■

4 PROOF OF THE THEOREM

Let $h = h(x)$ be a fixed smooth function on $[0,1]$ which assumes the value one in a neighborhood of zero and $h(1) = 0$. Define

$$\tilde{f}(x) = \left(\varphi(0) + \frac{\varphi'(0)}{2}x^2\right) h(x) + \left(\psi(0) + \frac{\psi'(0)}{2}(1-x)^2\right) h(1-x). \tag{4.1}$$

Let $w = w(x,t)$ denote the solution of (1.1)–(1.3) with $w(x,0) = \tilde{f}(x)$. From Ladyženskaya, Solonnikov and Ural'ceva [4, Chapter IV, Theorem 5.2 p. 320] we see that $w(x,t)$ exists, is unique, and belongs to $H^{2+\alpha,1+\alpha/2}(\overline{Q}_T)$ as the data φ, ψ and $\tilde{f}$ satisfy the compatibility conditions of the first order. Moreover,

$$\|w\|_T^{2+\alpha,1+\alpha/2} \le M_1\,, \tag{4.2}$$

where M_1 depends, upon M, T, and the upper bounds on the norm $\|\ \|_T^{\alpha,\alpha/2}$ of the data a, b, c and f. By the maximun principle, we see that

$$\begin{aligned} |w(x,t)| \ \le\ & e^{\lambda t}\{\|\varphi\|_t + \|\psi\|_t + \|f\|_t + \\ & +M_2\left(|\varphi(0)| + |\varphi'(0)| + |\psi(0)| + |\psi'(0)|\right)\} \end{aligned} \tag{4.3}$$

where M_2 depends upon h and where we select

$$\lambda = \|c\|_t + 1\,. \tag{4.4}$$

Thus, in order to estimate u it suffices to consider

$$v = u - w\,, \tag{4.5}$$

where v satisfies

$$\begin{aligned} v_t &= (1 + a(x,t))\,v_{xx} + b(x,t)v_x + c(x,t)v, \\ &\qquad 0 < x < 1, 0 \le t \le T\,, && (4.6) \\ v(0,t) &= 0, \quad 0 < t \le T\,, && (4.7) \\ v(1,t) &= 0, \quad 0 < t \le T\,, && (4.8) \\ v(\xi,t) &= \tilde{g}(t), \quad 0 < t \le T_m\,, && (4.9) \end{aligned}$$

and

$$\tilde{g}(t) = g(t) - w(\xi,t)\,. \tag{4.10}$$

By assumption A4 and (4.2), it follows that

$$\|v\|_T^{2+\alpha,1+\alpha/2} \le M_3\,, \tag{4.11}$$

where

$$M_3 = M + M_1\,.$$

Next, we write

$$v_t = v_{xx} + \phi(x,t)\,,$$

where

$$\phi(x,t) = a(x,t)v_{xx} + b(x,t)v_x + c(x,t)v\,. \tag{4.12}$$

Utilizing the representation for the inhomogeneous heat equation [2, Theorem 19.3.4, p. 339] we see that

$$\begin{aligned} v(x,t) &= \sum_{n=1}^{\infty} a_n e^{-n^2\pi^2 t} \sin n\pi x + \\ &+ \int_0^t \int_0^1 \{\Theta(x-y,t-\tau) - \Theta(x+y,t-\tau)\}\,\phi(y,\tau)dy\,d\tau, \end{aligned} \tag{4.13}$$

where

$$a_n = 2\int_0^1 v(x,0)\sin n\pi x\,dx\,, \tag{4.14}$$

As $|v_{xx}(x,0)| \leq M_3$ it follows from (4.7), (4.8) and integration by parts that

$$|a_n| \leq \frac{2M_3}{(\pi n)^2}\,. \tag{4.15}$$

An estimate of v will follow from the estimation of the a_n via consideration of the equation

$$\tilde{g}(t) = \sum_{n=1}^{\infty} a_n e^{-n^2\pi^2 t} \sin n\pi\xi + P(t) \tag{4.16}$$

where

$$P(t) = \int_0^t \int_0^1 \{\Theta(\xi-y,t-\tau) - \Theta(\xi+y,t-\tau)\}\,\phi(y,t)dy\,d\tau\,. \tag{4.17}$$

From Hölder's inequality, we see that

$$|P(t)| \leq \|\Theta(\xi-\cdot,t-\cdot) - \Theta(\xi+\cdot,t-\cdot)\|_{L^p(Q_t)}\,\|\phi\|_{L^q(Q_t)}\,. \tag{4.18}$$

Using Lemma 1, (4.11) and (4.12) it follows from (4.18) that

$$|P(t)| \leq M_5 t^{\alpha_p}\mathcal{E}(t) \tag{4.19}$$

where M_5 depends only upon M, T, p and upper bounds on the $\|\ \|_T^{\alpha,\alpha/2}$ norms of the data a, b, c, and f, and where $\mathcal{E}(t)$ is defined in (2.10).

Recalling (4.16), we see that from (4.3) and (4.10) it follows that

$$\left|\sum_{n=1}^{\infty} a_n \exp\left\{-n^2\pi^2 t\right\} \sin n\pi\xi\right| \leq |\tilde{g}(t)| + |P(t)| \leq \mu(t)\,, \tag{4.20}$$

where

$$\begin{aligned} \mu(t) &= \|g\|_t + e^{\lambda t}\{\|\varphi\|_t + \|\psi\|_t + \|f\|_t + \\ &\quad M_2\left(|\varphi(0)| + |\varphi'(0)| + |\psi(0)| + |\psi'(0)|\right)\} + M_5 t^{\alpha_p}\mathcal{E}(t)\,, \end{aligned} \tag{4.21}$$

$\lambda = \|c\|_T + 1$, M_2 depends on $h(x)$ (see (4.1)), and M_5 depends only upon M, T, p and upper bounds on the $\| \ \|_T^{\alpha,\alpha/2}$ norms of the data a, b, c, and f. Thus, we are led to study the analytic function

$$F(\eta + i\tau) = \sum_{n=1}^{\infty} a_n \exp\left\{-n^2\pi^2(\eta + i\tau)\right\} \sin n\pi\xi \tag{4.22}$$

for $\eta > 0$ and $-\infty < \tau < \infty$.

First, we note that $F(\eta + i\tau)$ is periodic with period $\frac{2i}{\pi}$ and we follow the analysis in [1]. Also, $F(\eta + i\tau)$ is bounded since

$$\begin{aligned} |F(\eta + i\tau)| &= \sum_{n=1}^{\infty} |a_n| \exp\left\{-n^2\pi^2\eta\right\} \\ &\leq \frac{2M_3}{\pi^2}\sum_{n=1}^{\infty}\frac{1}{n^2} < \frac{4M_3}{\pi^2} = M_6 \end{aligned} \tag{4.23}$$

Next, we fix t in $0 < t \leq T_m$. From (4.20) we see that

$$|F(\eta)| \leq \mu(t)\,, \qquad \frac{1}{4}t \leq \eta \leq \frac{3}{4}t\,, \tag{4.24}$$

as $\mu(t)$ is a monotone non-decreasing function.

Applying a trivial modification of Lemma 10.6.2 of [2], it follows that for $-\frac{1}{\pi} \leq \tau \leq \frac{1}{\pi}$,

$$\begin{aligned} \left|F\left(\frac{t}{2} + i\tau\right)\right| &\leq M_6^{1-\exp\{-\frac{\pi\tau}{t}\}} \quad \mu(t)^{\exp\{-\frac{\pi\tau}{t}\}} \\ &\leq M_6\ \mu(t)^{\exp\{-\frac{1}{t}\}}\,. \end{aligned} \tag{4.25}$$

Considering

$$G(\zeta) = \sum_{n=1}^{\infty} (a_n \sin n\pi\xi)\, \zeta^{n^2}\,, \tag{4.26}$$

we see that for

$$\zeta = \exp\left\{-\pi^2\left(\frac{t}{2} + i\tau\right)\right\}, \qquad -\frac{1}{\pi} \leq \tau \leq \frac{1}{\pi}\,,$$

$$G(\zeta) = F\left(1/2 + i\tau\right)\,.$$

Hence,

$$|G(\zeta)| < M_6\mu(t)^{\exp\{-\frac{1}{t}\}} \tag{4.27}$$

for all ζ on $|\zeta| = \exp\left\{-\frac{\pi^2 t}{2}\right\}$.

As

$$a_n \sin n\pi\xi = \frac{G^{(n^2)}(0)}{n^2!} \tag{4.28}$$

$$= \frac{1}{2\pi i}\oint \frac{G(\zeta)}{\zeta^{n^2+1}}\, d\zeta \qquad |\zeta| = \exp\left\{-\frac{\pi^2 t}{2}\right\}, \tag{4.29}$$

it follows from (4.25) that

$$|a_n| \leq M_6 |\sin n\pi\xi|^{-1} \exp\left\{\frac{n^2\pi^2 t}{2}\right\} \mu(t)^{\exp\{-\frac{1}{t}\}}\,. \tag{4.30}$$

We complete the estimate of $v(x,0)$ via (4.13), (4.15) and (4.30). Namely,

$$\begin{aligned}|v(x,0)| &\leq \sum_{n=1}^{N}|a_n| + \sum_{n=N+1}^{\infty}\frac{2M_3}{(\pi n)^2}\\ &\leq M_6\left\{\sum_{n=1}^{N}|\sin n\pi\xi|^{-1}\exp\left\{\frac{n^2\pi^2 t}{2}\right\}\right\}\mu(t)^{\exp\{-\frac{1}{t}\}} + \frac{2M_3}{\pi^2 N}\\ &= C(t,N)\mu(t)^{\exp\{-\frac{1}{t}\}} + \frac{2M_3}{\pi^2 N},\end{aligned} \tag{4.31}$$

where

$$C(t,N) = M_6\left[\sum_{n=1}^{N}|\sin n\pi\xi|^{-1}\exp\left\{\frac{n^2\pi^2 t}{2}\right\}\right]. \tag{4.32}$$

Using the inequality $(a+b)^\beta \leq a^\beta + b^\beta$ for a and b positive numbers and $0<\beta<1$, we can separate $M_5 t^{\alpha_p}\mathcal{E}(t)$ from the $\mu(t)$ in (4.21) to obtain for $0 \leq x \leq 1$, $0 < t < T_m$, and each positive integer N,

$$\begin{aligned}|v(x,0)| \leq\ & C(t,N)\mu_1(t)^{\exp\{-\frac{1}{t}\}} + C(t,N)\left\{M_5 t^{\alpha_p}\mathcal{E}(t)\right\}^{\exp\{-\frac{1}{t}\}} +\\ & + \frac{2M_3}{\pi^2 N},\end{aligned} \tag{4.33}$$

where

$$\begin{aligned}\mu_1(t) =\ & \|g\|_t + e^{\lambda t}\{\|\varphi\|_t + \|\psi\|_t + \|f\|_t +\\ & + M_2\left(|\varphi(0)| + |\varphi'(0)| + |\psi(0)| + |\psi'(0)|\right)\}.\end{aligned} \tag{4.34}$$

By the maximun principle,

$$\begin{aligned}\|v\|_T \leq\ & e^{\lambda T}\left\{C(t,N)\mu_1(t)^{\exp\{-\frac{1}{t}\}} +\right.\\ & \left. + C(t,N)\left\{M_5 t^{\alpha_p}\mathcal{E}(t)\right\}^{\exp\{-\frac{1}{t}\}} + \frac{2M_3}{\pi^2 N}\right\}\end{aligned} \tag{4.35}$$

for each positive integer N and each positive t in $0<t<T_m$. As

$$\|u\|_T \leq \|v\|_t + \|w\|_T,$$

it follows from (4.3) that

$$\begin{aligned}\|u\|_T \leq\ & e^{\lambda T}\left[C(t,N)\mu_1(t)^{\exp\{-\frac{1}{t}\}}C(t,N)\left\{M_5 t^{\alpha_p}\mathcal{E}(t)\right\}^{\exp\{-\frac{1}{t}\}} +\right.\\ & \left. + \frac{2M_3}{\pi^2 N}\right] + \mu_2(T),\end{aligned} \tag{4.36}$$

where

$$\begin{aligned}\mu_2(T) =\ & e^{\lambda T}\{\|\varphi\|_T + \|\psi\|_T + \|f\|_T +\\ & + M_2\left(|\varphi(0)| + |\varphi'(0)| + |\psi(0)| + |\psi'(0)|\right)\}\end{aligned} \tag{4.37}$$

and where λ is defined by (4.4). This completes the proof of the main theorem. ■

5 CONTINUOUS DEPENDENCE

As a corollary of the main theorem we show that u depends continuously upon the data φ, ψ, f and g within the class of solutions v for which $|u|_T^{2+\alpha,1+\alpha/2}$, is bounded by $M/2$. To accomplish this we need only to show that $\|u\|_T$ tends to zero in (2.14) as the sup norms of the data φ, ψ, f and g and $|\varphi'(0)|$ and $|\psi'(0)|$ tend to zero since the difference of any two solutions u_1 and u_2 of the class above satisfies the bound $\|u_1 - u_2\|_T^{2+\alpha,1+\alpha/2} \leq M$.

Given an $\varepsilon > 0$, we can select N sufficiently large so that

$$\frac{2M_3}{\pi^2 N} < \varepsilon/4. \tag{5.1}$$

for such an N selected and fixed, we render $C(t,N)$ fixed by setting $t = T_m$. We consider the third term in (2.14) which is bounded by

$$\exp\{\lambda T\}C(T_m,N)M_5\left\{t^{\alpha_p}\mathcal{E}(t)\right\}^{\exp\{-\frac{1}{t}\}}.$$

As $\{t^{\alpha_p}\mathcal{E}(t)\}^{\exp\{-\frac{1}{t}\}}$ tends to zero as $t \to 0$ via assumption A5, we can select a $t_* > 0$ yet sufficiently small so that

$$\exp\{\lambda T\}\, C(T_m,N)M_5\left\{t_*^{\alpha_p}\mathcal{E}(t_*)\right\}^{\exp\{-\frac{1}{t_*}\}} < \frac{\varepsilon}{4}. \tag{5.2}$$

With $t_* > 0$ fixed the second term in (2.14) is bounded by

$$\exp\,\{\lambda T\}\, C(T_m,N)\mu_1(T_m)^{\exp\{-\frac{1}{t_*}\}}. \tag{5.3}$$

For the norms $\|g\|_{T_m}$, $\|\varphi\|_T$, $\|\psi\|_T$, and $\|f\|_T$ and the values $|\varphi(0)|$, $|\varphi'(0)|$, $|\psi(0)|$ and $|\psi'(0)|$ sufficiently small

$$\exp\,\{\lambda T\}\, C(T_m,N)\mu_1(T_m)^{\exp\{-\frac{1}{t_*}\}} < \frac{\varepsilon}{4} \tag{5.4}$$

and

$$\mu_2(T) < \frac{\varepsilon}{4}. \tag{5.5}$$

Consequently,

$$\|u\|_T < \varepsilon. \tag{5.6}$$

As ε is any arbitrary positive number it follows that $\|u\|_T$ tends to zero as the data g, φ, ψ, and f tend to zero uniformly and the values $|\varphi'(0)|$ and $|\psi'(0)|$ tend to zero provided that the coefficients a, b, and c tend to zero sufficiently fast in the $L^q(Q_t)$ norm as t tends to zero, where q is the Hölder conjugate to p, $1 \leq p < 3$, which implies the range for q as $\frac{3}{2} < q \leq \infty$. ■

REFERENCES

[1] J. R. Cannon and R. E. Klein, Optimal selection of measurement locations in a conductor for approximate determination of temperature distributions, *Journal of Dynamic Systems, Measurement and Control, Vol. 93* Series G (1971), 193–199.

[2] J. R. Cannon, *The One-Dimensional Heat Equation*, Encyclopedia of Mathematics Series, Volume 23, Addison-Wesley, Reading, MA, 1984.

[3] R. E. Klein, *An investigation of Distributed System Observability from a Continuum Physics Viewpoint*, Ph.D Dissertation, Purdue University 1969.

[4] O. A. Ladyženskaja, V. A. Solonikov and N. N. Ural'ceva, *Linear and Quasilinear Equations of Parabolic Type*, Volume 23, Translations of Mathematical Monographs, American Mathematical Society, Providence, Rhode Island, 1968.

8. Decomposability of Rectangular and Triangular Probability Distributions

Giorgio Dall'Aglio Dipartimento di Statistica, Università di Roma la Sapienza, Rome, Italy

0 Introduction

A chapter in Probability Theory deals with the decomposability of probability distributions. A relevant topic is constituted by infinitely decomposable (or divisible) distributions, introduced by de Finetti (1929) in order to characterize the class of the limit distributions in the general central limit problem. And is well known the property of some important types of distributions, as the normal the Poisson and the binomial ones, to be closed under decomposition.

Also interesting is the more basic problem of decomposability, namely the property of a distribution to be the convolution of two other distributions (see Linnik and Ostrovskii, 1977).

In this field, the paper deals with the problem of the decomposability of the rectangular and triangular distributions, both discrete and continuous.

As for the discrete rectangular distribution, for which the characterization of the decomposability is known, some result are given for the structure of the components. This allows to present and study an infinite decomposition of the continuous rectangular distribution, which generalizes the well known one arising from the coin tossing problem.

For the discrete triangular distributions, which are convolution of two rectangular distributions, the search is for components which are not components of the corresponding rectangular distribution, nor their composition. It is shown that, except for some specified small values of the length of the triangular distribution, such components always exist. The same is true for the continuous triangular distribution.

1 Generalities

We will consider only one-dimensional random variables and distribution functions. The distribution functions will be taken left-continuous, so

Research under contract with Ministero Universita' e Ricerca - Italy

that a d.f. F will be a real, non-decreasing, left continuous function F(x), $x \in R$, with $\lim_{x\to-\infty} F(x) = 0$ and $\lim_{x\to+\infty} F(x) = 1$, and the d.f. of a real uni-dimensional random variable X is given by $F(x) = P(X<x)$.

A d.f. F is said decomposable (or divisible) when there exist two non-degenerate d.f.'s F_1 and F_2 such that

$$F(x) = F_1 * F_2(x) = \int_R F_1(x-y)\, dF_2(y)$$

A r.v. X is said decomposable when its d.f. is decomposable. This is equivalent to say that there are two independent non-constant r.v.'s Y_1, Y_2 such that

$$X \stackrel{d}{=} Y_1 + Y_2$$

where "$\stackrel{d}{=}$" means "equal in distribution".

* * *

A discrete rectangular (or uniformly distributed) random variable is a r.v. assuming with equal probabilities a finite number of equi-intervalled values. We will consider r.v.'s having span equal to 1 and minimum value equal to 0, and denote such a r.v. by $U^{(n)}$, i.e.

$$P(U^{(n)} = j) = \frac{1}{n}, \quad j = 0, 1, 2, \ldots, n-1$$

We will denote by $P_X(z)$ the probability generating function of a r.v. X taking only non-negative integer values:

$$P_X(z) = E\, z^X = \sum_{r=0}^{+\infty} z^r\, P(X = r), \quad |z| \leq 1$$

As a consequence

$$P_{U^{(n)}}(z) = \sum_{r=0}^{n-1} z^r \frac{1}{n} = \frac{1}{n} \frac{z^n - 1}{z-1}$$

In order to simplify the formulas, we will usually take the probability generating function of a discrete rectangular distribution multiplied by the length of the distribution, and denote it by G_n:

$$G_n(z) = n\, P_{U^{(n)}}(z) = \frac{z^n-1}{z-1} \tag{1}$$

* * *

A continuous rectangular (or uniformly distributed) random variable is a r.v. taking values in an interval, with probabilities of sub-intervals proportional to their lengths. We will denote by U a continuous rectangular r.v. over (0,1). By an abuse of notation (which, given the context, should not create confusion) also the corresponding d.f. will be denoted by U :

$$U(x) = \begin{cases} 0 & x \le 0 \\ x & 0 < x \le 1 \\ 1 & 1 < x \end{cases}$$

For such a r.v. the probability generating function is not defined; we will occasionally use the characteristic function

$$H_U(z) = E\, e^{izU} = \int_R e^{izx}\, dU(x) = \frac{e^{iz} - 1}{iz}$$

In the sequel the r.v.'s appearing in a sum will always be taken independent, so that the distribution function of the sum is given by the convolution.

* * *

We remark that in the decomposition of (discrete or continuous) rectangular distributions the cancellation property holds, i.e.

$$U \overset{d}{=} X + Y \ , \ U \overset{d}{=} X + Z \quad \Rightarrow \quad Y \overset{d}{=} Z$$

(and the same for $U^{(n)}$). This follows from the form of the characteristic function: if

$$H_X(z)H_Y(z) = H_U(z) = \frac{e^{iz} - 1}{iz}$$

the right hand side vanishes only for countably many values, and the same is true for $H_X(z)$. Therefore except for these values, and, by continuity, for all z ,

$$H_Y(z) = \frac{e^{iz} - 1}{iz} \left(H_X(z)\right)^{-1}$$

is uniquely determined.

The same argument holds for the characteristic function (or the probability generating function) of $U^{(n)}$.

2 Decomposition of discrete rectangular distributions

If the r.v. $U^{(N)}$ is decomposable, i.e.

$$U^{(N)} \overset{d}{=} X + Y \ , \quad X \text{ and } Y \text{ independent}$$

the r.v.'s X and Y may assume only integer values, except for translations. Without loss of generality X and Y can be taken starting from zero, i.e. assuming 0 as minimum value. We may put

$$\alpha_r = P(X = r) \ ; \quad r = 0,1,\dots,m-1$$

$$\beta_s = P(Y = s) \ ; \quad s = 0,1,\dots,n-1$$

In these conditions instead of the convolution for d.f.'s it is simpler to adopt the convolution for discrete distributions; and the conditions of decomposability of $U^{(N)}$ is given by

$$\sum_{0,\, j-n+1}^{m-1,\, j} {}_r\, \alpha_r \beta_{j-1} = P\,(U^{(N)} = j) = \frac{1}{N} \;; \quad j = 0, 1, \ldots, N-1 \tag{2}$$

and it follows immediately

$$\begin{aligned} m + n &= N - 1 \\ \alpha_0 \beta_0 &= \frac{1}{N} \neq 0 \\ \alpha_{m-1} \beta_{n-1} &= \frac{1}{N} \neq 0 \end{aligned} \tag{3}$$

The decomposability condition becomes, through the probability generating functions,

$$P_{U^{(N)}}(z) = P_1(z) P_2(z) = \left[\sum_0^{m-1} {}_r\, \alpha_r z^r\right] \left[\sum_0^{n-1} {}_s\, \alpha_s z^s\right]$$

or, if $a_r = \alpha_r / \alpha_0$, $b_s = \beta_s / \beta_0$

$$G_N(z) = \frac{1}{\alpha_0 \beta_0} P_1(z)\, P_2(z) = \left[\sum_0^{m-1} {}_r\, a_r z^r\right] \left[\sum_0^{n-1} {}_s\, b_s z^s\right] \tag{4}$$

where G_X and G_Y are polynomials of degree resp. m-1 and n-1 ($m, n > 1$, $m + n = N - 1$) with non-negative coefficients and the constant term equal to 1.

* * *

Simple calculations show that

$$U_1^{(r)} + r\, U_2^{(s)} \stackrel{d}{=} U^{(rs)} \tag{5}$$

so that if N is not prime $U^{(N)}$ is decomposable. It is known that the converse is also true:

THEOREM 1

i) A discrete rectangular distribution of length N is decomposable iff N is not prime

ii) If $U^{(N)} = X + Y$, X and Y independent, then X and Y are symmetrical, i.e.

$$\alpha_r = P(X = r) = P(X = m - 1 - r) = \alpha_{m-1-r}$$

$$\beta_s = P(Y = s) = P(Y = n - 1 - s) = \beta_{n-1-s}$$

iii) $\alpha_r = 0 \text{ or } \alpha_0 \quad ; \quad \beta_s = 0 \text{ or } \beta_0$

The proof can be found in Linnik and Ostrovskii (1977),III. 3. ■

The simple relation (5) allows an interesting development. If N is the product of k natural numbers,

$$N = \prod_{1}^{k} {}_r\, n_r$$

iterated application of (5) gives

$$U^{(N)} \stackrel{d}{=} U^{(n_1)} + n_1 U_2^{(n_2)} + n_1 n_2 U_3^{(n_3)} + n_1 n_2 \ldots n_{k-1} U_k^{(n_k)} \qquad (6)$$

* * *

For the study of the decomposition of $U^{(N)}$ the following result is useful

LEMMA 1.

$$G_N(z) = (1+z) \prod_{1}^{\frac{N-2}{2}} {}_r\, (1 + c_r z + z^2) \qquad N \text{ even}$$

$$= \prod_{1}^{\frac{N-1}{2}} {}_r\, (1 + c_r z + z^2) \qquad N \text{ odd} \qquad (7)$$

where

$$c_r = -2\cos \frac{2r\pi}{N} \qquad (8)$$

Proof. By the representation (1) the zeros of G_N are

$$z_r = \cos \frac{2r\pi}{N} + i \sin \frac{2r\pi}{N} \quad ; \qquad r = 1, 2, \ldots, N-1$$

since for $r = 0$ we get $z_r = 1$, which accounts for the denominator in (1) (remember that G_N has degree $N - 1$). So

$$G_N(z) = \prod_{1}^{N-1} {}_r\, (z - z_r)$$

If N is odd the $N - 1$ zeros are complex conjugated by pair; more precisely $z_{N-r-1} = \bar{z}_r$. By multiplying the factors $(z - z_r),(z - z_{N-r-1})$ and remebering that $|z_r| = 1$ we obtain the result in (7).

If N is even, for $r = N/2$ it is $z_r = -1$, which gives the factor $(1 + z)$, and with the same argument as above the lemma is proved. ■

Each factor which appears in Lemma 1 gives a component of G_N iff $c_r \geq 0$ for all r and also the product of the remaining factors has non-negative coefficients. With $N > 4$, for $r = 1$ (8) gives an angle smaller than $\pi/2$, and thus a negative value for c_r ; so at least a factor in (7) is not a component of G_N.

EXAMPLE 1. When $N = 2^n$, starting from Lemma 1 we have

$$G_{2^n}(z) = (1+z) \prod_{r=1}^{2^{n-1}-1} \left(1 - 2z \cos\frac{r\pi}{2^{n-1}} + z^2\right)$$

$$= (1+z) \prod_{h=1}^{n-1} \prod_{k=0}^{2^{h-1}-1} \left(1 - 2z\cos \frac{(2k+1)\pi}{2^h} + z^2\right) = (1+z) \prod_{h=1}^{n-1} K_h(z)$$

where

$$K_h(z) = \prod_{k=0}^{2^{h-1}-1} \left(1 - 2z \cos \frac{(2k+1)\pi}{2^h} + z^2\right) =$$

$$= \prod_{k=0}^{2^{h-2}-1} \left(1 - 2z\cos \frac{(2k+1)\pi}{2^h} + z^2\right) \prod_{2^{h-2}}^{2^{h-1}-1} \left(1 - 2z\cos \frac{(2k+1)\pi}{2^h} + z^2\right) =$$

$$= \prod_{k=0}^{2^{h-2}-1} \left[\left(1 - 2z\cos \frac{(2k+1)\pi}{2^h} + z^2\right)\left(1 - 2z\cos \frac{(2^h-2k-1)\pi}{2^h} + z^2\right)\right] =$$

$$= \prod_{k=0}^{2^{h-2}-1} \left[\left(1 - 2z\cos \frac{(2k+1)\pi}{2^h} + z^2\right)\left(1 + 2z\cos \frac{(2k+1)\pi}{2^h} + z^2\right)\right] =$$

$$= \prod_{k=0}^{2^{h-2}-1} \left[\left(1 - 2z^2\cos \frac{(2k+1)\pi}{2^{h-1}} + z^4\right) = K_{h-1}(z^2)\right.$$

From that, since

$$K_1(z) = 1 + z^2$$

we get

$$K_h(z) = 1 + z^{2^h}$$

and finally

$$P_{2^n}(z) = 2^{-n}\, G_{2^n}(z) = \prod_{h=0}^{n-1} \frac{1 + z^{2^h}}{2}$$

The factor in the product on the right hand side is the probability generating function of $2^h U^{(2)}$; so the factorization we arrived to is the one given by (6) with $N = 2^n$ and all the n_r's equal to 2 . ■

3 The case of continuous rectangular distributions

We can arrive to the continuous rectangular distribution starting from the discrete one.

Dividing (6) by N and renumbering the indexes we get

$$\frac{U^{(N)}}{N} \stackrel{d}{=} \frac{U_1^{(n_1)}}{n_1} + \frac{U_2^{(n_2)}}{n_1 n_2} + \ldots + \frac{U_k^{(n_k)}}{n_1 n_2 \ldots n_k} \tag{9}$$

It is easily seen that passing from N to Nn_{k+1} amounts to add a term to the right hand side of (9). Now when N goes to infinity the r.v. on the left hand side converges in distribution to the continuous rectangular r.v. U. Thus we get

$$U \stackrel{d}{=} \sum_{1}^{+\infty}{}_r \frac{U_r^{(n_r)}}{n_1 n_2 \; n_r} \tag{10}$$

Formula (10) is a generalization of the well known decomposition

$$U = \sum_{1}^{+\infty}{}_r \frac{U_r^{(2)}}{2^r} \tag{11}$$

arising from the coin tossing problem.

It is interesting to remark that the right hand side of (11) represents a random number in the interval [0, 1] in binary numeration. In other words (11) states that a random number chosen uniformly in the interval [0, 1] can be obtained by independent random choices among 0 and 1 for every digit in the binary representation.

In this optic (10) could be seen as a random number in [0, 1] obtained again with independent choices of a sequence of digits; but each digit can be chosen on a different basis of numeration.

* * *

A surprising consequence of (11) is that the continuous rectangular distribution is the convolution of two Cantor-type singular distributions (i.e. continuous distributions concentrated on a set of zero Lebesgue measure). The decomposition arises considering separately the sums of even-indexed and odd-indexed terms (see Feller (1966), V, 4, Ex. d).

The same result can be obtained for the decomposition (10). More precisely:

THEOREM 2.

Let $U = X + Y$, with

$$X = \sum_{r \in I} \frac{U_r^{(n_r)}}{n_1 n_2 \ldots n_r}$$
$$Y = \sum_{r \in I^c} \frac{U_r^{(n_r)}}{n_1 n_2 \ldots n_r} \tag{12}$$

where I is a set of natural numbers and I^c its complement, I and I^c non-empty. Then

i) If I is finite, X takes a finite number of values, all with the same probability, and Y is absolutely continuous, with density constant on a finite set of intervals and zero elsewhere.

ii) if both I and I^c are infinite, X and Y are singular.

Proof. Take I finite, and $k = \max\{r: r \in I\}$. Then, by (9),

$$\sum_{1}^{k}{}_r \frac{U_r^{(n_r)}}{n_1 n_2 \ldots n_r} \stackrel{d}{=} \frac{U^{(n_1 n_2 \ldots n_k)}}{n_1 n_2 \ldots n_k} \tag{13}$$

and X, being a component of a discrete rectangular r.v., by Theorem 1 takes all its values with the same probability. On the other side

$$Y = \sum_{\substack{r \in I^c \\ r \leq k}} \frac{U_r^{(n_r)}}{n_1 n_2 \ldots n_r} + \sum_{k+1}^{\infty} \frac{U_r^{(n_r)}}{n_1 n_2 \ldots n_r} = Y_1 + Y_2$$

By the argument above, Y_1 takes some values $y_1, \ldots, y_m$ with the same probability $1/m$, and it is easily seen that these values are multiples of $c = 1/n_1 n_2 \ldots n_k$. Moreover, by (9),

$$Y_2 = \sum_{k+1}^{+\infty}{}_r \frac{U_r^{(n_r)}}{n_1 n_2 \ldots n_r} = \frac{1}{n_1 n_2 \ldots n_k} \sum_{k+1}^{+\infty}{}_r \frac{U_r^{(n_r)}}{n_{k+1} \ldots n_r} \stackrel{d}{=} \frac{U}{n_1 n_2 \ldots n_k}$$

with density

$$f_{Y_2}(x) = \frac{1}{c} I_{(o,c)}(x)$$

and the density of Y is given by

$$f_Y(x) = \int_R f_{Y_2}(x-y)\, dF_{Y_1}(y) = \frac{1}{m} \sum_{1}^{m}{}_j f_{Y_2}(x-y_j) =$$

$$= \frac{1}{mc} \sum_{1}^{m}{}_j I_{(0,c)}(x-y_j) = \frac{1}{mc} \sum_{1}^{m}{}_j I_{(y_j\,,\, y_j+c)}(x)$$

Since the values y_j are multiple of c, the intervals in the last sum are not overlapping, and i) is proved.

In the case of I and I^c infinite put, for a given k,

$$Y = Y_1 + Y_3$$

where Y_1 is defined as above and

$$Y_3 = \sum_{\substack{r \in I^c \\ r > k}} \frac{U_r^{(n_r)}}{n_1 n_2 \ldots n_r}$$

From the structure of Y_1 it follows that the number m of its values is given by

$$m = \prod_{\substack{r \in I^c \\ r \leq k}} n_r$$

In fact $U^{(n_r)}$ takes n_r values, the product gives m sums, which are all different because the length of each addend is less than or equal to the length of the intervals between the values of the sum of the preceeding addends.

Since Y_3 is a component of Y_2, its length is less than the length c of Y_2. Then the Lebesgue measure of the support of Y, say μ, equal to m times the length of Y_3, is less than mc, i.e.

$$\mu < mc \leq \Big(\prod_{\substack{r \in I^c \\ r \leq k}} n_r\Big)\Big(\prod_{1}^{k} n_r\Big)^{-1} = \Big(\prod_{\substack{r \in I \\ r \leq k}} n_r\Big)^{-1}$$

Moreover, by the properties of the convolution,

$$\sup_{x \in R} P(Y = x) \leq \sup_{x \in R} P(Y_1 = x) \leq \frac{1}{m} = \Big(\prod_{\substack{r \in I^c \\ r \leq k}} n_r\Big)^{-1}$$

Since I and I^c are infinite, and $n_r \geq 2$, both the products above tend to $+\infty$ when k increases; therefore Y is continuous and its support has measure zero. So also ii) is proved. ■

3 Triangular distributions

Triangular distributions arise naturally as convolution of two identically distributed rectangular distributions. The name reflects the form the density function in the continuous case and of the representation of probabilities in the discrete one. A classical example is given by the sum of two dice.

We will denote by $T^{(n)}$ the discrete triangular random variable which is sum of two independent r.v. $U_1^{(n)}$ and $U_2^{(n)}$. Its length is $2n - 1$, and its distribution is given by

$$P(T^{(n)} = j) = \begin{cases} \dfrac{j+1}{n^2} & ; \quad j = 0, 1, \ldots, n-1 \\[2ex] \dfrac{2n - j - 1}{n^2} & ; \quad j = n, n+1, \ldots, 2n-2 \end{cases}$$

In the continuous case the triangular r.v. T is the sum of two independent r.v. U_1 and U_2 rectangularly distributed on (0, 1). Its distribution function and density are

$$T(x) = \begin{cases} 0 & x \le 0 \\ \dfrac{x^2}{2} & 0 < x \le 1 \\ 1 - \dfrac{(2-x)^2}{2} & 1 < x \le 2 \\ 1 & 2 < x \end{cases}$$

$$t(x) = \begin{cases} 0 & x \le 0 \\ x & 0 < x \le 1 \\ 2 - x & 1 < x \le 2 \\ 1 & 2 < x \end{cases}$$

* * *

Following the study for the rectangular distributions, we will consider now the decomposability of $T^{(N)}$. Since $T^{(N)}$ has trivially as components $U^{(N)}$ and its subcomponents, we will search for the existence of components of $T^{(N)}$ which are neither $U^{(N)}$ nor components of $U^{(N)}$. We will exclude also components of $T^{(N)}$ which are obtained by components of $U^{(N)}$. If, e.g.

$$U^{(N)} \stackrel{d}{=} X + Y$$

then

$$T^{(N)} \stackrel{d}{=} U^{(N)} + U_1^{(N)} \stackrel{d}{=} X + Y + X_1 + Y_1$$

where X_1, Y_1 have the same distribution as X, Y (we recall that all the r.v. in a sum are taken independent). So $X + X_1$ is a component of $T^{(N)}$, while it may not be a component of $U^{(N)}$; such is the case, e.g. , when X is rectangular, since then $X + X_1$ is triangular, and by Theorem 1 cannot be a component of $U^{(N)}$.

Summing up, we will say that $T^{(N)}$ is non-trivially decomposable when it has a component which is not a composition of components of $U^{(N)}$, possibly replicated.

As in the decomposition of $U^{(N)}$, it is clear that X and Y can take, except for translations, only integer values, and we may choose that they start from zero. For the probability generating functions we have

$$P_{T^{(N)}}(z) = P_X(z)P_Y(z) = \left[\sum_0^{m-1}{}_r \, \alpha_r z^r \right] \left[\sum_0^{n-1}{}_s \, \beta_s z^s \right]$$

where $m + n = 2N$. From

$$\frac{1}{N^2} = P(T^{(n)} = 0) = P(X = 0) \, P(Y = 0) = \alpha_0 \beta_0$$

we get $\alpha_0\beta_0 = 1/N^2$; and putting , as in (4), $a_r = \alpha_r/\alpha_0$, $b_r = \beta_r/\beta_0$,

$$G_N^2(z) = N^2 P_{T^{(N)}}(z) = \frac{1}{\alpha_0\beta_0}\left[\sum_0^{m-1}{}_r\ \alpha_r z^r\right]\left[\sum_0^{n-1}{}_s\ \beta_s z^s\right]$$

$$= \left[\sum_0^{m-1}{}_r\ a_r z^r\right]\left[\sum_0^{n-1}{}_s\ b_s z^s\right]$$

Therefore, by (1), the decomposition of $T^{(n)}$ will obtain iff

$$\left(\frac{z^N - 1}{z - 1}\right)^2 = G_X(z)G_Y(z) = \left[\sum_0^{m-1}{}_r\ a_r z^r\right]\left[\sum_0^{n-1}{}_s\ b_s z^s\right] \tag{14}$$

where G_X and G_Y are polynomials with non-negative coefficients and the constant term equal to 1.

Of course G_X and G_Y are, except for multiplying constants, the probability generating functions of two r.v. X and Y . By Lemma 1 we have

$$P_{T^{(N)}}(z) = G_N^2(z) = (1+z)^2 \prod_1^{\frac{N-2}{2}}{}_r (1 + c_r z + z^2)^2 \qquad \text{N even}$$

$$= \prod_1^{\frac{N-1}{2}}{}_r (1 + c_r z + z^2)^2 \qquad \text{N odd} \tag{15}$$

whith c_r as in (8). It's clear that a component of $T^{(N)}$ must be a product of polynomials in (15).

EXAMPLE 2. Let us see the case N = 5 . From (15) and (8) we have

$$P_{T^{(5)}}(z) = \left(1 + \frac{1 + \sqrt{5}}{2} z + z^2\right)^2 \left(1 + \frac{1 - \sqrt{5}}{2} z + z^2\right)^2$$

If we take

$$G_X(z) = \left(1 + \frac{1 + \sqrt{5}}{2} z + z^2\right)$$

$$G_Y(z) = \left(1 + \frac{1 + \sqrt{5}}{2} z + z^2\right)\left(1 + \frac{1 - \sqrt{5}}{2} z + z^2\right)^2 =$$

$$= 1 + \frac{3 - \sqrt{5}}{2} z + \frac{5 - \sqrt{5}}{2} z^2 + \frac{5 - \sqrt{5}}{2} z^3 +$$

$$+ \frac{5 - \sqrt{5}}{2} z^4 + \frac{3 - \sqrt{5}}{2} z^5 + z^6$$

G_X and G_Y having non-negative coefficients give two components of $T^{(N)}$, and G_X by theorem 1 is not a component of $U^{(5)}$. So $T^{(5)}$ has a non-trivial component. ∎

EXAMPLE 3. Take $N = 6$. By (7)

$$G_6(z) = (1 + z)(1 - z + z^2)(1 + z + z^2) = $$
$$= (1 + z)(1 + z^2 + z^4) = (1 + z + z^2)(1 + z^3)$$

so that $(1 + z)$, $(1 + z^2 + z^4)$, $(1 + z + z^2)$, $(1 + z^3)$ give components of $U^{(N)}$. A component of $T^{(N)}$ must be made by factors of G_6 , possibly to the square. In order to be non-trivial a component should contain the factor $1 - z + z^2$. But if it is taken with power 1, when multiplied by the other factors gives component of $U^{(N)}$, and therefore furnishes trivial components; if it is taken with power 2 it is easily seen that all possible products are polynomes with some negative coefficients. Therefore $T^{(6)}$ has not non-trivial components. ■

* * *

It is easy now to establish

THEOREM 3

For $N \le 6$, $N \neq 5$, $T^{(N)}$ has not non-trivial components.

Proof. For $N = 2, 3, 4$ the factorization (7) contains only polynomes with non-negative coefficients, which give components of $U^{(N)}$; therefore any of their composition is trivial. For $N = 6$ the result is given by Example 3. ■

* * *

We prove now

THEOREM 4

For every $N > 6$ $T^{(N)}$ has non-trivial components.

Proof. We will show that one of the factors in (7), which is not a component of $U^{(N)}$, is a component of $T^{(N)}$.

Take $c = -2\cos\vartheta$, where ϑ is one of the angles in (8), and pose

$$G_{T^{(N)}}(z) = G_X(z) \sum_{0}^{2N-4} a_j z^j = (1 + cz + z^2) \sum_{0}^{2N-4} a_j z^j \qquad (16)$$

Note that the coefficients a_j exist and are uniquely determined, since they are obtained as product of factors in (15). The polynomial $1 + cz + z^2$ gives a component of $T^{(N)}$ iff c and the a_j's are non-negative; this component is non-trivial iff $0 < c < 1$, since then $1 + cz + z^2$ is not a component of $U^{(N)}$ by Theorem 1 and cannot be made of components of $U^{(N)}$ because it is irreducible in the real field.

Since by (15) and (8) $0 < \vartheta < \pi$, $1 + cz + z^2$ will be a non-trivial component iff

$$
\begin{aligned}
&a_r \geq 0 && \forall\ r \\
&\frac{\pi}{2} < \vartheta < \pi \quad ; && \vartheta \neq \frac{2\pi}{3}
\end{aligned}
\tag{17}
$$

Because of the distribution of $T^{(N)}$, (16) give

$$
a_j + ca_{j-1} + a_{j-2} = \begin{cases} j + 1 & j = 0,\ 1,\ \ldots,\ N-1 \\ 2N - j - 1 & j = N,\ N+1, \ldots, 2N-2 \end{cases}
$$

where a_{-1}, a_{-2}, a_{2N-3}, a_{2N-2}, introduced for sake of homogeneity, are all equal to 0 .

The polynomial G_X as a product of factors in (15) is symmetrical, i.e. $a_r = a_{2N-4-r}$. This allows to restrict to the unknowns a_0, $a_1, \ldots$, a_{N-2} and to the equations with $j = 0, 1, \ldots, N - 2$. The equation obtained for $j = N - 1$, which contains again a_{N-2} , is necessarily satisfied since, as said above, a solution exists.

Passing to

$$
t_j = a_j - a_{j-1} - \frac{1}{2+c}
$$

we obtain the system

$$
t_j + ct_{j-1} + t_{j-2} = 0 \ ; \qquad j = 0,\ 1, \ldots,\ N-2
$$

which has the solution

$$
t_j = \gamma_1 e^{i\vartheta j} + \gamma_2 e^{-i\vartheta j} \ ; \qquad j = 0,\ 1, \ldots,\ N-2
$$

with γ_1 , γ_2 complex conjugated, whence

$$
a_j = \sum_0^j {}_r\, t_r + \frac{j+1}{2+c} = \gamma_1 \frac{e^{i\vartheta(j+1)} - 1}{e^{i\vartheta} - 1} + \gamma_2 \frac{e^{-i\vartheta(j+1)} - 1}{e^{-i\vartheta} - 1} + \frac{j+1}{2+c}
$$

Simple if cumbersome calculations bring to

$$
\gamma_1 = \frac{e^{i\vartheta}(e^{i\vartheta} - 1)}{4\mathrm{sen}\vartheta(1 - \cos\vartheta)}
$$

$$
a_j = \frac{1}{2(1-\cos\vartheta)} \left[j + 2 - \frac{\mathrm{sen}[\vartheta(j+2)]}{\mathrm{sen}\vartheta} \right] ; \qquad j = 0,\ 1, \ldots, N-2
$$

It is easy to see that the a_j's are all positive for every angle ϑ . In fact for $j = 0$

$$
\frac{\mathrm{sen}2\vartheta}{\mathrm{sen}\vartheta} = 2\cos\vartheta < 2
$$

(giving, as we already knew, $a_0 = 1$), and by recurrence

$$\left|\frac{\operatorname{sen}(j+1)\vartheta}{\operatorname{sen} n\vartheta}\right| = \left|\frac{\operatorname{sen} j\vartheta \cos\vartheta + \operatorname{sen}\vartheta \cos j\vartheta}{\operatorname{sen} n\vartheta}\right| \le$$

$$\le \left|\frac{\operatorname{sen} j\vartheta}{\operatorname{sen}\vartheta}\right| |\cos\vartheta| + |\cos j\vartheta| < j + 1$$

and the theorem is proved. ■

Note that conditions (17) are not fulfilled for $N = 6$ (which, as we have seen in Example 3, doesn't admit non-trivial components) since the only angle, among those in (8), with $\pi/2 < \vartheta < \pi$, is $2\pi/3$. The proof holds also for $N = 5$ (Example 2), in agreement with Theorem 3.

* * *

Theorem 4 for N large identifies several non-trivial components, i.e. all those which are given by G_X as in (16), with $N/4 < r < N/2$, $r \neq N/3$. Theorem 4 makes no difference between N prime and N not prime; thus also when $U^{(N)}$ is indecomposable $T^{(N)}$ has non-trivial components.

There are of course other non-trivial components beyond those given by Theorem 4.

EXAMPLE 4. Put $N = 2^n$. From the decomposition of Example 1 we have

$$G_{2^n}(z) = \prod_{h=0}^{n-1} (1 + z^{2^h}) = (1+z) \prod_{h=1}^{n-1} (1 + z^{2^h}) =$$

$$= (1+z) \prod_{h=0}^{n-1} \left[1 + (z^2)^{2^h}\right] = (1+z)\, G_{2^{n-1}}(z^2)$$

Then, by the proof of Theorem 4, if $n \ge 4$, so that $N = 2^{n-1} \ge 8$, also

$$G_X(z) = (1 + cz^2 + z^4)$$

with c as in Theorem 4 gives a non-trivial components of $T^{(2^n)}$.

By the same argument for $n-j \ge 3$ the decomposition

$$G_{2^n}(z) = \prod_{h=0}^{j-1} (1 + z^{2^h})\, G_{2^{n-j}}(z^{2^j})$$

obtains other non-trivial components of $T^{(n)}$. ■

* * *

Given two (non-trivial) components of $T^{(n)}$, their composition may not be a component.

EXAMPLE 5. Take $N = 16$. According to Theorem 4 two non-trivial components of $T^{(16)}$ are given by

$$G_{X_1}(z) = (1 - 2z \cos\tfrac{5}{8}\pi + z^2)$$

$$G_{X_2}(z) = (1 - 2z\cos\frac{7}{8}\pi + z^2)$$

Now $X_1 + X_2$ is a component of $T^{(16)}$ iff there exists a polynomial G_Y with non-negative coefficients such that

$$G_{X_1}(z)\, G_{X_2}(z)\, G_Y(z) = G^2_{16}(z)$$

But this is impossible: it suffices to look at the cofficient a_1 of z in G_Y , for which the equation above gives

$$a_1 - 2\cos\frac{5}{8}\pi - 2\cos\frac{7}{8}\pi = 2$$

whence

$$\frac{1}{2}a_1 = 1 + (\cos\frac{5}{8}\pi + \cos\frac{7}{8}\pi) = 1 + 2\cos\frac{6}{8}\pi\cos\frac{1}{8}\pi =$$

$$= 1 - 2\cos\frac{1}{8}\pi < 1 - 2\cos\frac{1}{3}\pi = 0$$

Therefore $X_1 + X_2$ is not a component of $T^{(16)}$. ∎

* * *

Considering now the continuous triangular distribution, we can prove

THEOREM 5

The continuous rectangular distribution has non-trivial components.

Proof. Start from the decomposition $T^{(5)} = X + Y$ of Example 2, with

$$G_X(z) = 1 + \frac{1 + \sqrt{5}}{2} z + z^2$$

so that X takes the values $0, 1, 2$, with probabili t ies, obtained normalizing the coefficients of G_X ,

$$\frac{2}{5 + \sqrt{5}}, \quad \frac{1 + \sqrt{5}}{5 + \sqrt{5}}, \quad \frac{2}{5 + \sqrt{5}}$$

We have now

$$\frac{T^{(5N)}}{5N} \stackrel{d}{=} \frac{U_1^{(5N)} + U_2^{(5N)}}{5N} \stackrel{d}{=} \frac{U_1^{(N)} + NU_2^{(5)} + U_3^{(N)} + NU_4^{(5)}}{5N} \stackrel{d}{=}$$

$$\stackrel{d}{=} \frac{1}{5}\frac{T^{(N)}}{N} + \frac{1}{5}T^{(5)} \stackrel{d}{=} \frac{1}{5}\frac{T^{(N)}}{N} + \frac{1}{5}X + \frac{1}{5}Y \tag{18}$$

Since when n increases

$$\frac{T^{(n)}}{n} \stackrel{d}{=} \frac{U_1^{(n)}}{n} + \frac{U_2^{(n)}}{n} \xrightarrow{d} U_1 + U_2 = T$$

from (18), for $N \to \infty$, we obtain

$$T \stackrel{d}{=} \frac{1}{5} T + \frac{1}{5} X + \frac{1}{5} Y$$

and $X/5$ is a component of T .

It is easy to see that $X/5$ cannot be a component of U . Otherwise it should be $U = X/5 + Z$ and, denoting by F the distribution function of the r.v. Z ,

$$U(x) = \int_R F(x-y)\, dF_{X/5}(y) =$$
$$= \frac{2}{5+\sqrt{5}} F(x) + \frac{1+\sqrt{5}}{5+\sqrt{5}} F(x-\frac{1}{5}) + \frac{2}{5+\sqrt{5}} F(x-\frac{2}{5}) \tag{19}$$

But (19) would imply $F(x) = 0$ for $x \leq 0$; then for $0 < x \leq 1/5$

$$\frac{2}{5+\sqrt{5}} F(x) = U(x) = x$$

and as a consequence, again by (19), for $1/2 < x \leq 2/5$

$$\frac{2}{5+\sqrt{5}} F(x) = x - \frac{1+\sqrt{5}}{5+\sqrt{5}} F(x - \frac{1}{5}) =$$

$$= x - \frac{1+\sqrt{5}}{2} (x - \frac{1}{5}) = \frac{1-\sqrt{5}}{2} x + \frac{1+\sqrt{5}}{10}$$

giving a decreasing function, contrary to the properties of the distribution functions ■

REFERENCES

de Finetti, B. (1929). Sulle funzioni a incremento aleatorio, Rendiconti Acc. dei Lincei, 10: 163-168

Linnik, Ju.V. and Ostrovskii, I.V. (1977). Decomposition of Random Variables and Vectors, American Math. Soc., Providence, Rhode Island.

Feller, W. (1966). An Introduction to Probability Theory and its Applications, Vol. II , Wiley, New York.

9. Nonlinear Infinite Networks with Nonsymmetric Resistances

Leonede De-Michele Dipartimento di Matematica, Università di Milano, Milan, Italy

Paolo M. Soardi Dipartimento di Matematica, Università di Milano, Milan, Italy

Abstract. In this paper we introduce a theory of infinite nonlinear resistive networks whose resistances are not necessarily odd symmetric. The main results are: a version of Thomson's principle for such networks; the description of the properties of the energy of the currents generated by multiples of a fixed energizing vector; the description of the behavior of the network if one or more resistances are replaced by their symmetric.

1. Introduction.

A resistive electrical network $\mathcal{N}$ consists of a locally finite connected oriented graph Γ, whose edges B are assigned a function p_B in such a way that the relation between the voltage v_B and the current i_B flowing in B has the form

$$v_B = p_B(i_B). \tag{1}$$

The network is energized y internal current generators represented by an infinite vector $\mathbf{h}$ and the current–voltage distribution obeys Kirchhoff's laws. The resistance functions p_B are continuous, strictly increasing and vanish at 0.

A theory of nonlinear resistive infinite networks in the framework of modular sequence spaces was studied in [1]. This theory was subsequently extended to more general infinite networks in [10]. In both these works the network was supposed to be symmetric i.e., the resistance functions p_B were supposed to be odd symmetric; moreover the p_B were required to satisfy a uniform variation condition (for other

1991 *Mathematics Subject Classification.* Primary 94C15, 49J45. Secondary 46A45.

approaches to infinite symmetric networks either in the Hilbert spaces setting or without finite energy assumptions see e.g. [10], [2], [12] and [13]).

The odd symmetry assumption on the resistances is a strong restriction on the network, since there are several electrical two–terminal devices, such as the semiconductor diode, which do not satisfy this condition. The aim of this paper is to generalize the theory of [1] by dropping the odd symmetry assumption as well as the uniform variation condition on the p_B's.

The strategy of our work is similar to [1]: there exists a unique current generated by $\mathbf{h}$ and this current minimizes the "nonlinear energy functional" $W(\mathbf{f}+\mathbf{h}) = \sum_B \int_0^{f_B+h_B} p_B(t)dt$ on the set of all cycles $\mathbf{f}$. In the general situation described in this paper, the mathematical framework is provided by the space S of all infinite vectors with the topology of componentwise convergence. The minimizing cycle is obtained in §3 as a limit of currents in finite subnetworks $\mathcal{N}_n$ defined by shorting $\mathcal{N}$ everywhere outside a finite subgraph Γ_n. Notice that this method was used in the case of linear networks in [11] and, indipendently, in [4]; see also [5] for the case of odd resistances.

In §4 we assume the uniform variation (but not the odd symmetry) condition and suppose that the network is energized by a multiple $k\mathbf{h}$ of a fixed vector $\mathbf{h}$. Then, denoting by $\mathbf{i}(k)$ the corresponding current, we show that $w(k) = W(\mathbf{i}(k) + k\mathbf{h})$ is a convex differentiable function of k, whose derivative is k^{-1} times the total power dissipated in the resistors. In particular, in the case of a dipole between the nodes a and b, the network behaves as a single nonlinear resistor between a and b with resistance function equal to the derivative of $w(k)$.

The most distinctive feature of nonsymmetric networks consists in the fact that, unlike the symmetric case, currents and energies depend on the orientation of the graph i.e., on the choice of the polarities of the resistors. This fact is investigated in §5. In particular, we prove that there exists a choice of the polarities (in general not unique) such that the energy W of the respective current $\mathbf{i}$ is minimal among all possible choices.

If, for every edge, the resistance in one direction is higher than the resistance in the opposite direction, then the minimizing currents are all equal, and the minimizing orientation is unique up to the symmetries of the network.

2. Notation.

Let Γ denote a (finite or infinite) connected, locally finite graph. Γ may have multiple edges but not self–loops. We label the edges of Γ by the natural integers $\mathbb{N}$ and fix an orientation for every edge. The result presented in sections 3–4 do not depend on the labeling, but they do depend on the orientation; see §5.

Electrical currents will be represented by vectors $\mathbf{f} = (f_1, f_2, \dots)$ with real components f_j. The support of a finite vector $\mathbf{f}$ is the set of all edges j such that $f_j \neq 0$. We say that a vector is finite if its support is finite.

For every node x we let $J(x) \subseteq \mathbb{N}$ denote the finite set of all the edges j incident with x. Let $\mathbf{f}$ be a vector on Γ. We set, for every node x,

(2) $$\partial \mathbf{f}(x) = \sum_{j \in J(x)} \pm f_j,$$

where we chose in (2) the sign + if the edge j_r is directed toward x, the sign − otherwise. The quantity defined in (2) is also called the boundary of $\mathbf{f}$ at x.

A vector $\mathbf{f}$ is a called cycle on Γ if $\partial \mathbf{f}(x) = 0$ for every node x. In other words, the cycles satisfy Kirchhoff's node law at every node x.

The space $S = \mathbb{R}^\infty$ is defined as the space of all vectors $\mathbf{f}$ with the topology of componentwise convergence: a sequence of vectors $\mathbf{f}_n = (f_{1,n}, f_{2,n}, \ldots)$ converges to $\mathbf{f}$ as $n \to \infty$ if $f_{j,n}$ tends to f_j for every $j = 1, 2, \ldots$. The topology in S can be metrized e.g. by setting

$$d(\mathbf{f}, \mathbf{g}) = \sum_{j=1}^{\infty} 2^{-j} \frac{|f_j - g_j|}{1 + |f_j - g_j|}.$$

In particular, the linear subspace of all cycles on Γ will be denoted by Z.

Next, we assume that for every edge j there is assigned a function $p_j : \mathbb{R} \mapsto \mathbb{R}$ with the following properties

(3) $p_j(t)$ is continuous and strictly increasing.

(4) $p_j(0) = 0$.

Note that $p_j(t)$ has the same sign as t.
For every vector $\mathbf{f} = (f_1, f_2, \ldots)$ we set

(5) $$W(\mathbf{f}) = \sum_{j=1}^{\infty} \int_0^{f_j} p_j(t)dt.$$

Suppose that $\mathbf{h} \in S$ is a fixed vector such that

(6) $$W(\mathbf{h}) < \infty.$$

Then, we set

$$S_{\mathbf{h}} = \{\mathbf{f} \in S : W(\mathbf{f} + \mathbf{h}) < \infty\}.$$

Clearly, $S_{\mathbf{h}}$ is convex and, by (6), every finite vector is in $S_{\mathbf{h}}$. The functional W is lower semicontinuous on S. Furthermore, $W(\mathbf{f} + \mathbf{h})$ is nonnegative and strictly convex on $S_{\mathbf{h}}$.

To check that W is lower semicontinuous, let $\mathbf{f}_n = (f_{1,n}, f_{2,n}, \ldots)$ tend to $\mathbf{f} = (f_1, f_2, \ldots)$ in S. Then, $f_{j,n}$ tends to f_j for all j as $n \to \infty$. By Fatou's Lemma

(7) $$W(\mathbf{f} + \mathbf{h}) \leq \liminf_{n \to \infty} \sum_{j=1}^{\infty} \int_0^{f_{j,n} + h_j} p_j(t)dt = \liminf_{n \to \infty} W(\mathbf{f}_n).$$

We can now define the notion of a nonlinear resistive electrical network. We shall assume that the network is energized by internal current generators represented by a vector $\mathbf{h}$ satisfying (6).

Definition. *Suppose that Γ is a locally finite connected oriented graph, possibly with multiple edges but no self–loops. Let $\mathcal{P}$ be a family of functions p_j satisfying assumptions (3)–(4) and let $\mathbf{h}$ be a vector satisfying (6). A nonlinear resistive electrical network $\mathcal{N}$ is defined as the triplet $\mathcal{N} = (\Gamma, \mathcal{P}, \mathbf{h})$.*

The functions p_j are called the resistance functions of $\mathcal{N}$.

We introduce now Tellengen's equation. First of all we have to define the nonlinear resistance operator. In the linear case the resistance operator encompasses Ohm's law. In the nonlinear case the resistance operator $R : S \mapsto S$ is defined by

$$R(\mathbf{f}) = (p_1(f_1), p_2(f_2), \dots)$$

for all vectors $\mathbf{f} = (f_1, f_2, \dots)$. We say that a cycle $\mathbf{i} = (i_1, i_2, \dots)$ on Γ is a solution of the nonlinear Tellengen's equation if

$$\langle \mathbf{z}, R(\mathbf{i} + \mathbf{h}) \rangle = \sum_j z_j p_j(i_j + h_j) = 0 \tag{8}$$

for every *finite* cycle $\mathbf{z} = (z_1, z_2, \dots)$.

Definition. *Given a network $\mathcal{N}$, a cycle $\mathbf{i} \in Z$ is called a current generated by $\mathbf{h}$ if $\mathbf{i}$ satisfies Tellengen's equation and minimizes $W(\mathbf{f} + \mathbf{h})$ among all cycles $\mathbf{f}$ on Γ.*

REMARK. It is well known that Tellengen's equation may have more than one solution even in the linear case: cfr. [8, §8.B] and [7]; see also [6, §4] for the homogeneous symmetric case. However, the solution with minimal energy is unique for networks with odd symmetric resistances: see [1]. We will show in the next section that this is also the case for networks without odd symmetry.

3. A minimization principle.

Let $\mathcal{N} = (\Gamma, \mathcal{P}, \mathbf{h})$ be an infinite network. An exhaustion of Γ is an increasing sequence $\Gamma_1 \subseteq \Gamma_2 \subseteq \cdots$ of finite connected oriented subgraphs of $\Gamma_n \subseteq \Gamma$ (with the orientation induced by Γ) such that

$$\bigcup_{n=1}^{\infty} \Gamma_n = \Gamma.$$

For every n we define a new finite connected oriented graph Υ_n by shorting Γ everywhere outside Γ_n. Namely, the nodes of Υ_n are the nodes of Γ_n plus a new node o_n. The edge set of Υ_n is the union of the edge set of Γ_n and the set of all the edges of Γ having one endpoint in Γ_n and the other endpoint not in Γ_n. The endpoints not in Γ_n of the edges of the latter set are all identified with o_n, and the edges inherit the same orientation as in Γ. We label the edges in Υ_n by the same indices as in Γ; in this way the edge set of Υ_n is identified with a finite subset of the edge set of Γ.

Let $\mathcal{P}_n$ and $\mathbf{h}_n$ denote the restriction to the edges of Υ_n of the family $\mathcal{P}$ and of $\mathbf{h}$, respectively.

Definition. *The sequence of finite networks $\mathcal{N}_n = (\Upsilon_n, \mathcal{P}_n, \mathbf{h}_n)$ described above is called an exhaustion of $\mathcal{N}$.*

Let $\mathcal{N}_n$ be any exhaustion of $\mathcal{N}$ and let S_n denote the space of all vectors on Υ_n. Every $\mathbf{f} \in S_n$ can be identified with a finite vector on Γ: if the edges of Υ_n are indexed by $j_1, j_2, \ldots, j_{k_n}$, then it suffices to set $f_j = 0$ whenever $j \neq j_i$, $i = 1, \ldots, k_n$. Clearly, this identification does not change the value of $W(\mathbf{f})$.

Conversely, restricting a vector on Γ to the edge set of Υ_n decreases the value of W.

Lemma 1. *Let Γ and Υ_n be as above and let $\mathbf{f}$ be a cycle on Γ. If $\mathbf{f}_n$ denote the restriction of $\mathbf{f}$ to the edge set of Υ_n, then $\mathbf{f}_n$ is a cycle on Υ_n.*

Proof. Denote by V the vertex set of Γ and by V_n the vertex set of Υ_n. Let, as in (2), $\partial\mathbf{f}(x)$ denote the boundary of $\mathbf{f}$ at $x \in V$, and let $\partial_n\mathbf{f}_n(x)$ denote the boundary of $\mathbf{f}_n = (f_{1,n}, f_{2,n}, \ldots)$ at $x \in V_n$ i.e,

$$\partial\mathbf{f}_n(x) = \sum_{j \in J_n(x)} \pm f_j, \tag{9}$$

where $J_n(x)$ is the set of all edges j of Υ_n incident to x and the signs plus or minus are chosen with the same convention as in equation (2).

Assume that $\mathbf{f}$ is a cycle i.e., $\partial\mathbf{f}(x) = 0$ for all $x \in V$. If $x \in V_n$ is different from o_n, then $\partial_n\mathbf{f}_n(x) = \partial\mathbf{f}(x) = 0$, since $J(x) = J_n(x)$ for such nodes x. On the other hand, we clearly have $\sum_{x \in V_n} \partial_n\mathbf{f}_n(x) = 0$. It follows that $\partial\mathbf{f}_n(o_n) = 0$. Hence $\mathbf{f}_n$ is a cycle on Υ_n. □

We are now ready to prove the main result of this section.

Theorem 1. *Let $\mathcal{N} = (\Gamma, \mathcal{P}, \mathbf{h})$ be a nonlinear resistive network. Then, there exists a unique current $\mathbf{i}$ generated by $\mathbf{h}$ in $\mathcal{N}$, and $W(\mathbf{i} + \mathbf{h}) \leq W(\mathbf{h}) < \infty$.*

If $\mathcal{N}$ is infinite, then, for any exhaustion $\mathcal{N}_n = (\Upsilon_n, \mathcal{P}_n, \mathbf{h}_n)$ of $\mathcal{N}$, the unique current $\mathbf{i}_n$ generated by $\mathbf{h}_n$ in $\mathcal{N}_n$ converges in S to the current $\mathbf{i}$, and $W(\mathbf{i}_n)$ converges to $W(\mathbf{i})$ as $n \to \infty$.

Proof. First we suppose that $\mathcal{N}$ is finite. Then the topology in S is induced by the euclidean norm and W is continuous. Moreover, $W(\mathbf{f} + \mathbf{h}) \to +\infty$ as $\mathbf{f} \to \infty$, since the p_j's are increasing. Therefore, if

$$\alpha = \inf\{W(\mathbf{f} + \mathbf{h}) : \mathbf{f} \in Z\},$$

then there exists $\mathbf{i} \in Z$ such that $\alpha = W(\mathbf{i} + \mathbf{h})$. The minimizing cycle $\mathbf{i}$ is unique by the strict convexity of W.

For every cycle $\mathbf{z}$ we must have

$$\sum_j p_j(i_j + h_j)z_j = \lim_{t \to 0} \frac{W(\mathbf{i} + t\mathbf{z} + \mathbf{h}) - W(\mathbf{i} + \mathbf{h})}{t} = 0 \tag{10}$$

Hence $\mathbf{i}$ satisfies (8) and $\mathbf{i}$ is the unique current generated by $\mathbf{h}$ in $\mathcal{N}$.

Suppose now $\mathcal{N}$ infinite. Let $\mathcal{N}_n = (\Upsilon_n, \mathcal{P}_n, \mathbf{h}_n)$ be any exhaustion of $\mathcal{N}$ and let $\mathbf{i}_n = (i_{1,n}, i_{2,n}, \dots)$ be the current generated in $\mathcal{N}_n$ by the restriction $\mathbf{h}_n = (h_{1,n}, h_{2,n}, \dots)$ of $\mathbf{h} = (h_1, h_2, \dots)$. By the minimum property of $\mathbf{i}_n$ we have, for every j and every n,

$$\int_0^{i_{j,n}+h_{j,n}} p_j(t)dt \leq W(\mathbf{i}_n + \mathbf{h}_n) \leq W(\mathbf{h}_n) \leq W(\mathbf{h}).$$

Therefore, always by monotonicity, for every j there exists a constant a_j such that for every n

$$|i_{j,n}| \leq a_j + |h_{j,n}| \leq a_j + |h_j| = b_j$$

Since the set

$$K = \{\mathbf{f} : |f_j| \leq b_j \quad \text{for all } j = 1, 2, \dots\}$$

is compact in S by Tichonov theorem, we can find a subsequence $\mathbf{i}_{n_r}$ converging to a vector $\mathbf{i} \in K$.

The vector $\mathbf{i}$ is clearly a cycle, since for any fixed node x of Γ and large values of r, $\partial\mathbf{i}(x) = \partial_{n_r}\mathbf{i}_{n_r}$ (∂_{n_r} is as in (9)). Furthermore, $\mathbf{i}$ satisfies Tellengen's equation. To see this it is sufficient to observe that the support of a finite cycle $\mathbf{z}$ on Γ is contained in the edge set of Υ_n for n large enough. Therefore

$$\langle \mathbf{z}, R(\mathbf{i}_{n_r} + \mathbf{h}_{n_r})\rangle = 0.$$

Note that, by (7),

$$W(\mathbf{i} + \mathbf{h}) \leq \liminf_{r\to\infty} W(\mathbf{i}_{n_r} + \mathbf{h}_{n_r}). \tag{11}$$

Suppose now that we are given a cycle $\mathbf{f}$ on Γ. Let $\mathbf{f}_n$ be its restriction to the edge set of Υ_n. By Lemma 1, $\mathbf{f}_n$ is cycle on Υ_n. By the minimality of $\mathbf{i}_{n_r}$, we have

$$W(\mathbf{i}_{n_r} + \mathbf{h}_{n_r}) \leq W(\mathbf{f}_{n_r} + \mathbf{h}_{n_r}). \tag{12}$$

Letting r tend to infinity in (12), and observing that $W(\mathbf{f}_{n_r} + \mathbf{h}_{n_r}) \to W(\mathbf{f} + \mathbf{h})$ as $r \to \infty$, we have by (11) that $W(\mathbf{i} + \mathbf{h}) \leq W(\mathbf{f} + \mathbf{h})$.

Hence $\mathbf{i}$ minimizes $W(\mathbf{f} + \mathbf{h})$ over all cycles in $\mathcal{N}$, and, by strict convexity, $\mathbf{i}$ is the unique minimum point of $W(\mathbf{f} + \mathbf{h})$ on Z. This argument shows also that the whole sequence $\mathbf{i}_n$ converges to $\mathbf{i}$.

Let now $\mathbf{y}_n$ denote the restriction of $\mathbf{i}$ to the edges of Υ_n. Then, by the Lemma 1 again, $\mathbf{y}_n$ is a cycle on Υ_n. By (7) and the minimality of $\mathbf{i}_n$

$$\begin{aligned} W(\mathbf{i} + \mathbf{h}) \leq \liminf_{n\to\infty} W(\mathbf{i}_n + \mathbf{h}_n) &\leq \limsup_{n\to\infty} W(\mathbf{i}_n + \mathbf{h}_n) \\ &\leq \lim_{n\to\infty} W(\mathbf{y}_n + \mathbf{h}_n) = W(\mathbf{i} + \mathbf{h}). \end{aligned}$$

Hence $W(\mathbf{i}_n + \mathbf{h}_n) \to W(\mathbf{i} + \mathbf{h})$ as $n \to \infty$ and $W(\mathbf{i} + \mathbf{h}) \leq W(\mathbf{h}) < \infty$. $\square$

REMARK. The above theorem can be seen as the version of Thomson's minimization principle for nonsymmetric networks. See [1] for the nonlinear symmetric case.

Clearly, if the network $\mathcal{N}$ satifies the same conditions as in the paper [1], then the current $\mathbf{i}$ provided by Theorem 1 coincides with the current found in [1].

4. Currents generated by multiples of a vector.

Let Γ be a locally finite connected oriented graph and let $\mathcal{P}$ be a family of resistance functions satisfying assumptions (3) and (4) above. We introduce the notation

$$M_j(t) = \int_0^t p_j(s)ds \qquad j = 1, 2, \dots .$$

for every real t. For every j we denote by $c_j^+ > 0$ and $c_j^- < 0$ the unique positive, respectively negative, number such that

$$M_j(c_j^+) = 1 \qquad \text{and} \qquad M_j(c_j^-) = 1.$$

Throughout this section we shall assume that the family $\mathcal{P}$ satisfies also the following uniform variation condition:
there exist a number r (necessarily greater than 1) and a positive integer j_0 such that, for all $j \geq j_0$,

$$\begin{aligned} tp_j(t) &\leq rM_j(t) && \text{for all} \quad t \in [0, c_j^+) \\ tp_j(t) &\leq rM_j(t) && \text{for all} \quad t \in (c_j^-, 0]. \end{aligned} \tag{13}$$

If the functions p_j are odd symmetric, then $c_j^+ = c_j^-$ and (13) corresponds to the uniform Δ_2 condition in [9].

It is easily seen, as in the symmetric case, that (13) implies

$$\begin{aligned} (t/c_j^+)^r &\leq M_j(t) \quad \text{for } 0 \leq t < c_j^+ \\ (t/c_j^-)^r &\leq M_j(t) \quad \text{for } c_j^- < t \leq 0. \end{aligned} \tag{14}$$

REMARK Assume that (13) holds. If $\mathbf{f}$ is a vector such that $W(\mathbf{f}) = \sum_{j=1}^{\infty} M_j(f_j) < \infty$, then, by the definition of the numbers $c_j^{\pm}$, there exists $j_1 > j_0$ such that, for all $j \geq j_1$, $f_j/c_j^+ < 1$ if $f_j > 0$ and $f_j/c_j^- < 1$ if $f_j < 0$. Consequently $f_j p_j(f_j) \leq rM_j(f_j)$ for all $j \geq j_1$.

Definition. *Let $\mathbf{i}$ denote the current generated in a network $\mathcal{N}$ by $\mathbf{h}$. The total power dissipated in the resistors is defined as*

$$E(\mathbf{i} + \mathbf{h}) = \sum_{j=1}^{\infty}(i_j + h_j)p_j(i_j + h_j).$$

If the functions p_j satisfy (13), then, by the above Remark, the total power dissipated is finite.

Assuming the uniform variation condition (13), we can prove that the set of all vectors with finite energy is a convex cone.

Lemma 2. *Suppose that the functions p_j satisfy (3), (4) and the uniform variation condition (13). Let $\mathbf{f} = (f_1, f_2, \dots)$ be a vector such that $W(\mathbf{f}) < \infty$. Then, $W(\lambda\mathbf{f}) < \infty$ for every $\lambda \geq 0$.*

Proof. The lemma is obviously true if $0 \leq \lambda \leq 1$. So, let us assume $\lambda > 1$. By a result of Woo (see [8, Proposition 3.1], see also [9, Lemma 4.6–4]), for every $\lambda > 1$ there exist numbers $\alpha^{\pm}(\lambda) > 0$ and $K^{\pm}(\lambda) \geq 1$, such that for all $j \geq j_0$

$$(15)\qquad \begin{aligned} M_j(\lambda t) &\leq K^{+}(\lambda)M_j(t) \quad \text{for all } t \text{ such that } 0 \leq t/c_j^{+} < \alpha^{+}(\lambda) \\ M_j(\lambda t) &\leq K^{-}(\lambda)M_j(t) \quad \text{for all } t \text{ such that } 0 \leq t/c_j^{-} < \alpha^{-}(\lambda). \end{aligned}$$

Since $\sum_{j=1}^{\infty} M_j(f_j)$ converges by assumption, there exists a positive integer $j_1 \geq j_0$ such that $\sum_{j \geq j_1} M_j(f_j) < \mu^r$, where $\mu = \min(1, \alpha^{+}(\lambda), \alpha^{-}(\lambda))$. By the definition of $c_j^{\pm}$, we see that $f_j/c_j^{+} < 1$ for $f_j > 0$ and $f_j/c_j^{-} < 1$ for $f_j < 0$. Then, by (14),

$$f_j/c_j^{+} \leq \mu \quad \text{if } f_j \geq 0 \quad \text{and} \quad f_j/c_j^{-} \leq \mu \quad \text{if } f_j < 0.$$

for all $j \geq j_1$. By (15),

$$\sum_{j \geq j_1} M_j(\lambda f_j) \leq \kappa \sum_{j \geq j_1} M_j(f_j),$$

where $\kappa = \max(K^{+}(\lambda), K^{-}(\lambda))$. Therefore $W(\lambda\mathbf{f}) < \infty$. □

REMARK. It is easy to construct examples of families $\mathcal{P}$ of functions p_j satisfying assumptions (3), (4), (13) such that, for some vector $\mathbf{f}$, $W(\mathbf{f}) < \infty$, but $W(-\mathbf{f}) = \infty$.

Let now the functions p_j satisfy conditions (3), (4), (13). Suppose that the vector $\mathbf{h}$ is such that

$$(16)\qquad W(\mathbf{h}) < \infty \quad \text{and} \quad W(-\mathbf{h}) < \infty.$$

Then, for every real number k, $W(k\mathbf{h}) < \infty$, by Lemma 2. Hence we may define the network $\mathcal{N}(k) = (\Gamma, \mathcal{P}, k\mathbf{h})$ for every real k. Let $\mathbf{i}(k) = (i_1(k), i_2(k), \dots)$ be the current generated by $k\mathbf{h}$ in $\mathcal{N}(k)$ according to Theorem 1. Note that in general, unlike the linear case, $\mathbf{i}(k)$ does not coincide with the cycle $k\mathbf{i}(1)$.

We set

$$w(k) = W(\mathbf{i}(k) + k\mathbf{h}),$$

and we want to study the behaviour of the function $w(k)$. The relevant properties of $w(k)$ are described by the following Theorem.

Theorem 2. *Suppose that the resistance functions p_j satisfy assumptions (3), (4), and the uniform variation condition (13). Assume that the vector $\mathbf{h}$ satisfies (16). Let $\mathbf{i}(k)$ be the current generated in $\mathcal{N}(k) = (\Gamma, \mathcal{P}, k\mathbf{h})$ by $k\mathbf{h}$, $k \in \mathbb{R}$. Then, the nonnegative function $w(k) = W(\mathbf{i}(k) + k\mathbf{h})$ is strictly increasing for $k \geq 0$, strictly* *decreasing for $k \leq 0$ and $w(0) = 0$. Moreover $w(k)$ is strictly convex and everywhere differentiable. We have*

$$w'(k) = k^{-1}E(\mathbf{i}(k) + k\mathbf{h}) \quad \text{if } k \neq 0 \tag{17}$$

$$w'(0) = 0. \tag{18}$$

Proof. We first prove that w is strictly increasing for $k \geq 0$ (the argument to show that $w(k)$ decreases for $k \leq 0$ is similar). Suppose that $0 \leq k_1 < k_2$. Then, by the minimality of $\mathbf{i}(k)$ and the convexity of W,

$$w(k_1) \leq W(k_2^{-1}k_1(\mathbf{i}(k_2) + k_2\mathbf{h})) \leq k_2^{-1}k_1 w(k_2) < w(k_2).$$

It is also clear that $w(0) = 0$.

Suppose now that $k_1 \neq k_2$ are real numbers and that λ_1, λ_2 are positive numbers such that $\lambda_1 + \lambda_2 = 1$. If $w(k_1) = w(k_2)$, then k_1 and k_2 have opposite sign and, by what has been just proved, $w(\lambda_1 k_1 + \lambda_2 k_2) < \lambda_1 w(k_1) + \lambda_2 w(k_2)$. If $w(k_1) \neq w(k_2)$, then $\mathbf{i}(k_1) + k_1\mathbf{h} \neq \mathbf{i}(k_2) + k_2\mathbf{h}$. Then, by minimality of $\mathbf{i}(\mathrm{k})$, and the strict convexity of W we have

$$\begin{aligned} w(\lambda_1 k_1 + \lambda_2 k_2) &\leq W\big(\lambda_1(\mathbf{i}(k_1) + k_1\mathbf{h}) + \lambda_2(\mathbf{i}(k_2) + k_2\mathbf{h})\big) \\ &< \lambda_1 w(k_1) + \lambda_2 w(k_2). \end{aligned}$$

Hence w is strictly convex.

We will now prove (17) and (18). Suppose first $k > 0$ and $t > 0$. By the minimality of $\mathbf{i}(k)$ again, we have

$$\begin{aligned} \frac{w(k+t) - w(k)}{t} &\leq \frac{W((1 + k^{-1}t)(\mathbf{i}(k) + k\mathbf{h})) - W(\mathbf{i}(k) + k\mathbf{h})}{t} \\ &= \sum_{j=1}^{\infty} \frac{1}{t} \int_{i_j(k)+kh_j}^{(1+k^{-1}t)(i_j(k)+kh_j)} p_j(s)ds. \end{aligned} \tag{19}$$

Note that $W((1 + k^{-1}t)(\mathbf{i}(k) + k\mathbf{h})) < \infty$ by Lemma 2. Now, always by Lemma 2, $W(2\mathbf{i}(k) + 2k\mathbf{h}) < \infty$. By the definition of the numbers $c_j^{\pm}$ there exists $j_1 \geq j_0$ such that, for all $j > j_1$, $(i_j(k) + kh_j)/c_j^+ < 1/2$ for $i_j(k) + kh_j > 0$ and $(i_j(k) + kh_j)/c_j^- < 1/2$ for $i_j(k) + kh_j < 0$. Therefore, if $t < k$,

$$\begin{aligned} \sum_{j>j_1} \frac{1}{t} \int_{i_j(k)+kh_j}^{(1+k^{-1}t)(i_j(k)+kh_j)} p_j(s)ds &\leq \frac{1}{2k} \sum_{j>j_1} (2i_j(k) + 2h_j)p_j(2i_j(k) + 2kh_j) \\ &\leq \frac{r}{2k} \sum_{j>j_1} M_j(2i_j(k) + 2kh_j) \\ &< \infty. \end{aligned}$$

Hence we may apply the dominated convergence theorem in (19) as $t \to 0+$. We obtain for the right derivative $w'_+(k)$

$$w'_+(k) \leq k^{-1} E(\mathbf{i}(k) + k\mathbf{h}). \tag{20}$$

If $t < 0$, then

$$\begin{aligned} \frac{w(k+t) - w(k)}{t} &\geq \frac{W((1+k^{-1}t)(\mathbf{i}(k)+k\mathbf{h})) - W(\mathbf{i}(k)+k\mathbf{h})}{t} \\ &= \sum_{j=1}^{\infty} \frac{1}{t} \int_{i_j(k)+kh_j}^{(1+k^{-1}t)(i_j(k)+kh_j)} p_j(s)ds \\ &\geq \frac{1}{k} \sum_{j=1}^{\infty} (i_j(k)+kh_j) p_j(i_j(k)+kh_j). \end{aligned} \tag{21}$$

From (21), we get for the left derivative $w'_-(k)$

$$w'_-(k) \geq k^{-1} E(\mathbf{i}(k) + k\mathbf{h}). \tag{22}$$

Since for a convex function we always have $w'_- \leq w'_+$, (20) and (22) imply (17) for $k > 0$. Arguing in the analogous way, one obtains (17) also for $k < 0$.

Finally, we prove (18). Let $1 > t > 0$. Then

$$\begin{aligned} \frac{w(t)}{t} &\leq \frac{W(t(\mathbf{i}(1)+\mathbf{h}))}{t} \\ &\leq \sum_{j=1}^{\infty} (i_j(1)+h_j) p_j(t(i_j(1)+h_j)). \end{aligned} \tag{23}$$

As before, there exists some $j_1 > j_0$ such that $(i_j(1)+h_j)p_j(t(i_j(1)+h_j)) < rM_j(i_j(1)+h_j)$ for all $j > j_1$. Hence, we get $w'_+(0) = 0$ from by the dominated convergence theorem and (23). By the same argument as for $t < 0$ we obtain (18). □

Lemma 3. *Let $\mathcal{N} = (\Gamma, \mathcal{P}, \mathbf{h})$ be a nonlinear network such that the functions p_j satisfy the uniform variation condition (13). Assume that $\mathbf{h}$ is finite. Then, denoting by $\mathbf{i}$ the current generated by $\mathbf{h}$ in $\mathcal{N}$,*

$$E(\mathbf{i} + \mathbf{h}) = \sum_{j=1}^{\infty} h_j p_j(i_j + h_j). \tag{24}$$

Proof. Since $E(\mathbf{i} + \mathbf{h}) < \infty$, the series $\sum_{j=1}^{\infty} i_j p_j(i_j + h_j)$ converges absolutely. To prove (24) it is sufficient to show that

$$\sum_{j=1}^{\infty} i_j p_j(i_j + h_j) = 0. \tag{25}$$

This is trivial if $\mathcal{N}$ is finite, since $\mathbf{i}$ is a cycle and satisfies (8). In the infinite case we argue as follows.

Since $\mathbf{i}$ minimizes $W(\mathbf{f}+\mathbf{h})$ among all cycles, we have

$$
\begin{aligned}
0 &= \lim_{t\to 0} \frac{W(\mathbf{i}+t\mathbf{i}+\mathbf{h}) - W(\mathbf{i}+\mathbf{h})}{t} \\
&= \lim_{t\to 0} \sum_{j=1}^{\infty} \frac{1}{t} \int_{i_j+h_j}^{(1+t)i_j+h_j} p_j(s)ds \qquad (26)\\
&= \lim_{t\to 0} \sum_{j=1}^{\infty} i_j p_j((1+t\theta_j)i_j + h_j),
\end{aligned}
$$

where $0 < \theta_j < 1$. Let $j_1 \geq j_0$ (where j_0 is as in (13)) be such that $h_j = 0$ for $j \geq j_1$. Then, for $j \geq j_1$ and $|t| < 1$,

$$
\begin{aligned}
0 \leq i_j p_j((1+t\theta_j)i_j + h_j) &= (h_j + i_j)p_j((1+t\theta_j)(i_j+h_j)) \\
&\leq 2(i_j+h_j)p_j(2(i_j+h_j)),
\end{aligned} \qquad (27)
$$

since $(i_j)p_j((1+t\theta_j)i_j) \geq 0$. But $W(2(\mathbf{i}+\mathbf{h})) < \infty$, by Lemma 2. Therefore by (13) and (14) there exists $j_2 \geq j_1$ such that $2(i_j+h_j)p_j(2(i_j+h_j)) \leq rM_j(2(i_j+h_j))$ for all $j \geq j_2$. By the dominated convergence theorem, we obtain (25) from (26). □

Let $\mathbf{i}$ be the current generated in $\mathcal{N}$ by $\mathbf{h}$. For every j, let x_j and y_j denote the first and the second endpoint of the j–th edge. It is well known that if a vector $\mathbf{f}$ annhilates all the finite cycles (i.e. $\langle \mathbf{z}, \mathbf{f}\rangle = 0$ for all finite cycle $\mathbf{z}$), then $\mathbf{f}$ admits a potential; see e.g. [3]. Therefore, there exists a potential function u (unique up to an additive constant), defined on the nodes of Γ, such that

$$
u(y_j) - u(x_j) = p_j(i_j + h_j) \qquad \text{for all all } j \qquad (28)
$$

Theorem 3 below shows that in nonlinear networks (with $\mathbf{h}$ finite) potential and total power dissipated are related in the same way as in the linear case.

Theorem 3. *Let $\mathcal{N} = (\Gamma, \mathcal{P}, \mathbf{h})$ be a nonlinear network such that $\mathbf{h}$ is a finite vector. Let u be the potential of the current $\mathbf{i}$ generated in $\mathcal{N}$ bt $\mathbf{h}$. Then, denoting by V the node set of Γ,*

$$
\sum_{x\in V} u(x)\partial\mathbf{h}(x) = E(\mathbf{i}+\mathbf{h}).
$$

Proof. First suppose $\mathcal{N}$ finite. Then, by Lemma 3, (28) and the definition (2) of boundary

$$
\begin{aligned}
E(\mathbf{i}+\mathbf{h}) &= \sum_j h_j p_j(i_j+h_j) = \sum_j h_j u(y_j) - \sum_j h_j u(x_j) \\
&= \sum_{x\in V} u(x) \sum_{j\in J(x)} \pm h_j = \sum_{x\in V} u(x)\partial\mathbf{h}(x).
\end{aligned}
$$

Suppose now $\mathcal{N}$ infinite, and let $\mathcal{N}_n = (\Upsilon_n, \mathcal{P}_n, \mathbf{h}_n)$ be an exhaustion of $\mathcal{N}$. Denote by V_n the node set of Υ_n and by U_1 the node set of Γ_1. We may assume that the support of $\mathbf{h}$ is contained in the edge set of Γ_1.

Chose any node v of Γ_1 and let u_n denote the potential of the current $\mathbf{i}_n$ generated in $\mathcal{N}_n$ by $\mathbf{h}_n$ such that $u_n(v) = 0$. Then, by what has been just proved,

$$E(\mathbf{i}_n + \mathbf{h}_n) = \sum_{x \in V_n} u_n(x)\partial_n \mathbf{h}_n(x) = \sum_{x \in U_1} u_n(x)\partial \mathbf{h}(x), \tag{29}$$

where ∂_n denotes the boundary of $\mathbf{h}_n$ on Υ_n. Now, for every j, $u_n(y_j) - u_n(x_j) = p_j(i_{j,n} + h_{j,n})$ converges to $p_j(i_j + h_j)$ as $n \to \infty$. On the other hand $u_n(v) = 0$ for all n. Therefore, by connectedness, $u_n(x)$ converges for all $x \in \Gamma$ to a function $u(x)$ which satisfies (28).

By Lemma 3 we have, as $n \to \infty$,

$$E(\mathbf{i}_n + \mathbf{h}_n) = \sum_j h_j p_j(i_{j,n} + h_{j,n}) \to \sum_j h_j p_j(i_j + h_j) = E(\mathbf{i} + \mathbf{h}).$$

Therefore, by (29) and Lemma 3,

$$E(\mathbf{i} + \mathbf{h}) = \lim_{n \to \infty} \sum_{x \in U_1} u_n(x)\partial \mathbf{h}(x) = \sum_{x \in V} u(x)\partial \mathbf{h}(x)$$

□

Suppose now that $\mathbf{h}$ is a finite vector such that there exists two nodes x_1 and x_2 such that

$$\begin{aligned} \partial \mathbf{h}(x_1) &= -\partial \mathbf{h}(x_2) \neq 0 \\ \partial \mathbf{h}(x) &= 0 \quad \text{if } x \neq x_i,\ i = 1, 2. \end{aligned} \tag{30}$$

Such a vector is also called a dipole with poles x_1 and x_2. The following Corollary is an immediate consequence of Theorem 3 and Lemma 3.

Corollary. *Let, as above, $\mathcal{N}(k) = (\Gamma, \mathcal{P}, k\mathbf{h})$ and suppose that $\mathbf{h}$ is a dipole. Let $u^{(k)}$ be the potential of the current $\mathbf{i}(k)$ generated by $k\mathbf{h}$ in $\mathcal{N}(k)$. Then,*

$$u^{(k)}(x_1) - u^{(k)}(x_2) = k^{-1}E(\mathbf{i}(k) + k\mathbf{h}) = \sum_j h_j p_j(i_j(k) + kh_j). \tag{31}$$

REMARK. The Corollary shows that, as in the linear case, the effective resistance between two nodes is $w'(k)$ (see equations (17) and (18)). By Theorem 1, the resistance function w' has the same properties (3) and (4) as the individual resistances of the network. In this sense the network is equivalent to a single nonlinear resistor with endpoints x_1 and x_2.

5. Minimal orientations.

We have so far supposed that the graph Γ is given a fixed orientation. This corresponds to fixing the polarities of the resistors. In this section we investigate the behaviour of the network if one or more polarities are reversed i.e., if one or more edges are given the opposite orientation. From a mathematical point of view we leave fixed the orientation of the graph but replace one or more functions p_j by their symmetric. We proceed as follows.

Let Γ be as before and let the functions p_j satisfy (3) and (4). Let

$$\omega = (\omega_1, \omega_2, \dots) \tag{32}$$

be any infinite sequence such that either $\omega_j = 1$ or $\omega_j = -1$, for all $j = 1, 2, \dots$. The set of all such sequences is in fact a compact topological group under coordinatewise multiplication and coordinatewise convergence topology. Such a group is known as the Cantor group and will be denoted (as usual) by $\mathbb{D}$.

We set for all ω as in (32) and all j

$$p_j^{\omega}(t) = \omega_j p_j(\omega_j t) \qquad \text{for } t \in \mathbb{R}.$$

The family of all functions p_j^{ω} is denoted by $\mathcal{P}^{\omega}$. Clearly, for every j, p_j^{ω} satisfies (3) and (4). Let $\mathbf{h}$ be a vector such that

$$W(\mathbf{h}) + W(-\mathbf{h}) < \infty. \tag{33}$$

Then, for all $\omega \in \mathbb{D}$, $\sum_{j=1}^{\infty} \int_0^{h_j} \omega_j p_j(\omega_j t)dt < \infty$ so that the vector $\mathbf{h}$ generates a unique current $\mathbf{i}(\omega) = (i_1(\omega), i_2(\omega), \dots)$ in the network $\mathcal{N}^{\omega} = (\Gamma, \mathcal{P}^{\omega}, \mathbf{h})$.

Set $\omega\mathbf{f} = (\omega_1 f_1, \omega_2 f_2, \dots)$ for all vectors $\mathbf{f}$ and all sequences $\omega \in \mathbb{D}$. Then, the current $\mathbf{i}(\omega)$ minimizes the functional

$$W(\omega(\mathbf{f}+\mathbf{h})) = \sum_{j=1}^{\infty} \int_0^{\omega_j(f_j+h_j)} p_j(t)dt = \sum_{j=1}^{\infty} \int_0^{f_j+h_j} \omega_j p_j(\omega_j t)dt.$$

among all cycles $\mathbf{f}$. We will use the notation

$$w(\omega) = W(\omega(\mathbf{i}(\omega) + \mathbf{h})).$$

If the resistances are not odd symmetric, then it is easy to see that $\mathbf{i}(\omega)$ and $w(\omega)$ depend in general on ω. However, we can show that there exist elements $\omega \in \mathbb{D}$ which minimize $w(\omega)$.

Theorem 4. *Let $\mathcal{N}^{\omega} = (\Gamma, \mathcal{P}^{\omega}, \mathbf{h})$ be as above. Then, there exists a closed nonempty subset $H \in \mathbb{D}$ such that, for all $\overline{\omega} \in H$,*

$$w(\overline{\omega}) = \min\{w(\omega) : \omega \in \mathbb{D}\} = m.$$

Proof. We first prove that w is a lower semicontinuous function from $\mathbb{D}$ to $\mathbb{R}^+$. Let $\omega_n = (\omega_{1,n}, \omega_{2,n}, \dots)$ tend to $\omega = (\omega_1, \omega_2, \dots)$ in $\mathbb{D}$ as $n \to \infty$. Then $\omega_{j,n} \to \omega_j$ for all j as $n \to \infty$.

Set $\alpha = \liminf_{n\to\infty} w(\omega_n)$. Then, there exists a subsequence, which we still denote by ω_n, such that

$$w(\omega_n) \to \alpha \quad (\text{and } \omega_n \to \omega). \tag{34}$$

As in the proof of Theorem 1,

$$w(\omega_n) = \sum_{j=1}^{\infty} \int_0^{\omega_{j,n}(i_j(\omega_n)+h_j)} p_j(t)dt \le W(h) + W(-h) < \infty.$$

Hence, for every j there exists a constant a_j such that for every n

$$|i_j(\omega_n)| \le a_j + |h_j| \le a_j + |h_j| = b_j.$$

Since the set

$$K = \{\mathbf{f} : |f_j| \le b_j \quad \text{for all } j = 1, 2, \dots\},$$

is compact in S, we see that, passing if necessary to a further subsequence, $\mathbf{i}(\omega_n)$ converges to a vector $\mathbf{g} = (g_1, g_2, \dots)$.

Clearly $\mathbf{g}$ is a cycle and $\omega_j(g_j + h_j) = \lim_{n\to\infty} \omega_{j,n}(i_j(\omega_n) + h_j)$ for all j. By Fatou's lemma and (34)

$$\begin{aligned} W(\omega(\mathbf{g}+\mathbf{h})) &= \sum_{j=1}^{\infty} \int_0^{\omega_j(g_j+h_j)} p_j(t)dt \\ &\le \liminf_{n\to\infty} \sum_{j=1}^{\infty} \int_0^{\omega_{j,n}(i_j(\omega_n)+h_j)} p_j(t)dt \\ &= \alpha. \end{aligned}$$

On the other hand, by the definition of current,

$$w(\omega) \le W(\omega(\mathbf{g}+\mathbf{h})) \le \alpha. \tag{35}$$

Therefore w is lower semicontinuous on $\mathbb{D}$. Since $\mathbb{D}$ is compact, the set H is nonempty. We now show that H is closed.

Let $\omega_n \in H$ converge to some $\omega \in \mathbb{D}$. As in the first part of the proof, passing to a subsequence if necessary, $\mathbf{i}(\omega_n)$ converges in S to a cycle $\mathbf{g}$ which satisfies (35). In this case α is equal to m, the minimum value of the function w on $\mathbb{D}$. Then, $m \le w(\omega) \le W(\omega(\mathbf{g}+\mathbf{h})) \le m$, so that $\omega \in H$ (and $\mathbf{g} = \mathbf{i}(\omega)$). □

The current $\mathbf{i}(\overline{\omega})$, where $\overline{\omega} \in H$, generated by $\mathbf{h}$ in $\mathcal{N}^{\overline{\omega}}$ will be called a minimal current.

If $p_j(t)$ is comparable with $-p_j(-t)$ we can estimate $w(\omega)$ in termes of its minimum m.

Corollary. *Suppose that there exist positive constants a and b such that*

$$-ap_j(-t) \leq p_j(t) \leq -bp_j(-t)$$

for all j and all t such that $M_j(t) \leq W(\mathbf{h}) + W(-\mathbf{h})$. Then, for all $\omega \in \mathbb{D}$, $\mathcal{W}(\omega) \leq \kappa m$, where m is as in Theorem 4 and $\kappa = \max\{a^{-1}, b\}$.

Proof. For all $\omega \in \mathbb{D}$ and all j

$$M_j(\omega_j(i_j(\omega) + h_j) \leq \mathcal{W}(\omega) \leq W(\omega\mathbf{h}) \leq W(\mathbf{h}) + W(-\mathbf{h}),$$

by Theorem 1. If $\overline{\omega} \in H$, then, for all ω

$$\begin{aligned}\sum_{j=1}^{\infty} \int_0^{i_j(\omega)+h_j} \omega_j p_j(\omega_j t)dt &\leq \sum_{j=1}^{\infty} \int_0^{i_j(\overline{\omega})+h_j} \omega_j p_j(\omega_j t)dt \\ &\leq \kappa \sum_{j=1}^{\infty} \int_0^{i_j(\overline{\omega})+h_j} \overline{\omega}_j p_j(\overline{\omega}_j t)dt \\ &= \kappa m.\end{aligned}$$

□

Observe now that if, for some $\overline{\jmath}$,

$$p_{\overline{\jmath}}(t) < -p_{\overline{\jmath}}(-t) \qquad \text{for all } t > 0, \tag{36}$$

then, for every $\omega \in H$, $\omega_{\overline{\jmath}}(i_{\overline{\jmath}}(\omega) + h_{\overline{\jmath}}) \geq 0$. Otherwise, let $\overline{\omega}$ denote the element of $\mathbb{D}$ such that $\overline{\omega}_j = \omega_j$ if $j \neq \overline{\jmath}$, and $\overline{\omega}_{\overline{\jmath}} = -\omega_{\overline{\jmath}}$. Then, $W(\overline{\omega}(\mathbf{i}(\omega)+\mathbf{h})) < W(\omega(i(\omega)+\mathbf{h}))$, which is absurd.

This remark suggests that we can get more information on the set of all the ω which minimize $\mathcal{W}$ and on the corresponding minimal currents $\mathbf{i}(\omega)$, by requiring a condition similar to (36) for every edge. Namely, we assume that there exists $\omega^* = (\omega_1^*, \omega_2^*, \dots)$ such that, for every j,

$$\omega_j p_j(\omega_j t) \geq \omega_j^* p_j(\omega_j^* t), \quad \text{for all } t \geq 0 \text{ and all } \omega = (\omega_1, \omega_2, \dots) \in \mathbb{D}. \tag{37}$$

We then have

Theorem 5. *Let $\mathcal{N}^\omega = (\Gamma, \mathcal{P}^\omega, \mathbf{h})$ be as in Theorem 4 and assume that the functions p_j satisfy (37) for some $\omega^* \in \mathbb{D}$. If $H \subseteq \mathbb{D}$ is the minimizing set of Theorem 4, then there exists a cycle $\mathbf{g} = (g_1, g_2, \dots)$ such that $\mathbf{i}(\omega) = \mathbf{g}$ for all $\omega \in H$. Moreover, there exists a subgroup $G \subseteq \mathbb{D}$ such that H is a coset of G.*

Proof. We may assume, without loss of generality, that $\omega^* = (1, 1, \dots)$. Then, for all $\omega \in \mathbb{D}$,

$$\sum_{j=1}^{\infty} \int_0^{\omega_j(i_j(\omega)+h_j)} p_j(t)dt \geq \sum_{j=1}^{\infty} \int_0^{|i_j(\omega)+h_j|} p_j(t)dt. \tag{38}$$

Let $\tilde{p}_j$ be the function obtained by symmetrizing p_j i.e.,

$$\tilde{p}_j(t) = \begin{cases} p_j(t) & \text{if } t \geq 0 \\ -p_j(-t) & \text{if } t < 0 \end{cases}.$$

Then, for all j

$$\int_0^{|i_j(\omega)+h_j|} p_j(t)dt = \int_0^{i_j(\omega)+h_j} \tilde{p}_j(t)dt. \tag{39}$$

The functions $\tilde{p}_j$ satisfy assumptions (3) and (4), so that there exixts a unique current $\mathbf{g}$ generated in $\tilde{\mathcal{N}} = (\Gamma, \tilde{\mathcal{P}}, \mathbf{h})$ by $\mathbf{h}$ (here $\tilde{\mathcal{P}}$ is the family of functions $\tilde{p}_j$).

Let $\epsilon = (\epsilon_1, \epsilon_2, \dots)$ be such that $|g_j + h_j| = \epsilon_j(g_j + h_j)$. Then, by (38) and the minimality of $\mathbf{g}$, we have for all $\omega \in \mathbb{D}$

$$\begin{aligned} \mathcal{W}(\omega) &= \sum_{j=1}^{\infty} \int_0^{\omega_j(i_j(\omega)+h_j)} p_j(t)dt \\ &\geq \sum_{j=1}^{\infty} \int_0^{\epsilon_j(g_j+h_j)} p_j(t)dt \\ &= \mathcal{W}(\epsilon). \end{aligned}$$

Hence $\epsilon \in H$ and $\mathbf{g} = \mathbf{i}(\epsilon)$. Furthermore, assume that $\omega \in \mathbb{D}$ is such that $\mathbf{i}(\omega) \neq \mathbf{g}$. Then, by the uniqueness of the current in $\tilde{\mathcal{N}}$, (38) and (39)

$$\sum_{j=1}^{\infty} \int_0^{\omega_j(i_j(\omega)+h_j)} p_j(t)dt > \sum_{j=1}^{\infty} \int_0^{\epsilon_j(g_j+h_j)} p_j(t)dt.$$

It follows that $\omega \notin H$.

To prove the second assertion of the Theorem, let J denote the subset of all integers j such that $p_j(t) = -p_j(-t)$ for all $|t| \leq g_j + h_j$. Set

$$G = \{\omega \in \mathbb{D} : \omega_j = 1 \quad \text{if } j \notin J\}$$

Then, we claim that $H = \epsilon G$. If $\omega \in \epsilon G$, then clearly

$$W(\epsilon(\mathbf{g}+\mathbf{h})) = W(\omega(\mathbf{g}+\mathbf{h})). \tag{40}$$

Conversely, if (40) holds, then $\omega_j = \epsilon_j$ for all $j \notin J$. Otherwise,

$$\int_0^{\omega_j(g_j+h_j)} p_j(t)dt > \int_0^{\epsilon_j(g_j+h_j)} p_j(t)dt$$

for some $j \notin J$, whence $\mathcal{W}(\omega) > \mathcal{W}(\epsilon)$. □

REMARK. If condition (37) holds with strict inequality for every j, then the H contains only ω^* and the minimal current is unique.

It is easy to see that, if condition (37) is not satisfied, then the we may have different minimal currents. Let for instance Γ be the graph with two vertices and two edges. Suppose that the edges are both oriented from the vertex x to the vertex y, so that the cycles are the vectors $(\lambda, -\lambda)$, $\lambda \in \mathbb{R}$. Let $p_1(t) = 4t$ for all t and $p_2(t) = 8(\sqrt{10}-4)^{-1}(\sqrt{10}-2)t^2$ if $t > 0$ and $p_2(t) = 4t$ if $t \leq 0$. Let $\mathbf{h} = (1/2, 1/2)$.

Then, there are only two current vectors and they have the same energy. The current vectors are $(0,0)$ and $((\sqrt{10}-3)/2, -(\sqrt{10}-3)/2)$.

References

[1] De Michele L. and Soardi P.M., *A Thomson's principle for infinite, nonlinear resistive networks*, Proc.Amer.Math.Soc. **109** (1990), 461–468.

[2] Dolezal, V, *Nonlinear Networks*, Elsevier, New York, 1977.

[3] Flanders, H., *Infinite networks I-Resistive networks*, IEEE Trans. Circuit Theory **18** (1971), 326–331.

[4] Schlesinger, E., *Infinite networks and Markov chain*, Boll.Un.Mat.Ital. **6–B** (1992), 23–37.

[5] Soardi, P.M., *Morphisms and currents in infinite nonlinear networks*, preprint (1992).

[6] Soardi P.M. and Yamasaki M., *Classification of infinite networks and its applications*, Circuits, Systems and Signal Processing **12** (1993), 143–149.

[7] Thomassen, C., *Resistances and currents in infinite electrical networks*, J. Combinatorial Theory, Series B **49** (1990), 87–102.

[8] Woess, W., *Random walks on infinite graphs and groups – a survey on selected papers*, Bull.London.Math.Soc. (to appear).

[9] Woo, J.Y.T., *On modular sequence spaces*, Studia Math. **48** (1973), 271–289.

[10] Zemanian, A.H., *Infinite Electrical Networks*, Cambridge University Press, Cambridge, 1991.

[11] Zemanian, A.H., *An electrical gridlike structure excited at infinity*, Math. Control Signal Systems **4** (1991), 217–231.

[12] Zemanian, A.H., *The limb analysis of countably infinite electrical networks*, J. Combinatorial Theory, Series B **24** (1978), 76–93.

[13] Zemanian, A.H., *Countably infinite, nonlinear, time-varying, active electrical networks*, SIAM J. Math. Anal. **10** (1979), 1193–1198.

10. About a Singular Parabolic Equation Arising in Thin Film Dynamics and in the Ricci Flow for Complete $I\!R^2$

Emmanuele DiBenedetto Mathematics Department, Northwestern University, Evanston, Illinois

David J. Diller Department of Mathematics, Northwestern University, Evanston, Illinois

1. Introduction

Let $u(x,t)$ be the thickness at the location $x \equiv (x_1, x_2) \in I\!R^2$ and at time $t \in I\!R^+$, of a thin (100Å–1000Å) colloidal film spread over a flat surface. According to [1,9], if the van der Waals forces are repulsive, a model for the evolution of u is the Cauchy problem,

$$(1.1) \qquad u_t - \Delta \ln u = 0 \quad \text{in} \quad S_T \equiv I\!R^2 \times (0,T), \quad \text{and} \quad u(\cdot, 0) = u_o,$$

for some given initial distribution u_o of the film. A weak solution to (1.1) is a non negative measurable function u defined in S_T, such that u and $\ln u$ are in $L^1_{\text{loc}}(S_T)$, which satisfies equation in $\mathcal{D}'(S_T)$ and takes the initial datum u_o in the sense of $L^1_{\text{loc}}(I\!R^2)$. The following are a one-parameter and respectively two-parameters families of explicit solutions of (1.1).

$$(1.2) \qquad u(x,t) = \frac{C}{e^{4t/C} + |x|^2}, \quad C>0; \qquad u(x,t) = \frac{8\lambda(T-t)_+}{(\lambda + |x|^2)^2}, \quad \lambda, T > 0.$$

Let $\{\Sigma; ds\}$ be a 2-dimensional, orientable, simply connected, non compact Riemaniann surface with metric $ds^2 = g_{ij}dx_i dx_j$, $i,j = 1,2$, where g_{ij} is the first fundamental form of Σ. The Gauss curvature K can be computed only in terms of the g_{ij}, and in a system of rectangular coordinates $(g_{12} \equiv 0)$ takes the form ([8] p.139)

$$(1.3) \qquad K = -\frac{1}{2\sqrt{g_{11}g_{12}}} \left\{ \frac{\partial}{\partial x_1} \frac{1}{\sqrt{g_{11}g_{22}}} \frac{\partial}{\partial x_1} g_{11} + \frac{\partial}{\partial x_2} \frac{1}{\sqrt{g_{11}g_{22}}} \frac{\partial}{\partial x_2} g_{22} \right\}.$$

The Ricci flow for $\{\Sigma; ds\}$, consists of evolving the metric ds^2 by its scalar curvature $R = 2K$, i.e., [6,10]

$$(1.4) \qquad \frac{\partial}{\partial t} ds^2 = -R ds^2 \quad \text{or equivalently} \quad \frac{\partial}{\partial t} g_{ij} = -2K g_{ij}, \qquad t \in I\!R^+.$$

---

[1] Department of Mathematics Northwestern Univ. Evanston Ill. 60208 USA. Partially Supported by NSF Grant DMS-9404379

If the g_{ij} are sufficiently regular, there exists a local system of rectangular coordinates and a positive, smooth function u, defined on a subset of $\mathbb{R}^2$, such that ([6] p. 15)

$$ds^2 = u(x)d|x|^2, \qquad d|x| \equiv \sqrt{dx_1^2 + dx_2^2}. \tag{1.5}$$

From (1.3) with $g_{ii} = u$, $i = 1,2$ we compute $2K = -u^{-1}\Delta \ln u$. Therefore by (1.4), the Ricci flow evolves, locally, by the p.d.e. in (1.1). Let $dA = u dx_1 dx_2$ denote the element of area on Σ. Then $K dA = -\frac{1}{2}(\Delta \ln u) dx_1 dx_2$. Therefore if u in (1.5) is defined in the whole $\mathbb{R}^2$, the total curvature of $\{\Sigma; ds\}$ is calculated from

$$\int_\Sigma K dA = -\frac{1}{2}\int_{\mathbb{R}^2} \Delta \ln u(x) dx \quad \text{provided the integrals are convergent.} \tag{1.6}$$

The metric ds is complete if Σ, endowed with the topology generated by ds is a complete metric space. Equivalently ds is complete if every divergent path on Σ has infinite length ([8] p. 220). A sufficient condition on the function u in (1.5), for $\{\Sigma; ds\}$ to be complete is that ([7] p. 55)

$$\int_\sigma \sqrt{u(|x|)} d|x| = \infty \quad \text{for every divergent locally rectifiable path } \sigma \in \Sigma.$$

As an example we have ([7] p. 61)

$$ds^2 = \frac{|dx|^2}{[|x| \ln(1+|x|)]^2} \qquad \text{is complete since} \qquad \int_1^\infty \frac{d|x|}{|x| \ln(1+|x|)} = \infty. \tag{1.7}$$

Thus completeness of $\{\Sigma; ds\}$ is related to the asymptotic behavior of $u(x)$ as $|x| \to \infty$. Completeness is also related to the total curvature by the Cohn–Vossen Theorem [2],

$$\{\Sigma; ds\} \text{ complete} \quad \Longrightarrow \quad \{\text{total curvature}\} \le 2\pi\chi, \tag{1.8}$$

where χ is the Euler–Poincaré characteristic of Σ. If Σ is compact, it is homeomorphic to a sphere with a finite number of *"handles"*. Then the Euler–Poincaré characteristic of Σ is defined by

$$\chi(\Sigma) \equiv 2(1-n) \quad \text{where } n \text{ is the number of handles.}$$

Thus for a sphere $\chi = 2$ and for a torus $\chi = 0$. If Σ is homeomorphic to a compact surface Σ' from which a finite number of points have been removed, then the Euler–Poincaré characteristic of Σ is

$$\chi(\Sigma) = \chi(\Sigma') - m \quad \text{where } m \text{ is the number of points removed out of } \Sigma'.$$

As an example take $\Sigma \equiv \mathbb{R}^2$. The plane is homeomorphic to a sphere from which one point has been removed. Thus $\chi(\mathbb{R}^2) = 1$. For equivalent definitions of χ, via triangulations we refer to [7] pp. 211-217 and [6].

We now take $\Sigma \equiv \mathbb{R}^2$ and combine the previous remarks by interpreting the Cauchy problem (1.1) as follows. Given an initial metric $ds_o^2 = u_o d|x|^2$ we evolve it by the Ricci flow to find at each time $t \in \mathbb{R}^+$ a new metric $ds_t^2 \equiv u(\cdot,t)d|x|^2$. In view of (1.7) such a metric is complete if

$$u(x,t) \geq \frac{C_o(t)}{[|x| \ln |x|]^2} \quad \text{for} \quad |x| \geq C_1(t), \tag{1.9}$$

for two continuous positive functions $t \to C_i(t)$, $i=1,2$. Also, by combining (1.8) and (1.6) and using that $\chi(\mathbb{R}^2)=1$, we conclude that such a metric cannot be complete, if

$$-\int_{\mathbb{R}^2} \Delta \ln u(\cdot,t)dx = 4\pi\, s \quad \text{for some } s>1. \tag{1.10}$$

1.1. Main Results

In [4] we derived a number of a priori estimates for a possible solution to (1.1). One of the conclusions of [4] is that, if the space dimension $N \geq 3$, the Cauchy problem (1.1) is ill posed in the sense that it cannot be solved for any initial datum $u_o \in C_o^\infty(\mathbb{R}^N)$. Indeed it cannot be solved for any initial datum $u_o \in L^1(\mathbb{R}^N)$. The case of space dimension $N=2$, left open in [4], is the object of this note. One of the main results is that even though (1.1) has a solution for *any* $u_o \in L^1(\mathbb{R}^2)$, the problem is still ill–posed since flagrant non–uniqueness occurs. Precisely (see Theorem 5.1) for every $u_o \in L^1(\mathbb{R}^2)$ there exists a continuous family of solutions u^s parametrized with $s>1$. These solutions all satisfy (1.10) and therefore the corresponding metrics they generate in $\mathbb{R}^2$ via (1.5) are not complete. Of these, the one corresponding to $s=1$ is the largest and we call it *maximal*. Even though it satisfies (1.10) with $s=1$ this is not sufficient to conclude that it generates a complete metric in $\mathbb{R}^2$, since the Cohn–Vossen Theorem (1.8) only provides a necessary condition for completeness. We show however that such a maximal solution satisfies (1.9) and therefore it generates a family of complete metrics in $\mathbb{R}^2$. There are two points worth making. The first is that an initial metric $ds_o^2 = u_o d|x|^2$, whether complete or not, always generates a family of complete metrics in $\mathbb{R}^2$, along with a continuoum of families of non–complete metrics. The second is that these families, complete or not, have a finite time span, i.e., they become extinct at least by time $T^s \equiv \|u_o\|_{1,\mathbb{R}^2}/4\pi s$ (Proposition 4.1). In terms of the physical models of [1,9], a droplet of fluid, or even a distribution of fluid of finite mass $u_o \in L^1(\mathbb{R}^2)$, generates a family of distributions of mass $u^s(\cdot,t)$, $s \geq 1$. They all disappear after a time $T^s = \|u_o\|_{1,\mathbb{R}^2}/4\pi s$, and the largest of them, lasts the longest, i.e., up to time $T = \|u_o\|_{1,\mathbb{R}^2}/4\pi$. We are uncertain of the physical significance of such a phenomenon.

This note builds on the investigations of [4,5]. The solutions u^s, $s \geq 1$ are constructed by a semigroup technique of time–discretization(Section 6). This leads to the solvability of singular elliptic equations of the type $\lambda u - \Delta \ln u = f$. Such a preliminary study has been carried out in [5]. A key fact in constructing solutions of (1.1), as wells as deriving their main qualitative and/or quantitative properties is a discrete version of a Harnack–type estimate established in [4].

2. Maximal Solutions and the Maximum Principle

A solution u to (1.1) is *maximal* if for any solution v with $v(\cdot,0) \leq u(\cdot,0)$ a.e. in $\mathbb{R}^2$, then $v \leq u$ a.e. in S_T. Thus a maximal solution is the largest among all the solutions generated by the same initial datum. The next Proposition asserts that a solution u is maximal if it does not decay too fast as $|x| \to \infty$.

PROPOSITION 2.1. *Let u be a solution to (1.1) in S_T and assume that there are positive constants R and C such that*

$$u(x,t) \geq C/|x|^2 \ln|x| \qquad \forall |x| \geq R. \tag{2.1}$$

Then u is maximal.

PROOF: For $\rho > 0$ and $\sigma \in (0,1)$, let $\zeta(\cdot)$ denote a non negative smooth cutoff function in $B_{\rho(1+\sigma)}$, such that

$$\begin{cases} \zeta \equiv 1 \text{ on } B_\rho \text{ and vanishes on } \partial B_{\rho(1+\sigma)}. \text{ Moreover } |\zeta_{x_i}| \leq \gamma/\sigma\rho \text{ and,} \\ |\zeta_{x_i x_j}| \leq \gamma/\sigma^2\rho^2 \text{ for some positive constant } \gamma \text{ and for all } i,j = 1,2. \end{cases}$$

For $n \in I\!N$ let $S_n(\cdot)$ denote a monotone approximation to the sign$(\cdot)$ function. We write the weak formulation of (1.1) for both u and v for the testing function $S_n(\ln v - \ln u)_+\zeta$, take their difference and integrate by parts. After we let $n \to \infty$, we obtain

$$\int_{B_\rho} (v-u)_+(x,t)dx \leq \frac{\gamma}{(\sigma\rho)^2} \int_0^t \int_{B_{\rho(1+\sigma)} \setminus B_\rho} (\ln v - \ln u)_+ dx, \quad \forall t \in (0,T). \tag{2.2}$$

Next, for all $\alpha \in (0,1]$ there is a constant γ_α such that for all positive numbers k and h,

$$(\ln k - \ln h)_+ = \ln\left\{1 + \frac{(k-h)_+}{h}\right\} \leq \gamma_\alpha \frac{(k-h)_+^\alpha}{h^\alpha}.$$

We use this inequality in (2.2), make use of the assumption (2.1) and integrate the resulting inequality in dt over $(0,T)$. This gives,

$$\begin{aligned} M_\rho \equiv \int_0^T \int_{B_\rho} (v-u)_+(x,t)dxdt &\leq \frac{\gamma\gamma_\alpha T(\ln\rho)^\alpha}{C^\alpha\sigma^2\rho^{2(1-\alpha)}} \int_0^T \int_{B_{\rho(1+\sigma)}\setminus B_\rho} (v-u)_+^\alpha dxdt \\ &= \frac{\gamma\gamma_\alpha T(\ln\rho)^\alpha}{C^\alpha\sigma^2\rho^{2(1-\alpha)}} \left\{ \int_0^T \int_{B_{\rho(1+\sigma)}} (v-u)_+^\alpha dxdt - \int_0^T \int_{B_\rho} (v-u)_+^\alpha dxdt \right\}. \end{aligned} \tag{2.3}$$

For $\rho > 0$ fixed, define sequences ρ_n and $\sigma_n \in (0,1)$ by

$$\rho_n \equiv \rho(2 - 2^{-n}), \quad \rho_{n+1} = \rho_n(1+\sigma_n), \quad M_n \equiv M_{\rho_n}, \qquad n = 0,1,2,\ldots.$$

We write (2.3) over the pair of balls $B_{\rho_{n+1}}$ and B_{ρ_n}, for which $\sigma_n \geq 2^{-(n+2)}$. Then we disregard the last non positive term on the right hand side, and majorize the first term by the Hölder inequality. This yields,

$$\begin{aligned} M_n &\leq \tilde{\gamma}_{\alpha,T}(\ln 2\rho)^\alpha 4^n M_{n+1}^\alpha & \tilde{\gamma}_{\alpha,T} &\equiv 8\gamma\gamma_\alpha T C^{-\alpha}\pi^{1-\alpha} \\ &\leq \varepsilon M_{n+1} + \gamma(\varepsilon,\alpha,T)(\ln 2\rho)^{\frac{\alpha}{1-\alpha}} 4^{\frac{n}{1-\alpha}}; & \gamma(\varepsilon,\alpha,T) &\equiv \{\alpha^\alpha \gamma_{\alpha,T}/\varepsilon^\alpha\}^{1/1-\alpha}, \end{aligned}$$

where $\varepsilon > 0$ is to be chosen. From these inequalities, by iteration we find

$$M_\rho \equiv M_{\rho_o} \leq \varepsilon^n M_n + \gamma(\varepsilon, \alpha, T)(\ln 2\rho)^{\frac{\alpha}{1-\alpha}} \sum_{i=0}^{n} \left(\varepsilon 4^{\frac{1}{1-\alpha}}\right)^n .$$

We choose ε so small that $\varepsilon 4^{1/\alpha - 1} \leq \frac{1}{2}$ and let $n \to \infty$, to conclude that, for every $\alpha \in (0,1)$ there exists a constant $\gamma = \gamma(\alpha, T)$ such that

$$\int_0^T \int_{B_\rho} (v-u)_+ dxdt \leq \gamma(\alpha, T)(\ln \rho)^{\frac{\alpha}{1-\alpha}}, \qquad \text{for all } \rho > 0. \tag{2.4}$$

We now return to (2.3) and write it for $\alpha = 1$ and over a sequence of balls of radii $\rho_n = 2^n \rho$, $n = 0, 1, 2, \ldots$. For these we have $\sigma = 1$, and

$$M_{2^n \rho} \leq \frac{\overline{\gamma} n}{1 + \overline{\gamma} n} M_{2^{n+1}\rho}, \qquad \overline{\gamma} \equiv \gamma \gamma_1 T \ln 2\rho / C.$$

We iterate these inequalities and make use of (2.4) to obtain,

$$M_\rho \leq \prod_{i=1}^{n} \frac{\overline{\gamma} n}{1 + \overline{\gamma} n} M_{2^n \rho} \leq \gamma(\alpha, T) \prod_{i=1}^{n} \frac{\overline{\gamma} n}{1 + \overline{\gamma} n} \left(\ln 2^{n+1}\rho\right)^{\frac{\alpha}{1-\alpha}} . \tag{2.5}$$

We now observe that there are constants $\gamma_o \geq 1$ and $\eta \in (0,1)$ that can be determined a priori only in terms of $\overline{\gamma}$ such that

$$\prod_{i=1}^{n} \frac{\overline{\gamma} n}{1 + \overline{\gamma} n} \leq \gamma_o n^{-\eta}.$$

We stress that η is independent of α. Substituting this in (2.5) gives

$$\int_0^T \int_{B_\rho} (v-u)_+ dxdt \leq \gamma_o \gamma(\alpha, T)(\ln 2\rho)^{\frac{\alpha}{1-\alpha}} n^{\frac{\alpha}{1-\alpha} - \eta}.$$

We choose α so small that the exponent of n is negative and let $n \to \infty$. ∎

3. Maximality, Stability and Semiconvexity

PROPOSITION 3.1. **(i).** *If u is a solution of (1.1) and $u \geq \varepsilon > 0$, then u is maximal.*

(ii) *If v is a solution of (1.1) with initial datum u_o, then there is a maximal solution u in S_T associated with u_o.*

(iii) *Every maximal solution u satisfies the semi-convexity inequality*

$$u_t(x,t) \leq u(x,t)/t \quad \text{in} \quad \mathcal{D}'(S_T). \tag{3.1}$$

PROOF OF **(ii)**: Let u^ε be the solution of (1.1) corresponding to the initial datum $u_o+\varepsilon$. These solutions can be constructed as limits, as $n\to\infty$ of the boundary value problems

$$\begin{cases} \frac{\partial}{\partial\tau}u_n^\varepsilon - \Delta\ln u_n^\varepsilon = 0 & \text{in } \{|x|<n\}\times(0,T); \\ u_n^\varepsilon(\cdot,t)\Big|_{|x|=n} = \varepsilon, & u_n^\varepsilon(\cdot,0)=u_o+\varepsilon. \end{cases} \tag{3.2}$$

By the maximum principle,

$$\varepsilon\le u_n^\varepsilon\le u_{n+1}^\varepsilon \text{ in } \{|x|<n\}\times(0,T), \text{ and } u_n^{\varepsilon'}\le u_n^\varepsilon, \text{ for all } n\in\mathbb{N} \text{ and } \varepsilon'\le\varepsilon.$$

Fix $\rho>0$ and in (3.2) for $n>2\rho$, use the testing function $S_\eta\left[(u_n^\varepsilon-1)_+\right]\zeta$ where $S_\eta(\cdot)$ is a monotone approximation to the sign$(\cdot)$ function, and $\zeta(\cdot)$ is the standard cutoff function in $B_{\rho(1+\sigma)}$ that equals one in B_ρ. By standard calculations and letting $\eta\searrow 0$,

$$\int_{B_\rho}\max\{u_n^\varepsilon(x,t);1\}dx\le\|u_o\|_{1,\mathbb{R}^2}+\varepsilon|B_{2\rho}|+\frac{\gamma}{\sigma^2\rho^2}\int_0^t\int_{B_{\rho(1+\sigma)}}\ln\max\{u_n^\varepsilon(x,\tau);1\}dxd\tau.$$

We integrate in dt over $(0,T)$ and estimate $\ln\max\{u_n^\varepsilon;1\}\le\gamma\max\{u_n^\varepsilon;1\}^{\frac12}$. This gives

$$\begin{aligned} M_\rho\equiv\int_0^T\int_{B_\rho}\max\{u_n^\varepsilon;1\}dxdt &\le T\left\{\|u_o\|_{1,\mathbb{R}^2}+\varepsilon|B_{2\rho}|\right\}+\frac{\tilde\gamma T^{3/2}}{\sigma^2\rho}\left(\int_0^T\int_{B_{\rho(1+\sigma)}}\max\{u_n^\varepsilon;1\}dxdt\right)^{1/2} \\ &\le T\left\{\|u_o\|_{1,\mathbb{R}^2}+\varepsilon|B_{2\rho}|\right\}+\frac{\tilde\gamma T^{3/2}}{\sigma^2\rho}\left(M_{\rho(1+\sigma)}\right)^{1/2}. \end{aligned}$$

Interpolating over the parameter σ, we conclude that for every $\rho>0$, there exists a constant $\gamma(\rho)$ independent of ε, such that

$$\int_0^T\int_{B_\rho}u_n^\varepsilon dxdt\le\gamma(\rho)\left\{\|u_o\|_{1,\mathbb{R}^2}+1\right\},\qquad \text{for all}\quad n\ge 2\rho. \tag{3.3}$$

These remarks permit to let $n\nearrow\infty$ in (3.2) to find a function $\varepsilon\le u^\varepsilon\in L^1_{\text{loc}}(S_T)$ solution of (1.1) with initial datum $u^\varepsilon(\cdot,0)=u_o+\varepsilon$. Since $u^\varepsilon\ge\varepsilon$, by **(i)** all the u^ε are maximal and $u^\varepsilon\ge v$ for all $\varepsilon\in(0,1)$. As $\varepsilon\searrow 0$ the net $\{u^\varepsilon\}$ decreases to a function $u\ge v$ in $L^1_{\text{loc}}(S_T)$. Since v is a solution to (1.1), we have $|\ln v|\in L^1_{\text{loc}}(S_T)$. Therefore also $|\ln u|\in L^1_{\text{loc}}(S_T)$, and u is a solution to (1.1). It remains to show that such a function takes the initial datum u_o in the sense of $L^1_{\text{loc}}(\mathbb{R}^2)$. For this fix a compact subset $\mathcal{K}$ of $\mathbb{R}^2$, and write

$$\begin{aligned} \int_{\mathcal{K}}|u(x,t)-u_o(x)|dx &\le\int_{\mathcal{K}}\{u(x,t)-u_o(x)\}_+dx+\int_{\mathcal{K}}\{u_o(x)-u(x,t)\}_+dx \\ &\le\int_{\mathcal{K}}\{u^\varepsilon(x,t)-u_o(x)\}_+dx+\int_{\mathcal{K}}\{u_o(x)-v(x,t)\}_+dx. \end{aligned}$$

We recall that u^ε takes the initial datum $u_o+\varepsilon$, and let $t\searrow 0$, to obtain

$$\limsup_{t\searrow 0}\int_{\mathcal{K}}|u(x,t)-u_o(x)|dx\le\varepsilon|\mathcal{K}|.$$

Since u^ε are maximal and u is the decreasing limit of $\{u^\varepsilon\}$, also u is maximal. ∎

Remark 3.1. A consequence of Proposition 3.1 is that all maximal solutions are constructed as decreasing limits of the solutions u^ε, starting from a given solution v. Since $u^\varepsilon \geq v$ and v is a solution, the existence of such a solution v, guarantees that the u^ε have a non trivial limit.

PROOF OF **(iii)**: We may assume, without loss of generality that u_n^ε is a classical solution of (3.2). To prove (3.1) it suffices to show that the function $w \equiv (\ln u)_t - 1/t$ is non positive in $B_n \times (0,T)$. By direct calculation one verifies that w satisfies

$$w_t - u^{-1}\Delta w = -w(w+2/t) \quad \text{in} \quad B_n \times (0,T).$$

Clearly $w \leq 0$ for $|x| = n$ and for t sufficiently small. If w has a positive maximum at some (x,t) in the interior of $B_n \times (0,T)$, then

$$0 \leq (w_t - u^{-1}\Delta w)\,\big|_{(x,t)} = -w(x,t)\,\{w(x,t) + 2/t\} < 0. \qquad \blacksquare$$

PROPOSITION 3.2. *Let u and v be maximal solutions with initial data u_o and v_o respectively. If $(u_o - v_o) \in L^1(\mathbb{R}^2)$ then also $\{u(\cdot,t) - v(\cdot,t)\} \in L^1(\mathbb{R}^2)$ for a.e. $t \in (0,T)$ and*

$$(3.4) \qquad \|u(\cdot,t) - v(\cdot,t)\|_{1,\mathbb{R}^2} \leq \|u_o - v_o\|_{1,\mathbb{R}^2}, \qquad \text{a.e.} \quad t \in (0,T).$$

PROOF: In view of Remark 3.1, it suffices to prove (3.4) for the solutions u^ε and v^ε corresponding to the initial data $u_o + \varepsilon$ and $v_o + \varepsilon$ respectively. Since $u^\varepsilon \geq \varepsilon$, it satisfies the estimate (2.4). Therefore for all $\rho > 2$,

$$\begin{aligned}\int_{B_\rho} \{v^\varepsilon(x,t) - u^\varepsilon(x,t)\}_+ \,dx &\leq \int_{B_{2\rho}} (v_o - u_o)_+ dx + \frac{\gamma}{\rho^2}\int_0^T\!\!\int_{B_{2\rho}} \{v^\varepsilon(x,t) - u^\varepsilon(x,t)\}_+ \,dxdt \\ &\leq \|u_o - v_o\|_{1,\mathbb{R}^2} + \gamma(\alpha,T)\,\{\ln 4\rho\}^{\frac{\alpha}{1-\alpha}}\,/\rho^2.\end{aligned}$$

The estimate follows by letting $\rho \to \infty$, using Fatou's Lemma and interchanging the role of u^ε and v^ε. $\blacksquare$

Remark 3.2. Proposition 3.2 can be regarded as a density and stability argument for maximal solutions, if there are any, whose initial data are in $L^1(\mathbb{R}^2)$.

Remark 3.3. By Proposition 2.1, the first of the explicit solutions given in (1.2) is maximal. A consequence of (5.3) of Theorem 5.1, will be that the second of (1.2) is not maximal. the res

4. Extinction in Finite Time

Let u be a solution of (1.1) originating from some initial datum u_o. If $u_o \in L^1(\mathbb{R}^2)$ then u must necessarily vanish in finite time. We estimate such an extintion time and show that such estimate is the best possible. Similar results are in [3] by different techniques.

PROPOSITION 4.1. *Let $u_o \in L^1(\mathbb{R}^2)$. Then any solution whose initial datum is u_o vanishes at least by time $T \equiv \|u_o\|_{1,\mathbb{R}^2}/4\pi$.*

PROOF: It suffices to prove the Proposition for maximal solutions. Assume first that

$$(4.1) \qquad u_o(x) \leq M/(1+|x|^2) \quad \text{a.e. in} \quad \mathbb{R}^2 \quad \text{for some} \quad M > 0.$$

By Remark 3.3 the first of (1.2) is maximal. Therefore $u(x,t)\le M/(1+|x|^2)$, a.e. in S_T. Let $(x,y)\to\mathcal{G}_R(x,y)$ denote the Green's function for the Laplacian in the ball B_R, i.e., for $x\neq y$,

$$\mathcal{G}(x,y)\equiv\frac{1}{2\pi}\{\Phi(x,y)-\ln|x-y|\}\qquad\text{where}\qquad\begin{cases}\Delta_y\Phi(x,y)=0 & \text{in } B_R;\\ \Phi(x,y)=\ln|x-y| & \text{for } |y|=R.\end{cases}$$

By the Poisson representation formula, we have for a.e. $s\in(0,T)$

$$\ln u(x,s)=\int_{\partial B_R}u(y,s)\mathcal{P}_R(x,y)d\sigma(y)-\int_{B_R}\mathcal{G}_R(x,y)u_t(y,s)dy,$$

where $\mathcal{P}_R$ is the Poisson kernel in ∂B_R. Since $\mathcal{P}_R\ge 0$ and has mass one on ∂B_R, integrating in ds over $(0,t)$ for some $t\in(0,T)$, gives

$$\begin{aligned}\int_0^t\ln u(x,s)ds&\le t\ln\left(\frac{M}{1+R^2}\right)-\int_{B_R}\mathcal{G}_R(x,y)\{u(y,t)-u_o(y)\}\,dy\\ &\le t\ln\left(\frac{M}{1+|x|^2}\right)+\int_{B_R}\mathcal{G}_R(x,y)u_o(y)dy.\end{aligned}$$

By (3.1) $t\to\ln\{u(x,t)/t\}$ is a decreasing function. Therefore

$$\int_0^t\ln u(x,s)ds=\int_0^t\ln\frac{u(x,s)}{s}ds+\int_0^t\ln s ds\ge t\ln\frac{u(x,t)}{t}+\int_0^t\ln s ds.$$

Combining these estimates gives

$$\ln u(x,t)\le\gamma(t,M)-2\ln R+\frac{1}{t}\int_{B_R}\mathcal{G}_R(x,y)u_o(y)dy, \tag{4.2}$$

for a constant γ depending on the indicated quantities. The function $y\to\Phi(x,y)$ is harmonic in B_R and $\Phi(x,y)\,\big|_{|y|=R}=\ln|x-y|\le\ln 2R$. Therefore by the maximum principle $\Phi(x,y)\le\ln 2R$ for all $x,y\in B_R$. Using this fact we estimate the last integral in (4.2) by,

$$\begin{aligned}2\pi\int_{B_R}\mathcal{G}_R(x,y)u_o(y)dy&\le\int_{B_R}\Phi(x,y)u_o(y)dy-\int_{\{|x-y|<1\}}\ln|x-y|u_o(y)dy\\ &\le\gamma(M)+\ln 2R\|u_o\|_{1,\mathbb{R}^2}\le\tilde{\gamma}(M,\|u_o\|_{1,\mathbb{R}^2})+\ln R\,\|u_o\|_{1,\mathbb{R}^2}.\end{aligned}$$

Combining these estimates we deduce that there is a constant $\gamma(t,M,\|u_o\|_{1,\mathbb{R}^2})$, such that

$$\ln u(x,t)\le\gamma(t,M,\|u_o\|_{1,\mathbb{R}^2})-2\ln R+\ln R\,\|u_o\|_{1,\mathbb{R}^2}/2\pi t.$$

The Proposition follows by letting $R\to\infty$. To remove the assumption (4.1), given $u_o\in L^1(\mathbb{R}^2)$ let $u_{o,n}$ be an increasing sequence of compactly supported and bounded functions in $\mathbb{R}^2$ such that $u_{o,n}\to u_o$ in $L^1(\mathbb{R}^2)$. Each of these satisfies (4.1) for some constant M depending upon n. Therefore the corresponding maximal solutions u_n, are identically zero after the time $T=\|u_o\|_{1,\mathbb{R}^2}/4\pi$. For each n fixed, u_n can be constructed as the decreasing limit of $\{u_{n,\varepsilon}\}_{\varepsilon\in(0,1)}$, where $u_{n,\varepsilon}$ is the maximal solution with initial datum $u_{o,n}+\varepsilon$. Let also u_ε be the maximal solution originating from $u_o+\varepsilon$. By (3.4),

$$\|u_\varepsilon(\cdot,t)-u_{n,\varepsilon}(\cdot,t)\|_{1,\mathbb{R}^2}\le\|u_o-u_{o,n}\|_{1,\mathbb{R}^2},\qquad\text{for a.e.}\quad t\in(0,T).$$

Letting $\varepsilon\searrow 0$, this gives $\|u(\cdot,t)\|_{1,\mathbb{R}^2}\le\|u_n(\cdot,t)\|_{1,\mathbb{R}^2}+\|u_o-u_{o,n}\|_{1,\mathbb{R}^2}$. ■

5. Existence of Solutions and Maximal Solutions

THEOREM 5.1. *For each non negative $u_o \in L^1(\mathbb{R}^2)$ and each $s \geq 1$, there is a solution u^s to (1.1) in the strip S_{T^s} where $T^s \equiv \|u_o\|_{1,\mathbb{R}^2}/4\pi s$. Moreover for a.e. $t \in (0, T^s)$,*

$$(5.1) \qquad \|u^s(\cdot,t)\|_{1,\mathbb{R}^2} = \|u_o\|_{1,\mathbb{R}^2} - 4\pi st \ \text{ a.e. } t \in (0,T) \quad \text{and} \quad u_t^s \leq u^s/t \ \text{ in } \mathcal{D}'(S_{T^s}).$$

More generally for a.e. $t \in (0, T^s)$ and a.e. $h > 0$ such that $t + h < T^s$,

$$(5.1)' \qquad \|u^s(\cdot, t+h)\|_{1,\mathbb{R}^2} = \|u(\cdot,t)\|_{1,\mathbb{R}^2} - 4\pi\, sh.$$

If in addition $u_o \in L^r_{\text{loc}}(S_{T^s})$ for some $r > 1$, then the solutions u^s are classical in S_{T^s}. Moreover $\Delta \ln u^s \in L^1(\mathbb{R}^2)$, and for all $t \in (0, T^s)$,

$$(5.2) \qquad -\int_{\mathbb{R}^2} \Delta \ln u(x,t)dx = 4\pi\, s; \qquad \int_{\mathbb{R}^2} |\Delta \ln u^s(x,t)|\, dx \leq 2\|u_o\|_{1,\mathbb{R}^2}/t.$$

Given $u_o \in L^1(\mathbb{R}^2)$, there exists a maximal solution u with initial datum u_o, in the strip S_T where $T \equiv \|u_o\|_{1,\mathbb{R}^2}$. Moreover such a maximal solution is characterized by

$$(5.1)_1 \qquad \|u(\cdot,t)\|_{1,\mathbb{R}^2} = \|u_o\|_{1,\mathbb{R}^2} - 4\pi t \ \text{ a.e. } t \in (0,T) \quad \text{and} \quad u_t \leq u/t \ \text{ in } \mathcal{D}'(S_T),$$

that is the maximal solution is that corresponding to the value one of the parameter s. Finally, as $|x| \to \infty$ such maximal solutions decrease no faster than $1/\left(|x| \ln |x|\right)^2$. Precisely, there exist continuous functions $t \to C_i(t) : (0,T) \to \mathbb{R}^+$, $i = 0, 1$, such that

$$(5.3) \qquad u(x,t) \geq C_o(t)/\left(|x| \ln |x|\right)^2 \quad \text{for a.e.} \quad |x| \geq C_1(t).$$

We will prove first the Theorem assuming that $s > 1$ and that u_o is bounded and has compact support. In particular there exists a constant M depending upon s and the support of u_o such that

$$(5.4) \qquad u_o(x) \leq M/(1 + |x|^2)^s, \qquad \text{for all} \quad x \in \mathbb{R}^2.$$

6. Time Discretization and Approximating Solutions

For a positive integer n, construct approximating solutions by setting $u_o^{s,n} \equiv u_o$ and for $j = 1, 2, \ldots, (n-1)$, the function $x \to u_{j+1}(x)$ is the solution of

$$(6.1) \qquad \frac{u_{j+1}^{s,n}}{h} - \Delta \ln u_{j+1}^{s,n} = \frac{u_j^{s,n}}{h}; \qquad h \equiv T/n, \quad T \equiv \|u_o\|_{1,\mathbb{R}^2}/4\pi\, s.$$

Such a solution is constructed in Section 16 of [5] and satisfies

$$(6.2) \qquad \int_{\mathbb{R}^2} u_{j+1}^{s,n}(x)dx = \int_{\mathbb{R}^2} u_j^{s,n}(x)dx - 4\pi s\, h \qquad \text{and} \qquad u_j^{s,n}(x) \leq \frac{M}{(1 + |x|^2)^s},$$

for all $j=0,1,2,\dots,(n-1)$. Set $t_j \equiv jh$ and construct the approximating solutions $(x,t)\to u^{s,n}(x,t)$ by a time interpolation of the $x\to u^{s,n}(x)$ within the time interval (t_j,t_{j+1}) for $j=0,1,\dots,(n-1)$, i.e.,

$$(6.3)\qquad u^{s,n}(x,t)\equiv \frac{t_{j+1}-t}{t_{j+1}-t_j}u_j^{s,n}(x)+\frac{t-t_j}{t_{j+1}-t_j}u_{j+1}^{s,n}(x); \qquad t\in[t_j,t_{j+1}].$$

LEMMA 6.1. *For all $j=0,1,2,\dots,(n-1)$ there holds*

$$\frac{u_{j+1}^{s,n}}{(j+1)h}\le\frac{u_j^{s,n}}{jh} \qquad \text{or equivalently} \qquad \frac{u_{j+1}^{s,n}-u_j^{s,n}}{h}\le\frac{u_{j+1}^{s,n}}{(j+1)h},$$

where in the first of these the right hand side for $j=0$ is defined as infinity.

Remark 6.1. These recursive inequalities can be regarded as a discrete version of the semiconvexity inequality (3.1). In particular, by taking into account the definition (6.3) of $u^{s,n}$ and the second of (6.2) they imply

$$(6.4)\qquad u_t^{s,n}\le M/t \quad \text{in} \quad \mathcal{D}'(S_T) \quad \text{for all} \quad n\in I\!N.$$

PROOF OF LEMMA 6.1: The inequality holds trivially for $j=0$. We will show that if it holds for some $j=1,2,\dots,(n-2)$, then it continues to holds for $(j+1)$. Each $u_{j+1}^{s,n}$ is constructed as the locally decreasing limit as $k\to\infty$ of the solutions of,

$$(6.5)\qquad \begin{cases} \dfrac{u_{j+1;k}^{s,n}}{h}-\Delta\ln u_{j+1;k}^{s,n}=\dfrac{u_j^{s,n}}{h} & \text{in the ball } B_k; \\ u_{j+1;k}^{s,n}\big|_{\partial B_k}=\dfrac{M}{(1+k^2)^s}; & k=1,2,\dots. \end{cases}$$

By Remark 16.1 of [5] this procedure yields the solutions $u_{j+1}^{s,n}$ of (6.1), provided

$$\frac{u_j^{s,n}(x)}{h}\le\frac{1}{h}\frac{M}{(1+|x|^2)^s}+\frac{4s}{(1+|x|^2)^2}.$$

Such an inequality is in turn guaranteed by the second of (6.2). We write (6.5) for j and $(j+1)$ and obtain by difference

$$(6.6)\qquad \begin{cases} \dfrac{u_{j+2;k}^{s,n}-u_{j+1;k}^{s,n}}{h}-\Delta\left(\ln\dfrac{u_{j+2;k}^{s,n}}{(j+2)h}-\ln\dfrac{u_{j+1;k}^{s,n}}{(j+1)h}\right)=\dfrac{u_{j+1}^{s,n}-u_j^{s,n}}{h}; \\ \left(\dfrac{u_{j+2;k}^{s,n}}{(j+2)h}-\dfrac{u_{j+1;k}^{s,n}}{(j+1)h}\right)\Bigg|_{\partial B_k}=-\dfrac{1}{(j+2)(j+1)h}\dfrac{M}{(1+k^2)^s}. \end{cases}$$

Next we transform the difference on the left hand side as

$$\begin{aligned}\frac{u_{j+2;k}^{s,n}-u_{j+1;k}^{s,n}}{h}&=\left(\frac{u_{j+2;k}^{s,n}}{(j+2)h}-\frac{u_{j+1;k}^{s,n}}{(j+1)h}\right)(j+2)+\frac{u_{j+1;k}^{s,n}}{(j+1)h}\\ &=(v_{j+2}-v_{j+1})(j+2)+\frac{u_{j+1;k}^{s,n}}{(j+1)h}; \qquad \text{where} \quad v_j\equiv\frac{u_{j;k}^{s,n}}{jh}.\end{aligned}$$

With this notation, making use of the induction hypothesis, we deduce from (6.6),

$$(v_{j+2}-v_{j+1})(j+2)-\Delta(\ln v_{j+2}-\ln v_{j+1})=\frac{u^{s,n}_{j+1}-u^{s,n}_{j}}{h}-\frac{u^{s,n}_{j+1;k}}{(j+1)h}$$
$$\leq\frac{u^{s,n}_{j+1}}{(j+1)h}-\frac{u^{s,n}_{j+1;k}}{(j+1)h}\leq 0.$$

Since $(v_{j+2}-v_{j+1})$ is negative on ∂B_k it is also negative in the whole B_k. Thus

$$\frac{u^{s,n}_{j+2;k}}{(j+2)h}\leq\frac{u^{s,n}_{j+1;k}}{(j+1)h}\qquad\text{in}\qquad B_k$$

and the Lemma follows by letting $k\to\infty$. ■

7. The Sequence $\{u^{s,n}\}_{n\in\mathbb{N}}$ Is Compact in $L^1(S_T)$

By the stability estimate of Proposition 12.1 of [5] we also have

$$\left\|u^{s,n}_{j+1}(x+y)-u^{s,n}_{j+1}(x)\right\|_{1,\mathbb{R}^2}\leq\left\|u^{s,n}_{j}(x+y)-u^{s,n}_{j}(x)\right\|_{1,\mathbb{R}^2},\qquad\forall\, y\in\mathbb{R}^2.$$

From this, by iteration and recalling the definition (6.3) of $u^{s,n}$,

$$\|u^{s,n}(x+y,t)-u^{s,n}(x,t)\|_{1,\mathbb{R}^2}\leq\|u_o(x+y)-u_o(x)\|_{1,\mathbb{R}^2},\qquad\forall\, y\in\mathbb{R}^2. \tag{7.1}$$

Also iterating the first of (6.2) gives $\|u^{s,n}(\cdot,t)\|_{1,\mathbb{R}^2}\leq\|u_o\|_{1,\mathbb{R}^2}$, $\forall t\in(0,T)$. From this

$$\int_{t_1}^{t_2}\int_{B_\rho}u^{s,n}(x,t)dxdt\leq(t_2-t_1)\|u_o\|_{1,\mathbb{R}^2},\qquad\forall\, 0\leq t_1<t_2<T\ \text{and}\ \forall\,\rho>0. \tag{7.2}$$

Fix $y\in\mathbb{R}^2$ and $\tau\in(0,T)$. Then for $(t_1,t_2)\subset(0,T-\tau)$ we estimate

$$\begin{aligned}\int_{t_1}^{t_2}\int_{B_\rho}|u^{s,n}(x+y,t+\tau)-u^{s,n}(x,t)|\,dxdt&\leq\int_{t_1}^{t_2}\int_{B_\rho}|u^{s,n}(x+y,t+\tau)-u^{s,n}(x,t+\tau)|\,dxdt\\&\quad+\int_{t_1}^{t_2}\int_{B_\rho}|u^{s,n}(x,t+\tau)-u^{s,n}(x,t)|\,dxdt\\&\equiv J_1+J_2.\end{aligned}\tag{7.3}$$

The first integral is estimated by (7.1) and gives $J_1\leq(t_2-t_1)\|u_o(x+y)-u_o(x)\|_{1,\mathbb{R}^2}$. To estimate the second we use the semiconvexity inequality in the form

$$\frac{d}{dt}\left\{u^{s,n}(x,t)-\frac{Mt}{t_1}\right\}\leq 0\quad\text{for}\quad t\geq t_1.$$

Therefore the function $t \to \{u^{s,n}(\cdot,t) - Mt/t_1\}$ is non increasing and,

$$\begin{aligned}|u^{s,n}(x,t+\tau)-u^{s,n}(x,t)| &= \left|\left\{u^{s,n}(x,t)-\frac{Mt}{t_1}\right\} - \left\{u^{s,n}(x,t+\tau)-\frac{M(t+\tau)}{t_1}\right\} - \frac{M\tau}{\tau_1}\right| \\ &\leq \{u^{s,n}(x,t)-u^{s,n}(x,t+\tau)\} + \frac{2M\tau}{t_1}.\end{aligned}$$

We may now estimate J_2 using these remarks and (7.2), as

$$\begin{aligned}J_2 &\leq \int_{t_1}^{t_2}\int_{B_\rho} \{u^{s,n}(x,t)-u^{s,n}(x,t+\tau)\}\,dxdt + 2(t_2-t_1)M\tau/t_1 \\ &\leq \int_{t_1}^{t_1+\tau}\int_{B_\rho} u^{s,n}(x,\sigma)dxd\sigma + 2(t_2-t_1)M\tau/t_1 \\ &\leq \tau\left(\|u_o\|_{1,\mathbb{R}^2} + 2TM/t_1\right).\end{aligned}$$

Combining these estimations in (7.3) yields,

$$\begin{aligned}\int_{t_1}^{t_2}\int_{B_\rho} |u^{s,n}(x+y,t+\tau)-u^{s,n}(x,t)|\,dxdt &\leq T\,\|u_o(x+y)-u_o(x)\|_{1,\mathbb{R}^2} \\ &\quad + \tau\left(\|u_o\|_{1,\mathbb{R}^2}+2TM/t_1\right).\end{aligned}$$ ■

8. The Sequence $\{\ln u^{s,n}\}_{n\in\mathbb{N}}$ Is Compact in $L^1_{\mathrm{loc}}(S_T)$

The compactness of $\{\ln u^{s,n}\}$ is established by a discrete version of the Harnack inequality. To derive it, we introduce the Green's type function

$$\mathcal{H}_\rho(x) = \frac{1}{2\pi}\left\{\ln\rho - \ln|x| + \frac{1}{2}\left(|x|^2-\rho^2\right)\right\} \qquad x\in B_\rho; \quad \rho>0 \text{ fixed.}$$

This function is non negative and vanishes on ∂B_ρ. Since also its normal derivative vanishes on ∂B_ρ, any $w\in C^2(\overline{B}_\rho)$ satisfies,

$$\fint_{B_\rho} w(x)dx = w(0) + \int_{B_\rho}\mathcal{H}_\rho(x)\Delta w(x)dx.$$

In particular for some $x_o\in\mathbb{R}^2$ and $w(x)\equiv \ln u^{s,n}_{j+1}(x-x_o)$,

$$\fint_{B_\rho(x_o)} \ln u^{s,n}_{j+1}(x)dx = \ln u^{s,n}_{j+1}(x_o) + \frac{1}{h}\int_{B_\rho(x_o)} \mathcal{H}_\rho(x-x_o)\left\{u^{s,n}_{j+1}(x)-u^{s,n}_j(x)\right\}dx.$$

Fix two integers $1<j_o<j_1\le n$ and add these equalities for $j=j_o,(j_o+1),\dots,j_1$,

$$\sum_{j=j_o}^{j_1-1} \fint_{B_\rho(x_o)} \ln u_{j+1}^{s,n}(x)dx = \sum_{j=j_o}^{j_1-1} \ln u_{j+1}^{s,n}(x_o) + \sum_{j=j_o}^{j_1-1} \frac{1}{h}\int_{B_\rho(x_o)} \mathcal{H}_\rho(x-x_o)\left\{u_{j+1}^{s,n}(x)-u_j^{s,n}(x)\right\}dx. \tag{8.1}$$

By the discrete semiconvexity inequality of Lemma 6.1,

$$\begin{aligned}\sum_{j=j_o}^{j_1-1} \fint_{B_\rho(x_o)} \ln u_{j+1}^{s,n}(x)dx &= \sum_{j=j_o}^{j_1-1} \fint_{B_\rho(x_o)} \ln \frac{u_{j+1}^{s,n}(x)}{(j+1)h}dx + \sum_{j=j_o}^{j_1-1} \ln(j+1)h \\ &\le (j_1-j_o)\fint_{B_\rho(x_o)} \ln\frac{u_{j_o}^{s,n}(x)}{j_oh}dx + \sum_{j=j_o}^{j_1-1}\ln(j+1)h,\end{aligned}$$

and

$$\sum_{j=j_o}^{j_1-1} \ln u_{j+1}^{s,n}(x_o) \ge (j_1-j_o)\ln\frac{u_{j_1}^{s,n}(x_o)}{j_1h} + \sum_{j=j_o}^{j_1-1}\ln(j+1)h.$$

Combining these estimates in (8.1) yields,

$$\fint_{B_\rho(x_o)} \ln u_{j_o}^{s,n}(x)dx \ge \ln u_{j_1}^{s,n}(x_o) - \ln\frac{j_1}{j_o} - \frac{1}{h(j_1-j_o)}\int_{B_\rho(x_o)} \mathcal{H}_\rho(x-x_o)u_{j_o}^{s,n}(x)dx. \tag{8.2}$$

The semiconvexity inequality of Lemma 6.1 and (6.1) also imply

$$\Delta \ln u_j^{s,n} \le u_j^{s,n}/jh \quad \text{for all} \quad j=1,2,\dots,n. \tag{8.3}$$

Therefore, by Proposition 10.1 of [4], there exists a constant γ independent of s, n, h, j, such that

$$-\inf_{B_{\rho/2}(x_o)} \ln u_j^{s,n}(x) \le \gamma \fint_{B_\rho(x_o)} \left(\ln u_j^{s,n}\right)_- dx + 2\ln\frac{\rho^2}{jh}.$$

This and (8.2) now imply

$$\begin{aligned}\inf_{B_{\rho/2}(x_o)} \ln u_{j_o}^{s,n}(x) &\ge -\gamma \fint_{B_\rho(x_o)} \left(\ln u_{j_o}^{s,n}\right)_+ dx + \gamma \ln u_{j_1}^{s,n}(x_o) - \gamma\ln\frac{j_1}{j_o} \\ &\quad - \frac{\gamma}{h(j_1-j_o)}\int_{B_\rho(x_o)} \mathcal{H}_\rho(x-x_o)u_{j_o}^{s,n}(x)dx - 2\ln\frac{\rho^2}{j_oh} \\ &\ge \ln\left(u_{j_1}^{s,n}(x_o)\right)^\gamma - \gamma\ln\{M\wedge 1\} - \gamma\ln\frac{j_1}{j_o} \\ &\quad - \frac{\gamma M\wedge 1}{h(j_1-j_o)}\rho^2\ln\rho - 2\ln\left(\frac{\rho^2}{j_oh}\right),\end{aligned} \tag{8.4}$$

where in the last estimate we have used the second of (6.2) and the definition of $\mathcal{H}_\rho$. We conclude that there exists an absolute constant γ such that for all $x_o \in \mathbb{R}^2$, every ball $B_\rho(x_o)$ and every $0<j_o<j_1<n$,

$$\inf_{B_\rho(x_o)} u_{j_o}^{s,n}(x) \geq \left(u_{j_1}^{s,n}(x_o)\right)^\gamma C(j_oh, j_1h, \rho, M)e^{\mathcal{I}(j_oh, j_1h, \rho, M)}, \tag{8.5}$$

where

$$C(j_oh, j_1h, \rho, M) \equiv (j_o/j_1)^\gamma (j_oh/\rho^2)^2, \qquad \mathcal{I}(j_oh, j_1h, \rho, M) \equiv \frac{\gamma M}{h(j_1-j_o)}\rho^2 \ln \rho.$$

Inequality (8.5) is a discrete version of the global Harnack–type inequality given in [4]. Essentially, whence $u^{s,n}(x_o, t_1)$ is bounded away from zero, uniformly in n, for some $(x_o, t_1) \in S_T$, then $u^{s,n}$ is uniformly bounded away from zero on sets of the form $B_\rho(x_o)\times(t_o, t_1)$. Let $\mathcal{K}$ be a compact subset of $\mathbb{R}^2$. Without loss of generality we may assume that $\mathcal{K}$ is of the form $\overline{B}_{\rho_o}\times[\tilde{t}_o, \tilde{t}_1]$ for some $\rho>0$ and $0<\tilde{t}_o<\tilde{t}_1<T$. Fix some t_1 such that $\tilde{t}_1<t_1<T$. Then, by virtue of (6.2) and (6.3), there exists $\rho \geq \rho_o$ so large that

$$\int_{B_\rho} u^{s,n}(x, t_1)dx \geq \frac{1}{2}\left\{\|u_o\|_{1,\mathbb{R}^2} - 4\pi\, t\, s\right\} > 0 \quad \text{for all} \quad n \geq n_1.$$

It follows that there exists some point $x_o \in B_\rho$ and some $n_o \in \mathbb{N}$, such that

$$u^{s,n}(x_o, t_1) > \frac{\|u_o\|_{1,\mathbb{R}^2} - 4\pi, s\, t}{4\,\text{meas}\, B_\rho} > 0, \quad \text{for all} \quad n \geq n_o.$$

This implies that $\{u^{s,n}\}$ is uniformly bounded away from zero in $\mathcal{K}$, and establishes the compactness of $\{\ln u^{s,n}\}$. ■.

9. Proof of Theorem 5.1 For $s>1$ and u_o as in (5.4)

Letting $n \to \infty$ yields a non negative function $u^s \in L^1(\mathbb{R}^2)$ satisfying

$$u_t^s - \Delta \ln u^s = 0 \ \text{ in } \ \mathcal{D}'(S_{T^s}) \qquad \text{and} \qquad u^s(x,t) \leq M/\left(1+|x|^2\right)^s \ \text{ in } \ S_{T^s}. \tag{9.1}$$

By the Harnack inequality (8.4)–(8.5), the function u^s is bounded away from zero on compact subsets of S_{T^s}. In view of (5.4) and the second of (6.2) it is also bounded above in S_{T^s}. Therefore by classical parabolic theory, u^s is a classical solution of (1.1) in S_{T^s}.

LEMMA 9.1. *The solution u^s takes the initial datum u_o in the sense of $L^1_{\text{loc}}(\mathbb{R}^2)$.*

PROOF: We first show that $u^s(\cdot, t) \to u_o$ weakly in $L^1_{\text{loc}}(\mathbb{R}^2)$. From (6.1) we deduce that for every $\varphi \in C_o^\infty(\mathbb{R}^2)$ and all $j = 0, 1, \ldots, (n-1)$,

$$\int_{\mathbb{R}^2} \left\{u_{j+1}^{s,n}(x) - u_j^{s,n}(x)\right\} \varphi dx = h\int_{\mathbb{R}^2} \ln u_{j+1}^{s,n}(x) \Delta\varphi dx. \tag{9.2}$$

Fix $j_1 \le (n-1)$. Then, again by the Harnack estimates (8.4)-(8.5), we have

$$\left| \inf_{\mathrm{supp}\{\varphi\}} \ln u_j^{s,n}(x) \right| \ge \gamma(\mathrm{supp}\{\varphi\}, T) \left| \ln(jh) \right| \qquad \text{for all } 1 \le j < j_1.$$

We take $n = 2^m$ for $m \in \mathbb{N}$ and apply the last two estimates for a fixed time $\bar{t}$ of the form

$$\text{(9.3)} \qquad \bar{t} \equiv j_m T^s / 2^m, \quad h = \|u_o\|_{1,\mathbb{R}^2} / 4\pi s 2^m \quad \text{for some positive integer } 1 \le j_m < 2^m.$$

We now add (9.2) for $j = 0, 1, \dots, (j_m - 1)$ to obtain

$$\left| \int_{\mathbb{R}^2} \{ u_{j_m}^{s,n}(x) - u_o(x) \} \varphi dx \right| \le C(\varphi, T) \, \bar{t} \left| \ln \bar{t} \right|.$$

For each diadic $n = 2^m$ there is j_m such that $\bar{t} = j_m h$ so that $u_{j_m}^{s,n}(x) = u^{s,n}(x, \bar{t})$. Therefore letting $m \to \infty$,

$$\text{(9.4)} \qquad \left| \int_{\mathbb{R}^2} \{ u^s(x, \bar{t}) - u_o(x) \} \varphi dx \right| \le C(\varphi, T) \, \bar{t} \left| \ln \bar{t} \right|.$$

Such inequality established for a dense subset of times in $(0, T^s)$ actually holds for all $t \in (0, T^s)$, since u^s is a classical solution. Inequality (9.4) written for all such t, shows that $u^s(\cdot, t) \to u_o$ weakly in $L^1_{\mathrm{loc}}(\mathbb{R}^2)$. To show that such a convergence is strong, we return to (7.1) written for diadic $n = 2^m$. Letting $m \to \infty$ yields,

$$\text{(9.5)} \qquad \int_{\mathbb{R}^2} |u^s(x+y, t) - u^s(x, t)| \, dx \le \int_{\mathbb{R}^2} |u_o(x+y) - u_o(x)| \, dx, \qquad \forall \, y \in \mathbb{R}^2.$$

This is first established for a dense subset of times in $(0, T^s)$ and then by continuity for all $t \in (0, T^s)$. A consequence of (9.5) and the second of (6.2) is that the net $\{u^s(\cdot, t)\}_{t \in (0, T^s)}$ is precompact in $L^1(\mathbb{R}^2)$. Thus for a subnet which we continue to index with t,

$$\{u^s(\cdot, t)\}_{t \in (0, T^s)} \quad \longrightarrow \quad u_o \quad \text{strongly in} \quad L^1_{\mathrm{loc}}(\mathbb{R}^2).$$

Whence the limit is identified, the selection of a subnet is unnecessary, and the convergence takes place along the whole net. ■

LEMMA 9.2. *$\Delta \ln u^s(\cdot, t) \in L^1(\mathbb{R}^2)$ and $\|\Delta \ln u^s(\cdot, t)\|_{1,\mathbb{R}^2} \le 2 \|u_o\|_{1,\mathbb{R}^2} / t$.*

PROOF: We write (8.3) and (6.2) for diadic $n = 2^m$ and for $(j+1) = j_m$. Then for a fixed $\bar{t}$ as indicated in (9.3),

$$\int_{\mathbb{R}^2} \{ \Delta \ln u^{s,n}(x, \bar{t}) \}_+ \, dx \le \frac{1}{\bar{t}} \int_{\mathbb{R}^2} u^{s,n}(x, \bar{t}) dx \quad \text{and} \quad - \int_{\mathbb{R}^2} \Delta \ln u^{s,n}(x, \bar{t}) dx = 4\pi s.$$

Therefore

$$\int_{\mathbb{R}^2} |\Delta \ln u^{s,n}(x, \bar{t})| \, dx \le \frac{2}{\bar{t}} \int_{\mathbb{R}^2} u^{s,n}(x, \bar{t}) dx + 4\pi s \le \frac{2}{\bar{t}} \|u_o\|_{1,\mathbb{R}^2}.$$

Moreover, since u^s is classical in S_{T^s}, for all $\varphi \in C_o^\infty(I\!R^2)$ such that $|\varphi| \le 1$, we have

$$\left|\int_{I\!R^2} \Delta \ln u^s(x,\bar{t})\varphi dx\right| = \lim_{n\to\infty}\left|\int_{I\!R^2} \ln u^{s,n}(x,\bar{t})\Delta\varphi dx\right|$$
$$= \lim_{n\to\infty}\left|\int_{I\!R^2} \Delta \ln u^{s,n}(x,\bar{t})\varphi dx\right| \le 2\|u_o\|_{1,I\!R^2}/\bar{t}.$$

Since φ is arbitrary, the Lemma is established for a dense subset of times $\bar{t} \in (0, T^s)$. It continues to hold for all $t \in (0, T^s)$ since u^s is classical in S_{T^s}. ∎

10. Proof of Theorem 5.1 Completed

Given $u_o \in L^1(I\!R^2)$ and $s > 1$, there exists an increasing sequence of functions $\{u_{o,i}\}_{i\in I\!N}$ all bounded, of compact support in $I\!R^2$ and satisfying (5.4) for some $M = M_i$. In addition $u_{o,i} \nearrow u_o$ in $L^1(I\!R^2)$, as $i \to \infty$. If u_i^s are the corresponding solutions, constructed in the previous sections,

$$\int_{I\!R^2} u_i^s(x,t)dx = \|u_{o,i}\|_{1,I\!R^2} - 4\pi st, \quad \forall t \in (0, T_i^s), \quad T_i^s \equiv \|u_{o,i}\|_{1,I\!R^2}/4\pi s.$$

Since $\{u_i^s\}_{i\in I\!N}$ is an increasing sequence, a standard limiting process shows that its limit u^s is a solution to (1.1) corresponding to u_o, in the strip S_{T^s}, where $T^s \equiv \|u_o\|_{1,I\!R^2}/4\pi$. In addition such a u^s satisfies (5.1). Assume now that $u_o \in L^1(I\!R^2) \cap L^r_{\mathrm{loc}}(I\!R^2)$ for some $r > 1$. Then the sequence $\{u_{o,i}\}_{i\in I\!N}$ can constructed so that

$$u_{o,i} \in L^1_{\mathrm{loc}}(I\!R^2) \quad \text{uniformly in} \quad i \in I\!N.$$

By Proposition 2.1 of [4] the corresponding solutions u_i^s are locally equibounded in S_{T^s}. Since by the Harnack estimates (8.4)–(8.5) they are also uniformly bounded below locally in S_{T^s}. Therefore by classical parabolic theory they are classical in S_{T^s} and all their derivatives are uniformly bounded on compact subsets of S_{T^s}. Therefore also the limit u^s is classical in S_{T^s} and (5.2) holds.

Set now $T^1 \equiv T = \|u_o\|_{1,I\!R^2}/4\pi$. By construction $u^s \le u^{s'}$ whenever $s \ge s'$. Therefore the function $u^1 \equiv \lim_{s\searrow 1} u^s$ is locally bounded below in S_T. Moreover by (5.1) and Fatou's Lemma, $u^1(\cdot,t) \in L^1(I\!R^2)$ uniformly in $t \in (0,T)$. Thus

$$u^s(\cdot,t) \to u^1(\cdot,t) \ \text{in} \ L^1(I\!R^2), \quad \forall t \in (0,T); \qquad \ln u^s \to \ln u \ \text{in} \ L^1_{\mathrm{loc}}(S_T),$$

and u^1 is a solution to (1.1). Such a solution takes the initial datum u_o in the sense of $L^1_{\mathrm{loc}}(I\!R^2)$ and satisfies $(5.1)_1$. If in addition $u_o \in L^r_{\mathrm{loc}}(I\!R^2)$, such a solution is classical in S_T and (5.2) holds for $s = 1$.

Next we establish that u^1 is maximal. By **(ii)** of Proposition 3.1 there exists a maximal solution u corresponding to u_o in S_{T^s} for all $s > 1$. Since $u \ge u^s$,

$$\|u(\cdot,t)\|_{1,I\!R^2} \ge \|u_o\|_{1,I\!R^2} - 4\pi t, \quad \forall t \in (0,T) \quad T \equiv \|u_o\|_{1,I\!R^2}/4\pi. \tag{10.1}$$

We now claim that (10.1) must hold with equality for all $t\in(0,T)$. Indeed if

$$\|u(\cdot,t_*)\|_{1,\mathbb{R}^2} > \|u_o\|_{1,\mathbb{R}^2} - 4\pi t_* \quad \text{for some} \quad t_*\in(0,T),$$

by using $u(\cdot,t_*)$ as a new initial datum, (10.1) would imply

$$\|u(\cdot,t)\|_{1,\mathbb{R}^2} \geq \|u(\cdot,t_*)\|_{1,\mathbb{R}^2} - 4\pi(t-t_*), \qquad \forall t\in(t_*,T_*),$$

where $T_*\equiv t_* + \|u(\cdot,t_*)\|_{1,\mathbb{R}^2}/4\pi > T$. This however is a contradiction, since by Proposition 4.1, the maximal solution $u(\cdot,t)$ must vanish identically at least by time T. ■

10.1. Proof Of (5.3)

Since maximal solutions satisfy the comparison principle, it suffices to prove (5.3) for the maximal solution associated with the initial datum $\min\{u_o;1\}$. Without loss of generality we may assume that u_o is bounded. Therefore u is a classical solution in S_T and (5.2) holds for $s=1$. From the semiconvexity inequality,

$$\frac{1}{t}u(\cdot,t) - \Delta \ln u(\cdot,t) = \frac{1}{t}u(\cdot,t) - u_t(\cdot,t) \equiv f \geq 0.$$

Setting $v\equiv u(\cdot,t)$, this has the form $\lambda v - \Delta \ln v = f$, where $\lambda = 1/t$. By (5.2) we have $f\in L^1(\mathbb{R}^2)$ and by $(5.1)_1$, $\lambda\|v\|_{1,\mathbb{R}^2} = \|f\|_{1,\mathbb{R}^2} - 4\pi$. Thus by Theorem 18.1 of [5] there exist constants C_i, $i=0,1$ depending upon t, such that (5.3) holds. The continuity of the functions $t\to C_i(t)$ follows from Lemma 19.1 of [5] and the fact that u is classical. ■

Acknowledgement We would like to thank R. Schoen and Jing–Yi Chen for enlightening conversations on the geometrical aspects of the problem.

References

[1] J. P. Burelbach, S.G. Bankoff and S.H. Davis, Non Linear Stability of Evaporating/ Condensing Liquid Films, *J. Fluid Mech. # 195 (1988) pp. 463–494.*

[2] S. Cohn–Vossen, Kürzeste Wege und Totalkrümmung auf Flächen, *Compositio Math. Vol. 2 (1935), pp. 69–133.*

[3] P. Daskalopoulos and M. Del Pino, On Fast Diffusion Nonlinear Heat Equations and Related Singular Elliptic Problems, *Indiana Univ. Math. J. Vol. 43 # 2(1994), pp. 703–728.*

[4] S.H. Davis, E. DiBenedetto and D. J. Diller, Some A Priori Estimates For A Singular Evolution Equation Arising In Thin Film Dynamics, *SIAM J. on Math. Analysis, (to appear).*

[5] E. DiBenedetto and D. J. Diller, Semilinear Elliptic Equations in $L^1(\mathbb{R}^2)$, *(to appear).*

[6] R.S. Hamilton, The Ricci Flow on Surfaces, *Math. and Gen. Relativity, Contemporary Mathematics # 71 (1988) pp. 237–262.*

[7] A. Huber, On Subharmonic Functions and Differential Geometry in the Large, *Comm. Math. Helvetici, Vol. 32 (1957) pp. 13–72.*

[8] J.J. Stoker, *Differential Geometry*, Wiley-Interscience, New York, 1969.

[9] M.B. Williams and S.H. Davis, Non Linear Theory of Film Rupture, *Jour. of Colloidal and Interface Sc. Vol. 90 #1(1982) pp. 220–228.*

[10] Wu Lang-Fang, The Ricci Flow on Complete $\mathbb{R}^2$, *Preprint Austr. Natl. Univ. 1991.*

11. Alternating-Direction Iteration for the p-Version of the Finite Element Method

Jim Douglas, Jr. Department of Mathematics, Purdue University, West Lafayette, Indiana

Abstract. Alternating-direction iterative methods are developed for the algebraic equations associated with the p-version of the finite element method for elliptic equations with variable coefficients. The technique given herein is associated with a single, rectangular domain (the reference element); extensions by means of domain decomposition approaches are presented elsewhere. An efficient choice is made for the basis of the finite element space.

1 Introduction

The object of this paper is to develop alternating-direction iteration procedures applicable to the linear algebraic equations generated by the p-version of the finite element method. The development here will be limited to the equations related to a single, rectangular element in either two or three dimensions. The final iterative technique (on a single element) for variable coefficients will be a multistage procedure and will use the basic alternating-direction method for the Laplace equation as a preconditioner in a fashion essentially equivalent to the procedures developed by Gunn [10, 11] and D'yakonov [8, 9] for finite differences and by Douglas and Dupont [5] for Galerkin methods. These procedures can be embedded in a domain decomposition algorithm to treat more general problems in relatively standard ways (see [2] and the references therein).

First, two model problems, both based on the Laplace equation but with different boundary conditions, will be treated. Each will be applicable in a domain decomposition process when the domain of actual interest is partitioned into a number of rectangles or images of rectangles. Problems in both two and three space variables will be discussed. Finally, a multistage procedure will be described for problems with variable coefficients. All p-method problems arising in this paper will be based on either of two particular bases that require fewer arithmetic operations to construct the stiffness and mass matrices associated with variable coefficient elliptic operators when numerical quadrature is essential than more commonly used bases. Both bases are related to Gauss-Lobatto quadrature; details of the analysis of the convergence of the method will be given only in the slightly more complicated case associated with one of the two bases.

Specifically, the Dirichlet problem for the Laplace equation on a unit square will be treated in detail in §2, followed by a brief discussion of a Robin problem for the Laplacian in §3. In §4, similar problems are treated in three dimensions. In §5, the general equation is handled by a two-stage Gunn iteration. Convergence analyses are given in all cases, along with estimates of the number of arithmetic operations required to achieve convergence to the level of the inherent truncation error associated with the finite element solution. In §6, the result of an experimental calculation to determine the asymptotic behavior of the condition number of a matrix arising in the alternating-direction process is reported.

2 The First Model Problem in Two Dimensions

Consider the Dirichlet problem on the unit square in $\mathbf{R}^2$:

$$-\Delta u = f, \qquad x \in \Omega \in [0,1]^2, \tag{1a}$$

$$u = g, \qquad x \in \partial\Omega, \tag{1b}$$

and consider its approximation by the p-version of the finite element method, as based on particular choices, to be discussed below, for a tensor-product basis for which the one-dimensional basis has a certain sparse structure.

It will be necessary to employ quadrature when treating equations with variable coefficients; the convenient choice will be a Gauss-Lobatto rule. Recall that it is necessary that the quadrature procedure be correct for tensor-product polynomials of degree $2p-1$ ([1], p. 205) in order to ensure nonsingularity of the stiffness matrix and maintenance of the optimal accuracy of the approximation; under certain circumstances, it is advisable to use a rule that is correct for tensor-products of degree $2p$. Both cases will be indicated below. Since the analysis is slightly more complicated when correctness is required for degree $2p$, the analysis below will be presented in that case; a similar development can be given in the simpler case of correctness for degree $2p-1$.

If correctness through degree $2p-1$ suffices, then consider the tensor product of $(p+1)$-point, one-dimensional Gauss-Lobatto quadrature rules (in either two or three space variables), and take as basis the tensor product of the one-dimensional Lagrangian polynomials of degree p over the Gauss-Lobatto points. This basis will lead to a diagonal mass matrix, as contrasted with the tensor product of tridiagonal matrices that will result from the second basis, as described below. However, the stiffness matrix will have essentially the same structure as that coming from the second basis, in that the "differentiated" term in the tensor product will again be full. Throughout the remainder of the paper, all of the analysis will be presented for the second basis; an entirely equivalent development can be carried out for the above basis.

Next, consider the tensor product of $(p+2)$-point Gauss–Lobatto quadrature rules, which is correct for polynomials of degree $2p+1$. Denote these quadrature points by

$$0 = \eta_0 < \eta_1 < \cdots < \eta_{p+1} = 1. \tag{2}$$

Define a basis for $P_p([0,1])$ such that each basis function ϕ_k, $k = 0, \ldots, p$, vanishes at all but two of the quadrature points; it is easy to see that the following constraints define such a basis for polynomials (in a single variable) of degree p:

$$\begin{aligned} \phi_k(\eta_\ell) &= 0, \qquad \ell \notin \{k, k+1\},\ k = 0, \ldots, p, &\text{(3a)}\\ \phi_k(\eta_k) &= 1, \qquad k \le (p+1)/2, &\text{(3b)}\\ \phi_k(\eta_{k+1}) &= 1, \qquad k > (p+1)/2. &\text{(3c)} \end{aligned}$$

The two-dimensional, tensor-product basis for $P_p([0,1]) \otimes P_p([0,1])$ will be taken in the form

$$\mathcal{M}_p = \text{Span}\{\psi_{ij}(x,y) = \phi_i(x)\phi_j(y) | i,j = 0, \ldots, p\}; \tag{4}$$

for later use, let

$$\mathcal{M}_p^0 = \{\psi \in \mathcal{M}_p \ : \ \psi = 0 \text{ on } \partial\Omega\}. \tag{5}$$

Define one-dimensional mass and stiffness matrices C_x and A_x, respectively, by

$$(C)_{k,\ell} = (\phi_k, \phi_\ell)_{GL}, \quad (A)_{k,\ell} = (\phi_k', \phi_\ell')_{GL}, \qquad k, \ell = 1, \ldots, p-1, \tag{6}$$

where the subscript GL indicates Gauss-Lobatto quadrature over the interval $[0,1]$. Note that C is tridiagonal and that the quadratures for its entries are over small support with respect to the quadrature points; however, ϕ_k' cannot vanish at any quadrature point, so that A is full. Below, the notation C_α and A_α, $\alpha = x, y$, will be used to identify directions. Then, let

$$M = C_x \otimes C_y \quad \text{and} \quad K = A_x \otimes C_y + C_x \otimes A_y \tag{7}$$

be the two-dimensional mass and stiffness matrices associated with the p–method for the tensor product basis for the Dirichlet problem (1) above. Thus, the discrete problem has the form

$$K\, U_p = \Phi, \tag{8}$$

where Φ contains the information related to f and g. The boundary data g is essential and is assumed to be applied strongly, up to GL-quadrature.

Consider the following analogue of the classical alternating–direction iterative method in two space variables (see [3, 7, 13]) for (8):

$$\begin{aligned} &x\text{-}sweep\text{:} \quad \rho_n C_x \otimes C_y(U^{n-\frac{1}{2}} - U^{n-1}) + A_x \otimes C_y U^{n-\frac{1}{2}} + C_x \otimes A_y U^{n-1} = \Phi, &\text{(9a)}\\ &y\text{-}sweep\text{:} \quad \rho_n C_x \otimes C_y(U^{n} - U^{n-\frac{1}{2}}) + A_x \otimes C_y U^{n-\frac{1}{2}} + C_x \otimes A_y U^{n} = \Phi. &\text{(9b)} \end{aligned}$$

One feasible computational procedure is given by the following:

- **x-sweep:**

i) Find $(I \otimes C_y)U^{n-\frac{1}{2}}$ such that

$$(\rho_n C_x + A_x) \otimes C_y U^{n-\frac{1}{2}} = \Phi + C_x \otimes (\rho_n C_y - A_y)U^{n-1}. \tag{10}$$

Thus,

$$(I \otimes C_y)U^{n-\frac{1}{2}} = \Big((\rho_n C_x + A_x)^{-1} \otimes I\Big)\Big(\Phi + C_x \otimes (\rho_n C_y - A_y)U^{n-1}\Big). \tag{11}$$

ii) Solve for $U^{n-\frac{1}{2}}$:

$$U^{n-\frac{1}{2}} = (I \otimes C_y^{-1})(I \otimes C_y)U^{n-\frac{1}{2}}. \tag{12}$$

- **y-sweep:**

i) Find $(C_x \otimes I)U^n$ such that

$$C_x \otimes (\rho_n C_y + A_y)U^n = \Phi + (\rho_n C_x - A_x) \otimes C_y U^{n-\frac{1}{2}}. \tag{13}$$

So,

$$(C_x \otimes I)U^n = \Big(I \otimes (\rho_n C_y + A_y)^{-1}\Big)\Big(\Phi + (\rho_n C_x - A_x) \otimes C_y U^{n-\frac{1}{2}}\Big). \tag{14}$$

ii) Solve for U^n:

$$U^n = (C_x^{-1} \otimes I)(C_x \otimes I)U^n. \tag{15}$$

Note that each sweep has two parts. The x–sweep first solves equations related to $(\rho_n C_x + A_y)z = b$ along lines of quadrature points in the x–direction and then solves the tridiagonal (or, for the first basis, diagonal) equations related to $C_y w = c$ along lines of quadrature points in the y-direction. The matrix $\rho_n C_x + A_x$ is a full $(p+1) \times (p+1)$ matrix; C_y is tridiagonal. Each can be factored just once into LU-form, with the factors being retained. Of course, the factorization of $\rho_n C_x + A_x$ must be carried out for each choice of the parameter ρ_n. The intermediate values $C_x \otimes I\, U^{n-\frac{1}{2}}$ and $I \otimes C_y U^n$ should be kept for use in the y-sweep of the current step and the x-sweep of the next step, respectively.

There are some possible simplifications in the algorithm. First, multiply (10) by $C_x^{-1} \otimes I$ and then $I \otimes C_y^{-1}$:

$$(\rho_n I + C_x^{-1} A_x) \otimes I\, U^{n-\frac{1}{2}} = (C_x^{-1} \otimes C_y^{-1})\Phi + I \otimes (\rho_n I - C_y^{-1} A_y)U^{n-1}. \tag{16}$$

Then,

$$U^{n-\frac{1}{2}} = (\rho_n I + C_x^{-1} A_x)^{-1} \otimes I\Big(C_x^{-1} \otimes C_y^{-1}\Phi + I \otimes (\rho_n I - C_y^{-1} A_y)U^{n-1}\Big), \tag{17a}$$

$$U^n = I \otimes (\rho_n I + C_y^{-1} A_y)^{-1}\Big(C_x^{-1} \otimes C_y^{-1}\Phi + (\rho_n I - C_x^{-1} A_x) \otimes I\, U^{n-\frac{1}{2}}\Big). \tag{17b}$$

The algorithm for an iterative step can be detailed as follows. Set

$$B_x = C_x^{-1} A_x, \qquad B_y = C_y^{-1} A_y; \tag{18}$$

each can be found by solving the trivial equations

$$C_x B_x = A_x \quad \text{or} \quad C_y B_y = A_y.$$

If the degree in the p–method is the same in the x– and y–directions, then $A_x = A_y$, $C_x = C_y$, and $B_x = B_y$. If different p_x and p_y are chosen, then they are all different. In any event, this is a one–time operation. Note that, if B_x and B_y are constructed, there is a slight reduction in computation required for the two–sweep iteration, since the tridiagonal solutions are not required; however, this is an almost negligible savings.

Next, set $\Psi = C_x^{-1} \otimes C_y^{-1}\Phi$. Clearly, Ψ can be computed trivially as an initialization. Now, rewrite (17) in the following form:

$$U^{n-\frac{1}{2}} = (\rho_n I + B_x)^{-1} \otimes I\Big(\Psi + I \otimes (\rho_n I - B_y)U^{n-1}\Big), \tag{19a}$$

$$U^n = I \otimes (\rho_n I + B_y)^{-1}\Big(\Psi + (\rho_n I - B_x) \otimes I\, U^{n-\frac{1}{2}}\Big). \tag{19b}$$

Let us turn to the choice of the parameter sequence $\{\rho_n\}$. We have to construct the matrices C_x and A_x; for simplicity, assume that $p_x = p_y$ from now on. Then, the eigenvalues associated with the generalized eigenvalue problem

$$A_x z = \lambda C_x z, \qquad \text{or, equivalently,} \qquad C_x^{-\frac{1}{2}} A_x C_x^{-\frac{1}{2}} w = \lambda w, \tag{20}$$

are important in the determination of an effective parameter sequence and in the analysis of the convergence of the iterative process; the eigenfunctions appear in the analysis, but it is not necessary to evaluate them. Neither eigenvalues nor eigenfunctions are obvious, as they are in the finite difference and piecewise-linear h–method cases, though it is obvious that $\lambda > 0$. The eigenvalues can be computed with sufficient accuracy using routines from EISPACK or LAPACK. Denote the eigenpairs for (20) by $\{\lambda_k, \xi_k\}$, $k = 1, \ldots, p-1$, with the ordering

$$0 < \lambda_1 < \cdots < \lambda_{p-1}.$$

Let $v^n = U^n - U$, where U is the solution of (8), U^0 is the initial guess at U, satisfying the discrete boundary condition but otherwise arbitrary, and U^n is the nth iterate as determined by (9). Then, the sequence $\{v^n\}$ satisfies the homogeneous form of the alternating-direction iteration equations, and v^n can be expanded in the form

$$v^n = \sum_{k,\ell=1}^{p-1} a_{k,\ell}^n \Xi_{k,\ell}, \tag{21}$$

where

$$\Xi_{k,\ell}(x,y) = \xi_k(x)\xi_\ell(y), \qquad k, \ell = 1, \ldots, p-1.$$

Let

$$R(\zeta, \lambda) = \frac{\zeta - \lambda}{\zeta + \lambda}. \tag{22}$$

Then, it is a standard result of alternating-direction analyses that

$$a_{k,\ell}^M = a_{k,\ell}^0 \prod_{m=1}^{M} R(\zeta_m, \lambda_k) R(\zeta_m, \lambda_\ell). \tag{23}$$

The simplest choice for a parameter cycle is given by taking

$$\rho_n = \lambda_n, \qquad n = 1, \ldots, p-1; \tag{24}$$

theoretically, the error after these $p-1$ iterations vanishes. As a result of round-off, there would be a minor error remaining, but this cycle should not need repeating for even modestly large p, particularly if the alternating-direction procedure is being used either as a preconditioner or in a domain decomposition iterative algorithm.

As a result of the computation of the condition number associated with (20) for a Chebychev-Lagrangian basis [12], it can be conjectured that the condition number λ_{p-1}/λ_1 is $\mathcal{O}(p^4)$; an experimental calculation for $p = 4, 8, \ldots, 256, 512$ confirms this conjecture at least up to 512. Consequently, it will be assumed throughout the rest of the paper that

$$\lambda_{p-1}/\lambda_1 = \mathcal{O}(p^4). \tag{25}$$

For $p = 512$, there are already more than 10^8 parameters on a single cell in the three-dimensional case; obviously, there is little interest in larger values of p in practical computation.

Under the assumption (25), it is possible for large p to define a more efficient cycle. Let

$$\rho_m = \left(\frac{1+\sqrt{\varepsilon}}{1-\sqrt{\varepsilon}}\right)^{2m-1} \lambda_1, \qquad m = 1, \ldots, Q, \tag{26}$$

where Q is the smallest integer such that

$$\frac{1+\sqrt{\varepsilon}}{1-\sqrt{\varepsilon}} \rho_Q \geq \lambda_{p-1}. \tag{27}$$

Then, by (23), it is easy to see that

$$|a^Q_{k,\ell}| \leq \varepsilon |a^0_{k,\ell}|, \qquad k, \ell = 1, \ldots, p-1, \tag{28}$$

so that the $L^2(\Omega)$ and $H^1(\Omega)$ norms of the function corresponding to v^Q satisfy the inequalities (confounding the notation for a vector and its corresponding function)

$$\|v^Q\|_s \leq \varepsilon \|v^0\|_s, \qquad s = 0, 1. \tag{29}$$

The optimal truncation error for the approximation is $\mathcal{O}(p^{-\alpha})$ if the solution of (1) is in H^α, $\alpha > 1$, and it seems to be uninteresting to solve the algebraic equations to any higher accuracy than that. Thus, if $\varepsilon < 1$ is held fixed, it suffices to carry out $\mathcal{O}(\log p)$ cycles of iteration. Moreover, by (25), $Q = \mathcal{O}(\log p)$, as well.

Now, we are in a position to estimate the total cost, in terms of arithmetic operations, for an error reduction by a factor $\mathcal{O}(p^{-\alpha})$. Consider first the cost of a single (double-sweep) iteration. Then, the number W_{fac} of arithmetic operations required to factor $(\rho_n I + B_x)$ into the product $T_\ell T_u$ of lower and upper triangular matrices is

$$W_{fac} = \mathcal{O}(p^3);$$

the same cost estimate holds if $\rho_n C_x + A_x$ is factored, instead. Solving the equations corresponding to one line of quadrature points costs $\mathcal{O}(p^2)$ operations, so that the solution cost W_{sol} for one iteration is also

$$W_{sol} = \mathcal{O}(p^3).$$

Thus, the total cost W_{step} for a single step is

$$W_{step} = \mathcal{O}(p^3). \tag{30}$$

If the geometric sequence is used for a parameter cycle, then the cost W for a reduction in the L^2- or H^1-norm by a factor $\mathcal{O}(p^{-\alpha})$ of the error is

$$W = \mathcal{O}(p^3(\log p)^2). \tag{31}$$

If the parameter cycle consists of all $p-1$ eigenvalues, then the work estimate becomes

$$W = \mathcal{O}(p^4). \tag{32}$$

Clearly, this estimate represents the maximum amount of work that could be required to find the solution of the algebraic equations.

If $N = (p-1)^2$ denotes the dimension of the finite element space, then the work estimate when (25) holds above can be expressed in terms of N as

$$W = \mathcal{O}(N^{\frac{3}{2}}(\log N)^2). \tag{33}$$

3 The Second Model Problem in Two Dimensions

Now, consider the Robin problem on the unit square in $\mathbf{R}^2$:

$$-\Delta u = f, \qquad x \in \Omega \in [0,1]^2, \tag{34a}$$

$$\frac{\partial u}{\partial \nu} + \beta u = g, \qquad x \in \partial\Omega, \tag{34b}$$

where ν denotes the outer normal to $\partial\Omega$ and $\beta > 0$. The variational form of this problem is given by seeking $u \in H^1(\Omega)$ such that

$$(\nabla u, \nabla v) + \beta\langle u, v\rangle = (f, v) + \langle g, v\rangle, \quad v \in H^1(\Omega), \tag{35}$$

where $\langle \cdot, \cdot \rangle$ denotes the L^2-inner product on $\partial\Omega$. Its approximation by the p-method based on either sparse basis of the preceding section leads again to a problem in the form (8); however, the dimension of the finite element space changes and the matrices A_x and A_y must be modified slightly to take the boundary integral into account. The functions ϕ_0 and ϕ_p must be included in the one-dimensional basis. This will induce some special cases; the mass matrix will have the same form as before:

$$(C_x)_{k,\ell} = (\phi_k, \phi_\ell)_{GL}, \qquad k, \ell = 0, \ldots, p, \tag{36}$$

while the stiffness matrix changes to

$$(A_x)_{k,\ell} = (\phi_k', \phi_\ell')_{GL} + \beta\Big(\phi_k(0)\phi_\ell(0) + \phi_k(1)\phi_\ell(1)\Big), \qquad k, \ell = 0, \ldots, p. \tag{37}$$

With these modifications, the algebraic problem and its alternating-direction iterative procedure have the same form, with different dimensions, as above. Again, use EISPACK or LAPACK to find the eigenvalues of the corresponding generalized eigenproblem and select the iteration parameters following the same rules as above. The entire analysis can be repeated without essential changes, and the work estimates are unchanged, except that the constants in the $\mathcal{O}$-terms will be a bit larger.

4 A Model Problem in Three Dimensions

Only the Dirichlet version of the model problem will be discussed in this section; the development for the Robin version can be treated with no more difficulty than arose in the two-dimensional case. So, consider the problem

$$-\Delta u = f, \qquad x \in \Omega = [0,1]^3, \tag{38a}$$

$$u = g, \qquad x \in \delta\Omega. \tag{38b}$$

Let

$$\mathcal{M}_p = P_p([0,1]) \otimes P_p([0,1]) \otimes P_p([0,1]), \quad \mathcal{M}_p^0 = \{\psi \in \mathcal{M}_p : \psi(x) = 0,\ x \in \partial\Omega\}.$$

Then, with an obvious extension of the notation employed above, the p-Galerkin problem can be expressed in the form of finding $U \in \mathcal{M}_p$ such that

$$KU = \Phi, \tag{39}$$

where U is the Gauss-Lobatto L^2-projection of g onto $P_p([0,1]) \otimes P_p([0,1])$ on each face of the cube and

$$K = A_x \otimes C_y \otimes C_z + C_x \otimes A_y \otimes C_z + C_x \otimes C_y \otimes A_z; \tag{40}$$

here, the matrices A_α and C_α have the same meaning as in (6).

This time, consider an analogue of the alternating-direction method introduced by the author in [4]; it was intended to be the alternating-direction modification of the standard Crank-Nicolson finite difference procedure for (38) (see also [6]). Each of three sweeps moves one of the second derivative terms forward, as follows:

$$\rho C_x \otimes C_y \otimes C_z(U^{n-\frac{2}{3}} - U^{n-1}) + \frac{1}{2}A_x \otimes C_y \otimes C_z(U^{n-\frac{2}{3}} + U^{n-1}) \tag{41a}$$
$$+ C_x \otimes A_y \otimes C_z U^{n-1} + C_x \otimes C_y \otimes A_z U^{n-1} = \Phi,$$

$$\rho C_x \otimes C_y \otimes C_z(U^{n-\frac{1}{3}} - U^{n-1}) + \frac{1}{2}A_x \otimes C_y \otimes C_z(U^{n-\frac{2}{3}} + U^{n-1}) \tag{41b}$$
$$+ \frac{1}{2}C_x \otimes A_y \otimes C_z(U^{n-\frac{1}{3}} + U^{n-1}) + C_x \otimes C_y \otimes A_z U^{n-1} = \Phi,$$

$$\rho C_x \otimes C_y \otimes C_z(U^{n} - U^{n-1}) + \frac{1}{2}A_x \otimes C_y \otimes C_z(U^{n-\frac{2}{3}} + U^{n-1}) \tag{41c}$$
$$+ \frac{1}{2}C_x \otimes A_y \otimes C_z(U^{n-\frac{1}{3}} + U^{n-1}) + \frac{1}{2}C_x \otimes C_y \otimes A_z(U^{n} + U^{n-1}) = \Phi.$$

Computationally, it is convenient to subtract the first equation from the second and the second from the third. Then, the system (41) can be written in the form

$$\rho C_x \otimes C_y \otimes C_z(U^{n-\frac{2}{3}} - U^{n-1}) + \frac{1}{2}A_x \otimes C_y \otimes C_z(U^{n-\frac{2}{3}} + U^{n-1})$$
$$+C_x \otimes A_y \otimes C_z U^{n-1} + C_x \otimes C_y \otimes A_z U^{n-1} = \Phi, \tag{42a}$$
$$\rho C_x \otimes C_y \otimes C_z(U^{n-\frac{1}{3}} - U^{n-\frac{2}{3}}) + \frac{1}{2}C_x \otimes A_y \otimes C_z(U^{n-\frac{1}{3}} - U^{n-1}) = 0, \tag{42b}$$
$$\rho C_x \otimes C_y \otimes C_z(U^{n} - U^{n-\frac{1}{3}}) + \frac{1}{2}C_x \otimes C_y \otimes A_z(U^{n} - U^{n-1}) = 0. \tag{42c}$$

Set

$$B_\alpha = \frac{1}{2}C_\alpha^{-1}A_\alpha, \quad \alpha = x, y, z; \quad I = I_x \otimes I_y \otimes I_z; \quad \mathcal{B}_x = B_x \otimes I_y \otimes I_z;$$

define $\mathcal{B}_y$ and $\mathcal{B}_z$ analogously. Then, (42) is equivalent to

$$(\rho I + \mathcal{B}_x)U^{n-\frac{2}{3}} = \Psi + (\rho I - \mathcal{B}_x - 2\mathcal{B}_y - 2\mathcal{B}_z)U^{n-1}, \tag{43a}$$
$$(\rho I + \mathcal{B}_y)U^{n-\frac{1}{3}} = \rho U^{n-\frac{2}{3}} + \mathcal{B}_y U^{n-1}, \tag{43b}$$
$$(\rho I + \mathcal{B}_z)U^{n} = \rho U^{n-\frac{1}{3}} + \mathcal{B}_z U^{n-1}, \tag{43c}$$

where $\Psi = C_x^{-1} \otimes C_y^{-1} \otimes C_z^{-1}\Phi$.

The computational algorithm can be summarized easily. After evaluating Ψ and the matrices $\mathcal{B}_\alpha$ as initialization, each iterative step consists of the following:

i) Evaluate $\mathcal{B}_\alpha U^{n-1}$, $\alpha = x, y, z$;

ii) With $\rho = \rho_n$, factor $\rho_n I + \mathcal{B}_\alpha$;

iii) Solve for $U^{n-\frac{2}{3}}$, $U^{n-\frac{1}{3}}$, and U^n.

The work per iterative step is $\mathcal{O}(p^4)$, since parts *i*) and *iii*) require that number of arithmetic operations, while the factorizations take $\mathcal{O}(p^3)$ operations.

In the three-dimensional case, there is no possibility of obtaining an exact solution unless complex arithmetic is used, since the ratio corresponding to $R(\zeta_m, \lambda_k)R(\zeta_m, \lambda_\ell)$ in (23) does not vanish for any real ρ_n. Consequently, it is appropriate to set up a geometric cycle of parameters; the choice of this cycle, which will be of length $\mathcal{O}(\log(\lambda_{\max}/\lambda_{\min}))$, was discussed in detail in [4] and will be indicated briefly here. Assume, for simplicity, that $A_x = A_y = A_z$ and $C_x = C_y = C_z$, and again consider the generalized eigenproblem (20). The error e^n in the iterative solution can be expanded as

$$e^n = \sum_{i,j,k=1}^{p-1} \varepsilon_{ijk}^n \Xi_{ijk}, \qquad \Xi_{ijk} = \xi_i(x)\xi_j(y)\xi_k(z). \tag{44}$$

A calculation (see [4]) shows that

$$\varepsilon_{ijk}^n = R_{ijk}(\rho)\varepsilon_{ijk}^{n-1}, \tag{45}$$

where

$$R_{ijk}(\rho) = 1 - \frac{\zeta_i + \zeta_j + \zeta_k}{(1+\zeta_i)(1+\zeta_j)(1+\zeta_k)}, \qquad \zeta_m = \zeta_m(\rho) = \frac{\lambda_m}{2\rho}. \tag{46}$$

Note that $0 < R_{ijk}(\rho) < 1$ for any positive ρ; hence, it is again the object to select a sequence $\{\rho_1, \ldots, \rho_N\}$ such that

$$\max_{ijk} \min_{n} R_{ijk}(\rho_n) \leq \gamma < 1. \tag{47}$$

This can be facilitated by applying Lemma 3 of [4], which states that, if $0 < \mu < 1 < \nu$, then

$$\hat{R}(\mu, \nu) = \max_{\mu \leq a \leq \nu; 0 \leq b, c \leq \nu} R(a, b, c) = \max\left\{1 - \frac{6\nu}{(1+\nu)^3}, 1 - \frac{2\mu}{1+\mu}\right\}. \tag{48}$$

If μ is chosen so that

$$\mu = \frac{3\nu}{1 + 3\nu^2 + \nu^3}, \tag{49}$$

the two terms in the maximum are equal; assume this relation to hold. An argument is given in [4] to indicate that the optimum choice for ν is about 1.78 to obtain a minimum constant in the work estimate to be derived below, with the corresponding value of μ being about 0.32. Now, take

$$\rho_n = \frac{\lambda_{\min}}{2\mu}\left(\frac{\nu}{\mu}\right)^{n-1}, \qquad n = 1, \ldots, Q, \tag{50}$$

where Q is chosen to be the smallest integer such that

$$2\nu\rho_Q \geq \lambda_{\max}. \tag{51}$$

Thus,

$$Q \sim \log \frac{\lambda_{\max}}{\lambda_{\min}} \Big/ \log \frac{\nu}{\mu}. \tag{52}$$

This choice of a parameter cycle implies that the error in the iterate is reduced by approximately one-half in either $L^2(\Omega)$ or $H^1(\Omega)$ after one alternating-direction cycle, as it can be seen easily from (47) that (48) holds. The factor one-half assumes the use of the optimal values for ν and μ.

In order to reduce the error by a factor $\mathcal{O}(p^{-\alpha})$, the total work will be

$$W = \mathcal{O}\left(p^4 \log p \log \frac{\lambda_{\max}}{\lambda_{\min}}\right), \tag{53}$$

where $\lambda_{\max}$ and $\lambda_{\min}$ are associated with the generalized eigenvalue problems $\frac{1}{2}A_\alpha\phi = \lambda C_\alpha\phi$. Thus, when (25) holds, the work estimate is

$$W = \mathcal{O}(p^4(\log p)^2) = \mathcal{O}(N^{\frac{4}{3}}(\log N)^2), \tag{54}$$

where $N = (p-1)^3$ is the dimension of the finite element space. This estimate is much better than the estimate $W = \mathcal{O}(N^{\frac{7}{3}})$ associated with solving the equations by Gaussian elimination.

5 The General Problem in Two or Three Dimensions

Let us consider the Dirichlet problem (the Robin problem can be handled in a fashion similar to the Dirichlet problem and will not be considered explicitly below) for the second-order elliptic equation

$$-\sum_{i,j=1}^{d} \frac{\partial}{\partial x_i}\left(a_{ij}(x)\frac{\partial u}{\partial x_j}\right) + c(x)u = f(x), \qquad x \in \Omega, \tag{55a}$$

$$u = g(x), \qquad x \in \partial\Omega, \tag{55b}$$

where Ω is again the unit square in $\mathbf{R}^d$, $d = 2$ or 3. Assume that $a_{ij} = a_{ji}$, that

$$\sigma_*|\gamma|^2 \le \sum_{i,j=1}^{d} a_{ij}(x)\gamma_i\gamma_j \le \sigma^*|\gamma|^2, \qquad x \in \Omega, \quad 0 < \sigma_* \le \sigma^* < \infty, \tag{56}$$

and that $0 \le c(x) \le c^* < \infty$. Set

$$\mathcal{A}(u,v) = \sum_{i,j=1}^{d}\left(a_{ij}\frac{\partial u}{\partial x_j}, \frac{\partial v}{\partial x_i}\right)_{GL} + (cu,v)_{GL}.$$

Then, it follows from (56) and the assumption on $c(x)$ that there exist constants K_* and K^*, independent of p, such that

$$K_* \le \frac{\mathcal{A}(u,u)}{(\nabla u, \nabla u)_{GL}} \le K^*, \quad u \in \mathcal{M}_p. \tag{57}$$

A two-stage iteration for the solution $U \in \mathcal{M}_p$ of the finite element equations

$$\mathcal{A}(U,V) = (f,V), \qquad V \in \mathcal{M}_p^0, \tag{58}$$

such that U coincides with the projection of g into P_p for $d = 2$ or $P_p \otimes P_p$ for $d = 3$ on each face of $\partial\Omega$ can be defined by the recursion

$$(\nabla(U^n - U^{n-1}), \nabla V)_{GL} + \omega(\mathcal{A}(U^{n-1},V) - (f,V)_{GL}) = 0, \quad V \in \mathcal{M}_p^0. \tag{59}$$

Let

$$e^n = U - U^n \in \mathcal{M}_p^0; \tag{60}$$

then,

$$(\nabla(e^n - e^{n-1}), \nabla v) + \omega\mathcal{A}(e^{n-1}, v) = 0, \qquad v \in \mathcal{M}_p^0. \tag{61}$$

(Note that the quadrature is exact for the product of the gradients.)

Two cases need be considered, the first applicable to the two-dimensional problem when the parameter set for the alternating-direction iteration for the Laplace problem consists of the $p-1$ eigenvalues of (20) and the second in either two or three dimensions when a geometric sequence is employed for the parameter cycle. The case in which the full parameter sequence is employed to achieve an exact solution of (59) will be treated first; in the case of finite differences, this procedure was first proposed by D'yakonov [8, 9].

There exist eigenpairs $\{\mu_k, \psi_k\}$, $k = 1, \ldots, (p-1)^2$, for the generalized eigenproblem

$$\mathcal{A}(\psi_k, v) = \mu_k(\nabla\psi_k, \nabla v), \quad v \in \mathcal{M}_p^0, \qquad (\nabla\psi_k, \nabla\psi_\ell) = \delta_{k\ell}. \tag{62}$$

Expand e^n in terms of $\{\psi_k\}$:

$$e^n = \sum_k \varepsilon_k^n \psi_k. \tag{63}$$

It is easy to see that

$$\varepsilon_k^n = (1 - \omega\mu_k)\varepsilon_k^{n-1}, \qquad k = 1, \ldots, (p-1)^2. \tag{64}$$

Thus,

$$|\varepsilon_k^n| \le \max(\omega\mu_{\max} - 1, 1 - \omega\mu_{\min}), \tag{65}$$

and convergence results if

$$0 < \omega < 2/\mu_{\max}. \tag{66}$$

While $\mu_{\max}$ and $\mu_{\min}$ are usually not known precisely, they can be estimated from the assumed ellipticity constants. It follows from (56) and the bounds on $c(x)$ that

$$\sigma_*\|\nabla u\|^2 \le \mathcal{A}(u,u) \le \sigma^*\|\nabla u\|^2 + c^*\|u\|^2 \le (\sigma^* + c^*/(d\pi^2))\|\nabla u\|^2, \tag{67}$$

where the Poincaré embedding constant $1/d\pi^2$ for the unit square has been applied. Hence,

$$\mu_{\min} \ge \sqrt{\sigma_*} = M_*, \quad \mu_{\max} \le \sqrt{\sigma^* + c^*/2\pi^2} = M^*; \tag{68}$$

note that these bounds are independent of p, which will imply that the convergence rate factor τ defined below in (70) is also independent of p. Choose

$$\omega = \frac{2}{M^* + M_*}; \tag{69}$$

then,

$$\|\nabla e^n\|^2 \le \frac{M^* - M_*}{M^* + M_*}\|\nabla e^{n-1}\|^2 = \tau\|\nabla e^{n-1}\|^2, \tag{70}$$

so that

$$\|\nabla e^n\|^2 \le \tau^n\|\nabla e^0\|^2. \tag{71}$$

Thus, in this case $\mathcal{O}(\log p)$ outer iterations are sufficient to reduce the error $U - U^n$ to the size $\mathcal{O}(p^{-\alpha+1})$ in $H^1(\Omega)$ of the truncation error associated with the finite element solution. Consequently, the total work required by the D'yakonov iteration procedure to produce the approximate solution U is

$$W = \mathcal{O}(p^4 \log p) = \mathcal{O}(N^2 \log N) \tag{72}$$

when $d = 2$.

Now, consider the case in which the finite element equations for the Laplace operator are incompletely solved by using one cycle of alternating-direction iteration (the analysis below, which is based on that by Gunn [10, 11], covers any iterator that is defined in advance and can be extended to cover some that are not, such as conjugate gradients). It will be convenient to give this analysis with the procedure being in operator form, rather than inner-product form. Let the equations for the p-Galerkin problem for the homogeneous Dirichlet problem (boundary values appear only on the right-hand side of the equation below) for Laplace's equation be given by

$$Dv = b, \tag{73}$$

and note that the inner product (Dv, v) is equal to the previously used inner product $(\nabla V, \nabla V) = (\nabla V, \nabla V)_{GL}$, where v is the vector of coefficients for $V \in M_p'$. Similarly, indicate the equations (58) by

$$Au = \Phi; \tag{74}$$

here again, the influence of the boundary values has been transferred to the right-hand side, so that u represents the $\mathcal{M}_p^0$-part of the solution U.

Denote by Λ the operator advancing the error in the solution of (73) as a result of one alternating-direction cycle; i.e., if the iterates are given by v^n, $n = 1, 2, \ldots$,

$$v - v^n = \Lambda(v - v^{n-1}). \tag{75}$$

It is important to remember that D and Λ commute. A short calculation shows that

$$(I - \Lambda)^{-1} D v^n = (I - \Lambda)^{-1} D v^{n-1} + b, \tag{76}$$

so that the single cycle iteration can be interpreted as solving equations (76) associated with $(I - \Lambda)^{-1} D$ exactly.

Gunn [10] proposed the following natural modification of the D'yakonov iteration:

$$Bu^{n+1} = Bu^n - \omega(Au^n + \Phi), \qquad \text{where} \quad B = (I - \Lambda)^{-1} D. \tag{77}$$

To analyze this iteration, it is helpful to obtain some operator comparisons. First, assume that

$$\|\Lambda\| \le \zeta < 1, \tag{78}$$

where ζ is chosen to be independent of p. A simple calculation shows that

$$(1 - \zeta)(Bz, z) \le (Dz, z) \le (1 + \zeta)(Bz, z). \tag{79}$$

Thus, it follows from (67) and (79) that

$$M_*(Bz, z) \le (Az, z) \le M^*(Bz, z), \tag{80}$$

where now

$$M_* = (1 - \zeta)\sigma_*, \qquad M^* = (1 + \zeta)\left(\sigma^* + \frac{c^*}{d\pi^2}\right). \tag{81}$$

If the parameter ω is again chosen by (69), then the remainder of the convergence argument is unchanged. The D'yakonov iteration is the special case of the Gunn iteration when $\zeta = 0$.

The work estimate is altered and improved, at least asymptotically for large p. The cycle length leading to fixed $\zeta < 1$ is of length $\mathcal{O}(\log(\lambda_{\max}/\lambda_{\min}))$ and the number of cycles required to obtain $\tau^n = \mathcal{O}(p^{-\alpha+1})$ remains $\mathcal{O}(\log p)$, so that the number of arithmetic operations required to obtain the solution to the desired accuracy is

$$W = \mathcal{O}(p^{d+1}(\log p)^2) = \mathcal{O}(N^{1+\frac{1}{d}}(\log N)^2), \tag{82}$$

under the assumption (25), as usual.

No known direct solution method compares favorably with this.

6 A Condition Number Estimate

The author wishes to thank Dr. Seongjai Kim for computing the eignevalues and condition numbers for the generalized eigenvalue problem (20) for the second basis. If C_p denotes the condition number corresponding to p, then the following table indicates the behavior of C_p.

p	C_p	$C_p/C_{p/2}$
4	1.03478e1	
8	8.90495e1	8.61
16	9.81944e2	11.03
32	1.29995e4	13.24
64	1.89226e5	14.56
128	2.88826e6	15.26
256	4.51387e7	15.63
512	7.13793e8	15.81

Table 1: Condition Numbers as a Function of p

The ratio $C_p/C_{p/2}$ is approaching the value 16; thus, the conjecture that $C_p = \mathcal{O}(p^4)$ seems to be correct. Recall that this asymptotic behavior has been proved for Chebychev interpolation [12]; the qualitative placement of the Chebychev interpolation points and the Gauss-Lobatto quadrature points is quite similar and was the basis for the conjecture.

References

[1] Ciarlet P. G., *Numerical Analysis of the Finite Element Method*, Les Presses de l'Université de Montréal, Montréal, 1976

[2] Quarteroni A., Valli A., *Numerical Approximation of Partial Differential Equations*, Springer-Verlag, Berlin Heidelberg, 1994

[3] Douglas J. Jr., *On the numerical integration of* $u_{xx} + u_{yy} = u_t$ *by implicit methods*, J. Soc. Indust. Appl. Math., **3** (1955), pp. 42–65

[4] Douglas J. Jr., *Alternating direction methods for three space variables* , Numerische Mathematik, **4** (1962), pp. 41–63

[5] Douglas J. Jr., Dupont T., *Alternating-direction Galerkin methods on rectangles*; in Numerical Solution of Partial Differential Equations II, (B. Hubbard, ed.), Academic Press, New York, 1971, pp. 133–214

[6] Douglas J. Jr., Gunn J. E., *A general formulation of alternating direction methods, I. Parabolic and hyperbolic problems*, Numerische Mathematik, **6** (1964), 428–453

[7] Douglas J. Jr., Peaceman D. W., *Numerical solution of two-dimensional heat flow problems*, A. I. Ch. E. Jour., **1** (1955), 505–512,

[8] D'yakonov E. G., *On an iterative method for the solution of a system of finite-difference equations*, Doklady Akad. Nauk USSR, **138** (1961), pp. 522–525,

[9] D'yakonov E. G., *On a method of solution of Poisson's equation*, Doklady Akad. Nauk USSR, **143** (1962), pp. 21–24

[10] Gunn J. E., *The numerical solution of* $\nabla \cdot a\nabla u = f$ *by a semi-explicit alternating-direction iterative method*, Numerische Mathematik, **6** (1964), pp. 181–184

[11] Gunn J. E., *The solution of elliptic difference equations by semi-explicit iterative techniques*, J. SIAM, Series B, Numerical Analysis, **2** (1965), pp. 24–45

[12] Olsen E. T., Douglas J. Jr., *Bounds on spectral condition numbers of matrices arising in the p–version of the finite element method*, Numerische Mathematik, to appear

[13] Peaceman D. W., Rachford H. H., *The numerical solution of parabolic and elliptic differential equations*, J. Soc. Ind. Appl. Math., **3** (1955), pp. 28-41

12. An Integrodifferential Analog of Semilinear Parabolic PDE's

Paul C. Fife Department of Mathematics, University of Utah, Salt Lake City, Utah

1 INTRODUCTION

The semilinear parabolic partial differential equation

$$\partial_t u = \Sigma \partial_i a_{ij}(x) \partial_j u - f(u,x), \quad x \in \Omega, \quad t \geq 0, \tag{1}$$

in which the bounded matrix $\{a_{ij}\}$ is symmetric and uniformly positive definite in a bounded domain $\Omega \subset R^N$, under one of the boundary conditions

$$\left\{ \begin{array}{ll} \Sigma a_{ij}(x) \nu_i \partial_j u(x,t) = 0 & \text{or} \\ u = 0 & \end{array} \right\} \quad x \in \partial\Omega, \quad \nu \text{ the unit outer normal,} \tag{2}$$

can be viewed as a gradient flow for the energy functional

$$E[u] = \int_\Omega \left[\frac{1}{2} \Sigma a_{ij}(x) \partial_i u \partial_j u + W(u,x) \right] dx, \tag{3}$$

$f(u,x) = \partial_u W(u,x)$, in the space $L_2(\Omega)$. Thus, solutions of (1), (2) evolve along curves of "steepest descent" for E in a dense domain $\mathcal{D}$ in that space. These curves, at each point, lie in the direction of the negative "gradient" $-\nabla E[u]$ [12, 13]. We take a_{ij}, Ω, and W to be smooth, and $\mathcal{D}$ to be the linear manifold of smooth functions with uniformly bounded derivatives of high enough order satisfying one or the other of (2). For each u in $\mathcal{D}$, $\nabla E[u]$ is defined to be a function $w[u] \in L_2$ such that for any smooth trajectory $h(t) \in \mathcal{D}$ with $h(0) = u$,

$$\frac{d}{dt} E[h(t)] \mid_{t=0} = \langle w, \frac{dh}{dt}(0) \rangle. \tag{4}$$

Here the L_2 scalar product is used. It is easily checked that for our energy functional (3), $\nabla E[u]$ is the negative of the right side of (1). Thus (1) is a gradient flow, in the sense that $\partial_t u = -\nabla E[u]$. Different scalar products produce different gradients and their associated evolutions.

In this sense, (1) evolves the state function u in a direction designed to optimize the rate of reduction of the energy, while constrained by (2). The portion of E (3) consisting of the integral of the quadratic form in derivatives can be considered an energy "penalty" resulting from spatial variation of the function u. If the matrix $\{a_{ij}\}$ is a scalar matrix, this penalty is isotropic; if it does not depend on x it is spatially homogeneous.

In place of such a quadratic form in the first derivatives of u, other possible penalties for the spatial variation of a function u come to mind. Possibly the simplest type is a weighted integral of $(u(x) - u(y))^2$. The corresponding energy $\bar{E}$ takes the following form, analogous to (3):

$$\bar{E}[u] = \frac{1}{4}\int_\Omega\int_\Omega J(x,y)(u(x) - u(y))^2 dxdy + \int_\Omega W(u(x), x)dx. \tag{5}$$

We do not require Ω to be bounded; in fact it could be the entire space.

We consider gradient flows for this functional, with smooth weight function J satisfying the following:

$$J(x,y) \geq 0; \quad J(x,y) = J(y,x); \quad \int_\Omega J(x,y)dy \equiv K(x) > 0; \quad J(x,y)|y|^4 \text{ bounded}. \tag{6}$$

The first two properties are analogous to the positive definiteness and symmetry of a_{ij}. Again, if J depends only on x and $|x-y|$, the penalty is isotropic; if it depends only on $x - y$, it is spatially homogeneous.

We now calculate $\nabla\bar{E}$ in $L_2(\Omega)$, using the notation $\dot{E}$, $\dot{h}$ for time derivatives:

$$\begin{aligned}
&\dot{\bar{E}}[h(\cdot,t)] \\
&= \tfrac{1}{2}\int\int J(x,y)(h(x,t) - h(y,t))(\dot{h}(x,t) - \dot{h}(y,t))dxdy + \int \partial_u W(h,x)\dot{h}(x,t)dx \\
&= \int\int J(x,y)(h(x,t) - h(y,t))\dot{h}(x,t)dxdy + \int \partial_u W(h,x)\dot{h}(x,t)dx \text{ [by (6b)]} \\
&= \int [\{h(x,t)\int J(x,y)dy - \int J(x,y)h(y,t)dy\} + \partial_u W(h(x,t),x)]\,\dot{h}(x,t)dx \\
&= \langle\{h(\cdot,t)K - \int J(\cdot,y)h(y,t)dy\} + \partial_u W(h(\cdot,t),x), \dot{h}(\cdot,t)\rangle.
\end{aligned} \tag{7}$$

Thus from (4) and the relation $f = \partial_u W$, we have

$$\nabla\bar{E}[u](x) = K(x)u(x) - \int_\Omega J(x,y)u(y)dy + f(u(x),x). \tag{8}$$

The associated gradient flow is

$$\tau\frac{\partial u}{\partial t}(x,t) = \int_\Omega J(x,y)u(y,t)dy - K(x)u(x,t) - f(u,x). \tag{9}$$

In this equation the arbitrary positive kinetic coefficient τ allows for changes in the time scale, which do not affect the gradient nature of the flow.

2 COMPARISON PRINCIPLES AND THE DIRECTION OF TIME

Evolution governed by (9) bears some similarity to that by (1). As a preliminary remark, notice that the following maximum principle holds: if a bounded measurable real-valued function $w(x)$ defined on $\bar{\Omega}$ attains a maximum at some finite point x_0, then it follows from properties (6) that $\int J(x_0,y)w(y)dy \le K(x_0)w(x_0)$, equality holding only if $w(x) = w(x_0)$ a.e. This latter is immediate if $J > 0$, and is seen by an iteration argument if J has compact support.

PROP. A. Let $f(u,x) \ge 0$ and let $u(x,t)$ be a solution of (9) in a closed cylinder $S = \{(x,t) : x \in \bar{\Omega},\ t_1 \le t \le t_2\}$, $t_1 < t_2$. If u attains a maximum value m at some point (x_0, t_2), $x_0 \in \bar{\Omega}$, then

$$u(x,t) = m$$

for all (x,t) in S.

Notice that unlike the strong maximum principle for (1), here the conclusion holds even if the maximum is attained for x on the boundary of Ω.

Proof. From the above maximum principle and the fact that $\partial_t u(x_0,t_2) \ge 0$, we obtain that $u(x,t_2) = m$ for all $x \in \bar{\Omega}$. We also obtain that $\partial_t u(x_0,t_2) = f(m,x) = 0$, so that $u(x,t) \equiv m$ is a solution of (9). By backward uniqueness of the initial value problem for (9) (see below), the conclusion follows.

Many other variants and consequences of this proposition hold, an example being the following comparison principle, given here without proof. (Its regularity conditions may be relaxed.) We denote the semilinear expression on the right of (9) by Lu. The cylinder S is as in Prop. A.

PROP. B. Let $\underline{u}(x,t)$ and $\bar{u}(x,t)$ be continuously differentiable functions in S satisfying

$$\tau\partial_t\underline{u}(x,t) - L\underline{u}(x,t) \le \tau\partial_t\bar{u}(x,t) - L\bar{u}(x,t), \quad (x,t) \in S,$$

and

$$\underline{u}(x,t_1) \le \bar{u}(x,t_1), \quad x \in \Omega,$$

Then

$$\underline{u}(x,t) \le \bar{u}(x,t), \quad (x,t) \in S.$$

This provides the possibility to use sub- and super-solutions to obtain estimates for solutions of the problem with general $f(u,x)$, and in fact to obtain the existence of stationary solutions in much the same manner that it is done in the case of (1).

Consider now the initial value problem for (9) with continuous bounded initial data

$$u(x,0) = \phi(x), \quad x \in \bar{\Omega}. \tag{10}$$

Unique local existence for both forward and backward time is straightforward by a Picard iteration, as is global existence forward and backward for bounded f (for example).

The backward existence here is in stark contract to (1), as is the fact that no boundary conditions on $\partial\Omega$ are required. However even in the case of (9), the solutions are better behaved in forward time than they are in backward time.

To illustrate this, consider the case $f = 0$, $\tau = 1$, and Ω bounded. The operator

$$\mathbf{J}w(x) = \int_\Omega J(x,y)w(y)dy \tag{11}$$

is (by (6)) compact and self-adjoint on $L_2(\Omega)$. Let $\mathbf{K}$ denote the operator of multiplication by the bounded positive function $K(x)$. By definition of this function, we find that

$$\begin{aligned}\langle \mathbf{J}w - \mathbf{K}w, w\rangle &\le \tfrac{1}{2}\int\int J(x,y)[w(x)^2 + w(y)^2]dxdy - \int K(x)w(x)^2dx \\ &= \int J(x,y)\int w(x)^2dxdy - \int K(x)w(x)^2dx = 0.\end{aligned} \tag{12}$$

Therefore the spectrum of the operator **J - K** is real, nonpositive and bounded. Constants are the only eigenfunctions with eigenvalue 0, by the above maximum principle, so there are spectral points on the negative real axis. In general, therefore, solutions of (9) will be unbounded exponentially as $t \to -\infty$. Of course Prop. B. prohibits any such growth of any solution in positive time if it is initially bounded.

3 TRAVELING WAVES IN THE BISTABLE CASE

From now on, we consider only functions $f(u,x) = f(u)$ which do not dependexplicitly on x. An especially important case arising in models in materials science [5, 6, 7] and population genetics [1] is that when the nonlinearity in (1) is bistable. This means that the antiderivative $W(u)$ of f has exactly two local minima, which we shall always take to be $u = \pm 1$. We now consider this bistable case in the context of (9).

I report in this section on some recent results of Bates, Ren, Wang and myself [3]. Another nonlocal evolution equation with bistable-like nonlinearity of a different type has been studied extensively [9, 22, 23, 24, 25]. It arises in a continuum analog of a dynamic Ising model.

We consider only the case that **J** is a convolution: $J(x,y) = \hat{J}(x-y)$, and the domain Ω is the whole real line (so space is one dimensional). In this case, $K(x)$ is a constant, which with no loss of generality we take to be unity. We shall assume that the function f is continuously differentiable and bistable, i.e. it has exactly three zeros at -1, 1, and an intermediate value $u = \alpha$, with $f'(\pm 1) > 0$, $f'(\alpha) < 0$.

The most important property of the equation (1) in the bistable autonomous case in 1D (in which (1) reduces to $u_t = au_{xx} - f(u)$) is the existence and stability of traveling waves, i.e. of solutions of the form

$$u(x,t) = U(x - ct)$$

for some velocity c. They exist, are unique, and satisfy a global stability property [21, 1, 16, 17]. Moreover the sign of c is the same as that of $\int_{-1}^{1} f(u)du$. The following results refer to traveling waves for solutions of (9). Much is the same as for (1); but the differences are nevertheless striking.

The 1D version of (9) on the real line with $\tau = K = 1$ is

$$\frac{\partial}{\partial t}u(x,t) = \hat{J} * u - g(u), \tag{13}$$

where the first term on the right is a 1D convolution and we define $g(u) = u + f(u)$. The function g satisfies

$$g(\pm 1) = \pm 1; \tag{14}$$

it may or may not be monotone.

We shall work with monotone "truncations", as follows:

DEFINITION. A *null truncation* of the function $g(u)$ is a continuous monotone function $\hat{g}(u)$ satisfying (14), $\hat{g}'(u) = 0$ whenever $\hat{g}(u) \neq g(u)$, and $\int_{-1}^{1} \hat{g}(u)du = 0$. If g is already monotone, then g has a null truncation only if $\int_{-1}^{1} f(u)du = 0$, and in that case $g = \hat{g}$. A null truncation is *proper* if there is an interval of positive length on which $\hat{g}'(u) = 0$.

In the following we shall assume throughout that at most one null truncation of g exists, and that it has at most one maximal interval of constancy. The results can be extended to other cases as well.

THEOREM 1. Under the above assumptions on f, J and Ω, there exists a unique traveling wave solution (U, c) of (13) (up to shifts in the independent variable). The velocity c is zero if and only if g has a null truncation, and the profile is discontinuous if and only if there exists a proper null truncation. (In particular if $c \neq 0$, then the profile is continuous; in fact it is smooth.) The function $U(z)$ is discontinuous for at most one value of z, which we take to be $z = 0$; at that discontinuity, U jumps from a to b, where (a, b) is the maximal open interval on which $\hat{g}'(u) = 0$.

THEOREM 2. The traveling wave $U(z)$ is stable in the L_∞ norm. Consider the solution $u(x, t)$ of (13) with bounded initial data (10) satisfying

$$\limsup_{x \to -\infty} \phi(x) < \alpha < \liminf_{x \to \infty} \phi(x). \tag{15}$$

If $c = 0$ and U is smooth, then for some "shift" ξ,

$$\lim_{t \to \infty} |u(x, t) - U(x - \xi)| = 0 \tag{16}$$

uniformly in x, exponentially in t. If $c = 0$, ϕ is monotone, and $U(x)$ is discontinuous (at $x = 0$, say), then (16) holds uniformly on the set $\{x : |x - \xi| > \delta\}$ for any $\delta > 0$.

In summary, the main point of difference with the theory of traveling waves for the spatially homogeneous equation (1) is simply that stationary profiles are not necessary continuous. The convergence results at this time are also weaker.

4 THE SPREAD OF STATES IN HIGHER DIMENSIONS.

In this and the following section, I outline recent results of Wang and myself [18]. We consider the integrodifferential equation

$$\partial_t u(x, t) = \hat{J} * u - u - f(u), \tag{17}$$

and initial condition

$$u(x, 0) = \phi(x) \tag{18}$$

in the case that **J** is a convolution, Ω is all of R^N, ϕ is bounded and continuous, and f is bistable, as in the previous section. In the present section we also assume isotropy of J, namely $J(x, y) = \tilde{J}(|x - y|)$.

The equation (17) has two stable constant solutions: $u \equiv 1$ and $u \equiv -1$. Moreover, it has time-dependent solutions $u(x, t)$ such that for any fixed time t, the function u takes on values near +1 in some open set $\mathcal{D}_+$, and values near -1 in another such set $\mathcal{D}_-$. We shall call these loosely defined regions the upper and lower state regions.

Traveling waves in 1D with nonzero velocity are a prototypical process by which one of the two state regions advances at the expense of the other. The conversion of one state into the other is accomplished this way through a propagation event with a definite speed. A similar result, somewhat less precise, can be proved in higher dimensions, where the state regions may be less regular in outline. That is the gist of the following theorem. It states that if at any one time the solution u of (17) surpasses the value α by some given amount over a large enough region, and if f is such that traveling waves result in the advance of the upper state into the lower in a 1D scenario, then in higher dimensions as well, the territory of the upper state will expand at roughly the speed of the 1D traveling wave. This result, and the following ones in this section, are similar in spirit to results for the bistable nonlinear diffusion equation established, for example, in [1, 2, 8, 10, 14, 15, 19, 28, 29, 30].

THEOREM 3. In addition to the above assumptions on J, f, ϕ and Ω, suppose $\int_{-1}^{1} f(u)du < 0$. For all $\eta > 0$ and small enough (depending on η) $\beta > 0$, there exists a positive constant $L(\eta, \beta)$) such that if $\phi(x) \geq \alpha + \eta$ for $|x| \leq L(\eta, \beta)$, then

$$u(x,t) \geq U(-|x| - ct - \xi(t)) - q(t), \quad x \in R^N, \quad t \geq 0, \tag{19}$$

where $c < 0$ is the velocity of 1D traveling waves, and the functions $\xi(t)$ and $q(t)$ depend on η and β and satisfy

$$\xi'(t) = \beta \text{ for } t \text{ large } ; \quad \lim_{t\to\infty} q(t) = 0. \tag{20}$$

The remaining results reported here have to do with situations when one expects the two state regions to be separated by a thin "internal layer". It is known, at least in the case when the partial differential operator on the right is ϵ^2 times the Laplacian, that (1) has such "layered" solutions of width $O(\epsilon)$, and that the layers generally will move in time. In the context of (9), the analogous scaling of space results in the kernel J on the right being replaced by $J_\epsilon = \frac{1}{\epsilon^N} J\left(\frac{x}{\epsilon}, \frac{y}{\epsilon}\right)$, N being the space dimension. We also scale time with ϵ.

Again, we consider the case that $\Omega = R^N$ and $J(x, y)$ depends only on $|xy|$, so that **J** is a convolution. Our equation now becomes

$$\epsilon\partial_t u(x,t) = \hat{J}_\epsilon * u - u - f(u), \tag{21}$$

The initial condition will remain (18), with ϕ independent of ϵ.

The following theorem shows that the partition into state regions divided by thin layers is a very naturally occurring situation: even if initially there is no such partition, then one quickly (in time $O(\epsilon|\ln\epsilon|)$) develops at the surface where the initial function crosses the value α.

THEOREM 4. Suppose that $||\phi||_{C^2}$ is finite, and that J satisfies the same assumptions as in Theorem 3. Let ϵ be a sufficiently small positive parameter. Let $u(x,t)$ be the solution of (21), (18). It exists for all positive ϵ and positive t. There exist constants $\tau_0 > 0$ (depending only on f) and $M_0 > 0$, $M_1 > 0$, both depending on f and $||\phi||_{C^2}$, such that

(i) for x such that $\phi(x) \geq \alpha + M_0\sqrt{\epsilon}|\ln\epsilon|$, we have $u(x, \tau_0\epsilon|\ln\epsilon|) \geq 1 - M_1\epsilon$, and

(ii) if $\phi(x) \leq \alpha - M_0\sqrt{\epsilon}|\ln\epsilon|$, then $u(x, \tau_0\epsilon|\ln\epsilon|) \leq -1 + M_1\epsilon$.

In somewhat rough form, one can trace the motion of these internal layers for small ϵ. The following limiting behavior is seen. It is analogous to the pioneering results of Gartner and Freidlin [19, 20].

THEOREM 5. Let J, f, ϕ satisfy the same assumptions as in the previous theorem. Let $u^\epsilon(x,t)$ be the solution of (21), (18). Let $\Omega_0^+ = \{x \in R^N \,|\, \phi(x) > \alpha\}$. Then the following is true:

(i) If the one-dimensional traveling wave $U(x - ct)$ has velocity $c < 0$, then for any fixed $t > 0$,

$$\lim_{\epsilon\to 0^+} u^\epsilon(x,t) = \begin{cases} 1, & \text{if } \operatorname{dist}(x, \Omega_0^+) < -ct, \\ -1, & \text{if } \operatorname{dist}(x, \Omega_0^+) > -ct, \end{cases} \tag{22}$$

A similar assertion holds if $c > 0$.

(ii) If the one-dimensional traveling wave has velocity $c = 0$, then for any $t > 0$,

$$\lim_{\epsilon\to 0^+} u^\epsilon(x,t) = \begin{cases} 1, & \text{if } \phi(x) > \alpha, \\ -1, & \text{if } \phi(x) < \alpha \end{cases} . \tag{23}$$

5 MOTION BY CURVATURE

Finally, there is more precise knowledge [26, 27, 8, 30, 4, 6, 7, 29, 10, 11] about the motion of internal layers of (1) when it takes the form $\epsilon^2\partial_t u = \epsilon^2\Delta u - f(u)$, where f is bistable with $\int_{-1}^{1} f(u)du = 0$. Time is scaled here with another factor ϵ, as compared to (21), to reflect the fact that the motion of interfaces in higher dimensions is slower in this case by that order of magnitude. The motion is then driven by the mean curvature of the interface. In the same vein, a proof of the existence of layered solutions moving by mean curvature in a nonlocal model deriving from an isotropic Ising model with Glauber dynamics was given in [23].

This same type of motion holds, by a formal asymptotic argument [18], in the case of (21) when the integral condition indicated above is imposed on the function f, and the coefficient of the time derivative on the left is changed from ϵ to ϵ^2. In this argument, we no longer require that J be isotropic, but still require that the integral operator be a convolution.

This argument yields that if an internal layer exists and is regular enough for the asymptotic arguments to be valid, then the normal velocity v of the interface in two space dimensions at each point is given by

$$v = \beta(\nu)\kappa,$$

where κ is the curvature of the interface and $\beta(\nu)$ is a constant which depends on the orientation of the interface at that point, i.e. on the unit outward normal vector ν. This constant can in principle be calculated in terms of the functions J and f. Its dependence on ν, of course, is a reflection of any anisotropy in the kernel J.

The sign convention chosen here is such that the signs of v and κ are taken to be the same if the motion of the interface is toward the center of curvature. In the isotropic case, β is a positive constant. But if J is anisotropic, it is possible that $\beta < 0$ for some (not all) normal vectors ν. In that case, this motion by curvature problem is ill-posed, and any assumption that there exists a layer with such a disallowed normal vector would be ill founded. Realistically, the interface would be faceted in this case.

6 CONCLUSION

The purpose of this account was to show that in many important respects, the evolution problems governed by (1) and (9) are similar. The basic existence theory for initial value problems is much easier in the latter case. No boundary conditions are needed for (9); this is reflected in corresponding differences in basic comparison principles.

When we specialize to bistable problems with J a convolution operator, then again most of the basic properties associated with the spread of states for (1) (with $a_{ij} = \gamma\delta_{ij}$) are also valid in the other case, but this time the proofs are considerably more involved. The theory of traveling waves is similar for the two evolution problems, but in the case of waves with zero speed, the profile (stationary solution of (13) in 1D) may be discontinuous.

References

[1] D. G. Aronson and H. F. Weinberger, Multidimensional nonlinear diffusion arising in population genetics, *Advances in Math. 30*, 33-76 (1978).

[2] G. Barles, H. M. Soner and P. E. Souganidis, Front propagation and phase field theory, *SIAM J. Control Optim., 31*, 439-469 (1993).

[3] Peter W. Bates, Paul C. Fife, Xiaofeng Ren, and Xuefeng Wang, Traveling waves in a convolution model for phase transitions, preprint.

[4] G. Caginalp and P. C. Fife, Dynamics of layered interfaces arising from phase boundaries, *SIAM J. Appl. Math. 48*, 506-518 (1988).

[5] J. W. Cahn, Theory of crystal growth and interface motion in crystalline materials, *Acta Metallurgica 8*, 554-562 (1960).

[6] J. W. Cahn and S. M. Allen, A microscopic theory for domain wall motion and its experimental verification in Fe-Al alloy domain growth kinetics, *J. de Physique 38 Colloque C7*, 51-54 (1977).

[7] J. W. Cahn and S. M. Allen, A microscopic theory for antiphase boundary motion and its application to antiphase domain coarsening, *Acta. Met. 27*, 1085-1095 (1979).

[8] Xinfu Chen, Generation and propagation of interface in reaction-diffusion equations, *J. Diff. Equations 96*, 116-141 (1992).

[9] R. Dal Passo and P. de Mottoni, The heat equation with a nonlocal density dependent advection term, preprint.

[10] L. C. Evans, H. M. Soner, and P. E. Souganidis, Phase transitions and generalized motion by mean curvature, *Comm. Pure Appl. Math. 45*, 1097-1123 (1992).

[11] P. C. Fife, Dynamics of Internal Layers and Diffusive Interfaces, *CBMS-NSF Regional Conference Series in Applied Mathematics No. 53*, Soc. Ind. Appl. Math, Philadelphia (1988).

[12] P. C. Fife, "Barrett Lecture Notes, Univ. of Tennessee," 1991.

[13] P. C. Fife, Models for phase separation and their mathematics, in *Nonlinear Partial Differential Equations with Applications to Patterns, Waves, and Interfaces*, (M. Mimura and T. Nishida, eds.), pp. 183-212, Proc. Conf. on Nonlinear Partial Differential Equations, Kyoto (1992).

[14] P. C. Fife, Long time behavior of solutions of bistable nonlinear diffusion equations, *Arch. Rat. Mech. Anal. 70*, 31-46 (1979).

[15] P. C. Fife and L. Hsiao, The generation and propagation of internal layers, *Nonlinear Anal. 70*, 31-46 (1988).

[16] P. C. Fife and J. B. McLeod, The approach of solutions of nonlinear diffusion equations to travelling front solutions, *Arch. Rational Mech. Anal. 65*, 335-361 (1977). Also announcement in *Bull. Amer. Math. Soc. 81*, 1075-1078 (1975).

[17] P. C. Fife and J. M. McLeod, A phase plane discussion of convergence to travelling fronts for nonlinear diffusion, *Arch. Rat. Mech. Anal. 75*, 281-314 (1981).

[18] Paul C. Fife and Xuefeng Wang, A convolution model for phase transitions: the generation and propagation of internal layers in higher space dimensions, preprint.

[19] J. Gartner, Bistable reaction-diffusion equations and excitable media, *Math. Nachr. 112*, 125-152 (1983).

[20] M.I. Freidlin, Geometric optics approach for reaction diffusion equations, *SIAM J. Appl. Math. 46*, 222-232 (1986).

[21] Ya. I. Kanel', On the stabilization of solutions of the Cauchy problem for the equations arising in the theory of combustion, *Mat. Sbornik 59*, 245-288 (1962).

[22] A. de Masi, T. Gobron, and E. Presutti, Travelling fronts in non local evolution equations, *Arch. Rat. Mech. Anal*, in press.

[23] A. de Masi, E. Orlandi, E. Presutti, and L. Triolo, Motion by curvature by scaling nonlocal evolution equations, *J. Stat. Physics 73*, 543-570 (1993).

[24] A. de Masi, E. Orlandi, E. Presutti, and L. Triolo, Stability of the interface in a model of phase separation, *Proc. Roy. Soc. Edinburgh 124A*, 1013-1022 (1994).

[25] A. de Masi, E. Orlandi, E. Presutti, and L. Triolo, Uniqueness of the instanton profile and global stability in non local evolution equations, *Rend. Math. 14*, 693-723 (1994).

[26] P. de Mottoni and M. Schatzman, Evolution geometrique d'interfaces, *C. R. Acad. Paris 309*, 453-458 (1989).

[27] P. de Mottoni and M. Schatzman, Geometrical evolution of developped interfaces, *Trans. Amer. Math. Soc.*, in press.

[28] P. de Mottoni and M. Schatzman, Development of interfaces in N-dimensional space, *Proc. Roy. Soc. Edinburgh 116A*, 207-220 (1990).

[29] J. Rubinstein, P. Sternberg, and J. B. Keller, Fast reaction, slow diffusion and curve shortening, *SIAM J. Appl. Math. 49*, 116-133 (1989).

[30] H. M. Soner, Ginzburg-Landau equation and motion by mean curvature, I: convergence; II: Development of the initial interface, *J. Geom. Anal*, in press.

13. On Solutions of Mean Curvature Type Inequalities

Robert Finn Department of Mathematics, Stanford University, Stanford, California

Abstract: New results of Bernstein type and related a-priori estimates are given for solutions of differential inequalities described in terms of a generalized form of the mean curvature operator. The estimates depend in an essential way on the particular nonlinearity of the operator.

Hiroyuki Usami considers in [14] differential inequalities of the form

$$\mathrm{M}\,u \equiv \operatorname{div}\left[\frac{\nabla u}{\left(1+|\nabla u|^2\right)^{\alpha}}\right] \geq p(x)f(u), \qquad x \in \mathbf{R}^N, \tag{1}$$

where $N \geq 2$, $\alpha \in \left[0,\frac{1}{2}\right]$ is a constant, and $p\colon \mathbf{R}^N \to (0,\infty)$ and $f\colon (0,\infty) \to (0,\infty)$ are continuous functions subject to specific hypotheses. Under those hypotheses, Usami proves the nonexistence of positive entire solutions of the inequality.

If $\alpha = \frac{1}{2}$ then Mu becomes the *mean curvature operator*. If in addition $N = 2$ and $p(x)f(u) \equiv \kappa u + \lambda$, with κ and λ both constant, then (1) with equality sign becomes the *capillarity equation*, and has a special physical interest.

Although mean curvature operators are comprised in the class considered in [14], the capillarity equation is excluded by the hypotheses. Nevertheless, for the capillarity equation it should be expected that much stronger results hold than are indicated in [14], cf. the material in the references [3,4,5,12,13].

In the present note, we introduce a class of differential inequalities that generalize mean curvature relations in a different way than is undertaken in [14]. Corresponding to the formats initiated in [13] and in [6], we consider a prescribed function F(s) defined for non-negative s, set $s = |\nabla u|$ and study inequalities of the form

$$\mathcal{M}u \equiv \sum_{i=0}^{N} \frac{\partial}{\partial x_i} \mathcal{F}_{u_{x_i}}(s) \geq f(x;u). \tag{2}$$

The choice F(s) = $\left(1+s^2\right)^{1-\alpha}$ yields the operator considered by Usami. In general, we introduce the requirements:

Hypothesis F : F(s) can be extended to be an even function of s, with $\mathcal{F}(s) \in C^{(2)}(-\infty,+\infty)$, $\mathcal{F}_{ss}(s) > 0$ and $|\mathcal{F}_s(s)| < T < \infty$.

The condition $\mathcal{F}_{ss}(s) > 0$ ensures the ellipticity of the differential operator in (2). The requirement $|\mathcal{F}_s(s)| < T < \infty$ imposes a nonlinearity analogous to that of the mean curvature operator. In such a case, the ellipticity cannot be uniform. We may assume that $\lim_{s\to\infty} \mathcal{F}_s(s) = T$.

With regard to the function $f(x;u)$, we introduce two hypotheses.

Hypothesis $\mathbf{f_0}$: We assume that $f(x;u)$ is continuous and positive on $\mathbf{R}^N \times (0,\infty)$ and that for each fixed x, $\liminf_{u\to+\infty} f(x;u) > 0$.

We note that any function $f(x;u)$ taking the separated form considered in [14] and satisfying the hypotheses of [14] satisfies Hypothesis f_0 but that the converse is not true. We note especially that monotonicity of the function in u is not required. For the capillarity equation, Hypothesis f_0 is satisfied if $\kappa\lambda \geq 0$, $\kappa + \lambda > 0$, and thus the physically important cases, of "positive" gravity without volume constraint and also zero gravity configurations with positive curvature, are included. Our hypotheses can be modified to encompass also the case $f(x;u) = \lambda u - u^q$ studied in [7,10,11], although we are restricted to different ranges for λ and q than are considered in those references.

From Hypothesis f_0 we may conclude that for each x and $u_0 > 0$, there holds $\operatorname{glb}_{u\geq u_0} f(x;u) > 0$. We will need a more specific requirement:

Hypothesis $\mathbf{f_\infty}$: We assume that for each $u_0 > 0$ there exists a continuous $p(x) > 0$ and an inequality of the form $f(x;u) \geq p(x)q(u_0) > 0$ for all $u \geq u_0$; here $p(x)$ is to have the following property: writing $p_*(r;a) = \min_{|x-a|=r} p(x)$, we require that $\lim_{r\to\infty} \frac{1}{r^{N-1}} \int_0^r t^{N-1} p_*(t;a)\,dt = \infty$, for all $a \in \mathbf{R}^N$.

This requirement is (necessarily for our purpose) in some ways more restrictive than the corresponding one in [14]. It clearly holds, however, for the capillarity equation under the conditions mentioned above. We obtain in this way a class of inequalities overlapping those of [14], which include the capillarity equation in physically important cases; perhaps more to the point, the capillarity equation is the conceptual model for the relations we have introduced.

In Hypothesis f_∞, we may choose for $q(u_0)$ the maximum of values for which the inequalities hold. This function will then be continuous and non-decreasing in u_0.

We intend to prove:

THEOREM 1: *Assume Hypotheses* F, f_0 *and* f_∞. *Then any function u(x) that satisfies* (2) *for all* $x \in \mathbf{R}^N$ *is nowhere positive.*

Proof: We suppose for contradiction that a solution $u(x)$ of (2) exists, for which $u(a) > 0$ at some $a \in \mathbf{R}^N$. For functions v with rotational symmetry about the point $x = a$ and satisfying (2), we obtain, with $r = |x - a|$

$$\mathcal{M} v \equiv \left(r^{N-1} \mathcal{F}_s\right)_r \geq r^{N-1} f(x;v) \geq r^{N-1} q(v_0) p_*(r;a) \tag{3}$$

whenever $v \geq v_0 > 0$. We consider the equation

$$\mathcal{M} v \equiv \left(r^{N-1} \mathcal{F}_s(v_r)\right)_r = (1-\varepsilon) r^{N-1} q(u(a)) p_*(r;a) \tag{4}$$

for fixed $\varepsilon < 1$ and observe that in consequence of Hypothesis F the unique solution for which $v(0) = v_0$, $v_r(0) = 0$ is given by

$$v(r) = v_0 + \int_0^r \mathcal{F}_s^{-1}\left\{(1-\varepsilon) q(u(a)) \frac{1}{\rho^{N-1}} \int_0^\rho t^{N-1} p_*(t;a)\, dt\right\} d\rho \tag{5}$$

(Actually, the condition $v_r(0) = 0$ is superfluous; because of the nonlinearity, every solution defined in a deleted neighborhood of $r = 0$ can be extended in that way.)

Note that in (5), $v(r)$ is increasing, so that the inequality $v \geq v_0$ persists throughout the traverse. In view of the hypotheses F and f_∞, the solution (5) can be extended only until the (finite) value $r = R_a$ for which $(1-\varepsilon) q(u(a)) \frac{1}{r^{N-1}} \int_0^r t^{N-1} p_*(t;a)\, dt = T$, and we find $\lim_{r \to R} v_r(r) = \infty$.

We note that in the particular case of the mean curvature operator with $p(x) \equiv$ const. (as occurs for the capillarity equation in "non-negative" gravity fields) the solution $v(r)$ describes a meridional section of a lower hemishere.

Since R_a is independent of v_0, it follows that when v_0 is sufficiently large, the rotation surface $v(r;u(a))$ lies above the graph of u throughout the ball $r < R_a$.

We now allow v_0 to decrease continuously until points (x_0, u_0) of contact first appear. In view of the assumption that $u(a) > 0$, there holds $u_0 \geq v_0 \geq u(a) > 0$ at all such points, and hence also $f(x_0;u_0) > (1-\varepsilon) q(u(a)) p_*(|x_0 - a|;a)$. Since each such point must be a point of tangency and since $|\nabla u|$ is bounded in the closed ball $|x-a| \le R_a$, there exists $\delta > 0$ such that $|x_0 - a| \le R_a - \delta$ for every contact point. We may thus select a particular such point for which $|x_0 - a|$ is as large as possible. By the continuity of u, there exists δ_0 with $0 < \delta_0 < \delta/2$, such that $f(x;u) > (1-\varepsilon)q(u(a))p_*(|x-a|;a)$ throughout a closed ball B_{δ_0} of radius δ_0, centered at a distance $|x_0 - a| + \delta_0$ from the origin on the radial segment from a through x_0. Interior to B_{δ_0} we therefore have $\mathrm{M}u \ge \mathrm{M}v$. Using the mean value theorem, one finds that the difference $w = u - v$ satisfies an inequality $Ew \ge 0$, for a non-degenerate positive elliptic operator E without zero-order terms. Interior to B_{δ_0} we have $w < 0$, while $w = 0$, $\nabla w = 0$ at the boundary point x_0, where the two surfaces contact. This configuration contradicts the E. Hopf boundary point lemma [7], according to which the outer normal derivative of w at x_0 must be positive. □

We obtain immediately from the identical reasoning applied both to u and to its negative:

Corollary 1.1: *Under the hypotheses for Theorem 1, suppose that $f(x;u)$ can be extended to $\mathbf{R}^N \times \mathbf{R}$ as an odd function of u. Then the only entire solution of* (2) *with equality sign is the function $u \equiv 0$.*

For the theorem as we have stated it, our result cannot be significantly improved. As an example, we set $\mathrm{F}(s) = \sqrt{1+s^2}$ and consider the resulting equation

$$\mathcal{M}u \equiv \sum_1^2 \frac{\partial}{\partial x_i}\left(\frac{u_{x_i}}{\sqrt{1+|\nabla u|^2}}\right) = f(x) = \frac{2+|x|^2}{\left(1+|x|^2\right)^{3/2}}. \tag{6}$$

Here

$$\lim_{r\to\infty} \frac{1}{r}\int_0^r \rho f(\rho;0)\,d\rho = 1 < \infty$$

and correspondingly (6) admits the entire solution $u(r) = \frac{1}{2}r^2 + \text{const.}$, for any constant. Nevertheless, our procedure does yield further information, of specific character. For example, if $f(x;u)$ does not depend on u, then the radius R_a is independent of $u(a)$. Let $u(x)$ satisfy (2) interior to a ball $B_R(a)$. If $R_a < R$ then the above procedure yields a contradiction. We conclude:

Corollary 1.2: *Under the hypotheses of Theorem 1, if $f(x;u)$ is independent of u and if $u(x)$* satisfies (2) in a ball $B_R(a)$ then $R \le R_a$.

This result generalizes a theorem going back to S. Bernstein [1], according to which no surface $u(x)$ of mean curvature $\ge c^2$ can be defined and regular in a ball of radius exceeding $1/c^2$ (for sharper forms of that theorem, see [8] and [2]). We note here the particular case $f(x;u) \equiv 0$, for which (2) with equality sign becomes a generalized form of the minimal surface equation. In this case $R_a = \infty$ for every a, and in fact the planes $u = u_0 + \sum_1^N a_i x_i$ provide particular entire solutions. But if there should hold $f(x;u) > Nc^2 > 0$, then the procedure for Corollary 1.2 yields $R < T/c^2$ as an upper bound for the radius of any ball in which (2) holds.

We can carry these ideas further. The procedure provides an upper bound for R_a depending on $q(u(a))$, and that bound becomes a non-increasing function of $u(a)$. If by increasing $u(a)$ we can achieve $R_a < R$ then the procedure yields a contradiction. In fact, this inequality can always be achieved when $\lim_{t\to\infty} q(t) = \infty$. In view of these remarks and noting the discussion just above, we find under our hypotheses restricted to a given point $a \in \mathbf{R}^N$:

THEOREM 2: *Let $u(x)$ satisfy* (2) *in $B_R(a)$ and let $Q^- = \operatorname{glb} q(t)$, $Q^+ = \operatorname{lub} q(t)$, $t \in (0,\infty)$. The behavior of the solution is limited according to the following case distinctions*:

i) $Q^- = Q^+ = 0$. *Then no information is provided by the method.*

ii) $Q^- = Q^+ > 0$. *Then there is a finite $R_a > 0$ such that $R \le R_a$.*

iii) $Q^- = 0$, $0 < Q^+ < \infty$. *Then there exists $R^- > 0$, such that if $R > R^-$ there holds $u(a) < M(R)$, with $\lim_{R\to\infty} M(R) = 0$.* If *$R \le R^-$, no information is provided.*

iv) $Q^- = 0$, $Q^+ = \infty$. *Then for every $R > 0$, there holds an inequality $u(a) < M(R)$, with $\lim_{R\to\infty} M(R) = 0$, $\lim_{R\to 0} M(R) = \infty$.*

v) $0 < Q^- < Q^+ < \infty$. *Then there exist R^-, R^+, with $0 < R^- < R^+ < \infty$, such that $R \le R^+$ and in the interval $R^- < R < R^+$ there holds an inequality $u(a) < M(R)$. If $R \le R^-$, no information is provided.*

vi) $Q^- > 0$, $Q^+ = \infty$. *Then $R \le R^+$ as in v), and an inequality $u(a) < M(R)$ holds for all $R < R^+$, with $\lim_{R\to 0} M(R) = \infty$.*

Theorem 2 qualitatively improves earlier results, in [4, Sec.5] and in [6,Theorem 1]. We illustrate the proof by considering Case v). Since $q(u(a)) \ge Q^-$, the radius R_a of definition for the comparison function $v(r;a)$ cannot exceed the value R_ε^+ of r for which

$$(1-\varepsilon)Q^- \frac{1}{r^{N-1}} \int_0^r t^{N-1} p_*(t;a)\,dt = T$$

If $R > R_\varepsilon^+$ then the domain for $v(r;a)$ is strictly included in $B_R(a)$ for any $u(a)$, and the procedure used for the proof of Theorem 1 yields a contradiction. We thus obtain $R \le R^+ = \lim_{\varepsilon\to 0} R_\varepsilon^+$.

Since $q(u(a)) \le Q^+$, the domain for v always includes the ball $B_{R_\varepsilon^-}(a)$, where R_ε^- is determined by

$$(1-\varepsilon)Q^+ \frac{1}{r^{N-1}} \int_0^r t^{N-1} p_*(t;a)\,dt = T.$$

Thus, the procedure can provide no information when $R < R^- = \lim_{\varepsilon \to 0} R_\varepsilon^-$.

Finally, observe that the radius R_a for $v(r;a)$ (in the limit as $\varepsilon \to 0$) varies monotonely from R_ε^- to R_ε^+ according to

$$q(u(a))\frac{1}{r^{N-1}}\int_0^r t^{N-1} p_*(t;a)\,dt = T$$

as the height $u(a)$ varies from ∞ to 0, and there holds necessarily $R \le R_a$, otherwise the above procedure leads to a contradiction. Thus, $u(a)$ is bounded above by the maximum value $M(R)$ of τ such that the value R_a determined from

$$q(\tau)\frac{1}{r^{N-1}}\int_0^r t^{N-1} p_*(t;a)\,dt = T$$

satisfies $R_a \ge R$. The proof is complete.

As an example, we consider the equation

$$\mathcal{M}u \equiv \sum_1^2 \frac{\partial}{\partial x_i}\left(\frac{u_{x_i}}{\sqrt{1+|\nabla u|^2}}\right) = q(u) = \begin{cases} 4, & u \ge 4 \\ u, & 2 \le u \le 4 \\ 2, & u \le 2 \end{cases}$$

which fits into Case v). We find $R^- = 1/2$, $R^+ = 1$. If $R = R^+$ a lower hemishere of radius 1, with $0 \le u(a) \le 1$, satisfies the equation. In the range $1/2 \le R \le 1$ we obtain $u(a) \le 2/R$, while if $R < 1/2$ every lower hemishere of radius 1/2 with $u(a) \ge 4$ provides a solution. Thus, we see that it is not a weakness of the method that no bound is provided when $R < R^-$; it cannot be expected that $u(a)$ will admit a bound depending only on R in that event.

This investigation was supported in part by a grant from the National Aeronautics and Space Administration, and in part by a grant from the National Science Foundation.

REFERENCES

1. Bernstein, S. N.: Sur les surfaces définies au moyen de leur courbure moyenne ou totale, *Ann. Ecole Norm. Sup. 27*: 233-256 (1910)
2. Finn, R., Remarks relevant to minimal surfaces and to surfaces of prescribed mean curvature, *J. d'Analyse Math. 14*: 139-160 (1965).
3. Finn, R., The inclination of an *H*-graph, *Springer-Verlag Lecture Notes, 1340*: 40-60 (1988).
4. Finn, R., Comparison methods in capillarity, *Springer-Verlag Lecture Notes 1357*: 156-197 (1988)
5. Finn, R., Gradient and Gauss-curvature bounds for *H*-graphs, in *Proc. Intl. Symp. Compar. Meth. Stability Theory, Waterloo 1993*, Marcel Dekker (1994), pp.137-150.
6. Finn, R., and E. Giusti, On the Touching Principle, in *Partial Differential Equations and the Calculus of Variations; Essays in honor of Ennio De Giorgi, Volume I*, Birkhaüser Verlag (1989), pp. 451-460.

7. Franchi, B., E. Lanconelli, and J. Serrin, Esistenza e unicitá degli stati fondamentali per equazioni ellitici quasilineari, *Atti Accad. Naz. Lincei Rend., 79*: 121-126 (1985)

8. Heinz, E.: Über Flächen mit eineindeutiger Projektion auf eine Ebene, deren Krümmungen durch Ungleichungen einbeschränkt sind, *Math. Annalen 129*: 451-454 (1955).

9. Hopf, E., Elementare Bemerkungen über die Lösungen partieller Differentialgleichungen zweiter Ordnung vom elliptischen Typus, *Sitzber. preuss. Akad. Wiss. Berlin, 19*: 147-152 (1927). See also: A remark on linear elliptic differential equations of second order, *Proc. Amer. Math. Soc. 3*: 791-793 (1952).

10. Ni, W.M. and J. Serrin, Non-existence theorems for quasilinear partial differential equations, *Rend. Circ. Mat. Palermo, (2) 8*: 171-185 (1985).

11. Peletier, L.A., and J. Serrin, Ground States for the Prescribed Mean Curvature Equation, *Proc. Amer. Math. Soc. 190*: 694-700 (1987).

12. Siegel, D., Height Estimates for Capillary Surfaces, *Pacific J. Math., 88*: 471-515 (1980).

13. Turkington, B., Height Estimates for Exterior Problems of Capillary Type, *Pacific J. Math., 88*: 517-540 (1980)

14. USAMI, H., Nonexistence of Positive Entire Solutions for Elliptic Inequalities of the Mean Curvature Type, *J. Diff. Eqns. 111*: 472-480 (1994).

14. An Application of the Calculus of Variations to the Study of Optimal Foraging

Stefano Focardi Istituto Nazionale per la Fauna Selvatica, Ozzano dell'Emilia, Italy

Paolo Marcellini Dipartimento di Matematica, Università di Firenze, Florence, Italy

1 Introduction

Classical mathematical models in *optimal foraging theory* have been often criticized by field and experimental biologists because too simple to reproduce the actual behavioural mechanisms of choice used by animals under natural conditions. In fact these models deal with *average* values of environment (food density, its spatial distribution, ecc.) but do not take into account the short term dynamics in food gathering experienced by an animal during a foraging excursion (see for example the discussion of Owen Smith 1993).

In this paper we try to show how calculus of variations can be used to modelize the foraging behaviour, in particular of herbivores, with less restrictive assumptions with respect to the previous models.

The first to use calculus of variations in this context have been Arditi and Dacorogna (1985, 1987, 1988), Botteron and Dacorogna (1990), who developed a mathematical model for the study of optimal foraging with arbitrary food distribution, by maximizing the total food gathered by the animal. Despite the theoretical importance of these studies, it is not possible to apply these models to the foraging behaviour of herbivores because the total food gathered by the animal in this case is fixed, equal to the maximum of gut content. Focardi and Marcellini (1995) have introduced a specific model for foraging of ruminants based on the minimization of foraging time $T(v,l)$, which results to be an integral of the calculus of variations of the type

$$T(v,l) = \int_0^l f(x, v'(x))\, dx \ , \tag{1.1}$$

related to a function of two real variables $f = f(x,\xi)$, which will be described later (cfr. (4.2)) where l is the length of the one-dimensional patch $[0,l]$ physically described by the animal, $v = v(x)$ (≥ 0) is the amount of food gained by the animal till the point x, and $v' = \frac{dv}{dx}$ is the density of acquisition of food.

We minimize $T(v,l)$ both with respect to $l \in I$ (I is an interval in $\mathbb{R}$, inferiorly bounded; see (4.5)) and with respect to $v \in W$ (W is a Sobolev class of functions defined in $[0,l]$, with pointwise constraints in the gradient v'; see (4.3)); note that W depends on l. Moreover v and l satisfy the *a-priori* constraint that, at the end of the excursion, the animal has filled his gut.

2 Biological background

The use of mathematical models have had a deep influence in our understanding of the foraging behaviour of animals. In particular, optimization models (Stephens and Krebs, 1986) have been used to embed the huge amount of available experimental data on foraging in the framework of evolutionary theory in that they allow scholars to explain animal behaviour in terms of fitness maximization, where fitness is a direct or indirect measure of the number of vital offsprings an animal may get. For most of animal species food is a limiting factor and the optimization of the strategy used to collect and process it is one of the main determinant of their reproductive fitness.

Among the first studies, an important result obtained within optimal foraging theory (OFT) was to show that an optimal-foraging predator should leave its pray before having consumed all available food. Ecologists were unable to understand this behaviour which, from the point of view of efficiency of the ecosystems, may appear quite negative. Charnov (1976) analyzed this problem and demonstrated the marginal value theorem (MVT): the predator should leave its prey when the foraging rate on the prey equals the long-term average rate of foraging in a given environment; in other word larger is the number of potential prey items higher is the fraction of prey left unconsumed. The validity of MVT for predators has been clearly assessed but the application of such a approach to herbivores, specifically to ruminants (i.e. cattle, deer, antelopes) is yet limited. There are some studies (Astrom, Lundberg and Danell 1990; Jiang and Hudson 1993) which successfully explain patch exploitation by ruminants on the basis of the MVT. However Focardi and Marcellini (1995) noted that, in general, the assumption of a patchy food distribution is not realistic at all for organisms, like ungulates, which exploit common and often of poor-quality food sources. On the basis of the studies of Arditi and Dacorogna (1985 and related papers) Focardi and Marcellini (1995) have modeled the foraging behaviour of ruminants, relaxing the assumption of patchy food distribution. The main assumptions of the model are that (i) animals are informed about some general parameters of food distribution in the environment, but have no specific knowledge of the position of favorable foraging stations in the habitat; (ii) search and foraging are intermingled activities and (iii) the search process is at short spatial scale, i.e. there is no long-range detection of food items. The model predicts the existence of a critical biomass value λ (*the threshold*) which separates selected and refused foraging stations. More precisely, let $\rho(x)$ be the food density at the point x; the animal forages at x if and only if $\rho(x) > \lambda$, otherwise it ignores the station. The biomass threshold depends on animal's functional response, speed, rumen content and length of the foraging path.

In both approaches (patchy or arbitrary food distribution) an interesting point is that animals should exhibit a non-linear foraging response to food density, while alternative theoretical approach to foraging behaviour, e.g. satisfacing (Myers 1983; Ward 1992) or hedyphagia (see Proveza and Balph, 1990 for a recent review) predict more gradual changes in the foraging behaviour. Even anatomo-physiological constraints are unlikely to predict thresholds. If diet choice in ruminants depends on body size (Demment and Van Soest 1985) or on the structure of the digestive system (Hofmann 1989) we expect to observe a smooth gradient of feeding styles rather than abrupt discontinuities.

Recent analyses of the foraging behaviour in fallow deer (*Dama dama*) (Focardi, Marcellini and Montanaro, in prep.) showed that the assumptions of Focardi and Marcellini's (1995) model are consistent with the actual behaviour of animals and that the predicted threshold is likely to occur, at least under semi-natural conditions. Even other authors (Astrom et al. 1990; Saether 1990) have described the process of partial food consumption in the moose (*Alces alces*).

3 Mathematical model

To introduce our mathematical model, let us consider a curve (physically described by the animal) in a two-dimensional domain, parametrized by a variable $x \in [0, l]$, with $x = 0$ and $x = l$ corresponding respectively to the initial and to the final points of the curve.

We denote by $\rho = \rho(x)$, the *density* of food at a generic point $x \in [0, l]$. This food density $\rho(x)$ is not necessarily continuous and, in the application, usually has discontinuities in the form of jumps in the interior of $[0, l]$.

To find how long an animal remains feeding at a point x we study the dynamics of food consumption at x in terms of the *functional response* Φ of the animal, i.e. its *velocity* of food acquisition. For this purpose, let us denote by τ the foraging time that the animal spends at a point x to acquire food; moreover, let us denote by δ the density of food eaten at x. Then the velocity of food acquisition at x is the derivative $\frac{d\delta}{d\tau}$ of δ with respect to τ; therefore we have

$$\frac{d\delta}{d\tau} = \Phi(\rho - \delta), \tag{3.1}$$

where $\rho - \delta$ is the density of remaining food and $\Phi: [0, +\infty) \to [0, +\infty)$ is a strictly increasing function with $\Phi(0) = 0$.

As examples let us mention the classical model of Lotka and Volterra, with a linear *functional response*

$$\Phi(\xi) = k \cdot \xi, \tag{3.2}$$

where k is a positive constant; or the models of *functional responses* proposed for sheep, rabbits and kangaroos by Short (1985), and for *Peromyscus maniculatus* by Real (1979), given (for some constants $k, \alpha, b > 0$) respectively by

$$\Phi(\xi) = k \cdot (1 - e^{-\alpha\xi}), \tag{3.3}$$

$$\Phi(\xi) = k \cdot \frac{\xi^\alpha}{b + \xi^\alpha}. \tag{3.4}$$

We can solve the ordinary differential equation (3.1)

$$d\tau = \frac{d\delta}{\Phi(\rho - \delta)} \tag{3.5}$$

together with the initial condition that, if the animal does not eat at the point x (i.e. $\delta = 0$), then at x it does not spend time acquiring food ($\tau = 0$). We obtain the time (per unit length) τ that the animal spends at the point x:

$$\tau = G(\rho - \delta) - G(\rho),$$

where G is a primitive of $-1/\Phi$, i.e.

$$\frac{dG}{d\xi} = -\frac{1}{\Phi(\xi)}, \qquad \forall \xi > 0. \tag{3.6}$$

Therefore the *total time to acquire food* is given by

$$\int_0^l \tau\, dx = \int_0^l \{G(\rho-\delta) - G(\rho)\} dx . \tag{3.7}$$

Moreover, we have to take into account the time used by the animal to move and to search for food. We assume that it moves at a point x with velocity $V_0(x)$, so that the *total time to move* in the path of length l is given by

$$\int_0^l dt = \int_0^l \frac{1}{V_0(x)}\, dx . \tag{3.8}$$

The mathematical model is based on the minimization of the total time used by the animal to search and to gather food. Therefore we sum (3.7) and (3.8) to obtain the *total time to acquire food and to move*:

$$T = \int_0^l \left\{\frac{1}{V_0} + G(\rho-\delta) - G(\rho)\right\} dx . \tag{3.9}$$

Let us denote by $v(x)$ the amount of food gathered by the animal *up* to the point x. Then, in particular, we have $v(0) = 0$ and $v(l) = A$, where A is the rumen content, i.e. the maximum amount of food which can be eaten by the animal during a meal. Then v is related to δ (density of consumed food) by the condition that v is the integral of δ over the interval $[0, x]$; therefore

$$v'(x) = \frac{dv}{dx} = \delta . \tag{3.10}$$

By (3.9) and (3.10) we are led to *minimize*, with respect to the variables v and l, the *total time to acquire food and to move*:

$$T(v,l) = \int_0^l \left\{\frac{1}{V_0(x)} + G(\rho(x) - v'(x)) - G(\rho(x))\right\} dx . \tag{3.11}$$

Since $0 \le \delta \le \rho$ and $\delta = v'$, the *class of functions* v where to look for a minimizer is given by

$$\{v\colon v(0) = 0,\ v(l) = A;\ 0 \le v'(x) \le \rho(x),\ \forall x \in [0,l]\}, \tag{3.12}$$

while the real variable l satisfies the constraint

$$\int_0^l \rho(x)\, dx > A , \tag{3.13}$$

imposed by the assumption that the animal finds in the excursion all the food it needs.

Thus we have shown that the total foraging time T is an integral of the calculus of variations depending on the food eaten v and on the length l of animal path. We will prove in the next section that there exists a *threshold* λ in the acquisition of food. More explicitly, it exists a positive real number λ (independent of the animal position along the foraging excursion x) such that, at any point x of the path, the animal either eats till the density of food is decreased to the value λ or, if the density of food at x is less than λ, there it does not eat.

4 The existence and representation results

Let $\Phi = \Phi(\xi)$ be a nonnegative, strictly increasing function, with $\Phi(0) = 0$. Let $G = G(\xi)$ be a primitive of the function $\xi: \to -1/\Phi(\xi)$, i.e. such that

$$\frac{d}{d\xi} G = \frac{-1}{\Phi(\xi)} . \tag{4.1}$$

Let $V_0 = V_0(x)$ be a positive bounded function such that $1/V_0(x)$ is summable. We consider the minimization of the integral functional

$$T(v,l) = \int_0^l \left\{ \frac{1}{V_0(x)} + G(\rho(x) - v'(x)) - G(\rho(x)) \right\} dx , \tag{4.2}$$

where $v \in W$, and W is the class of Lipschitz continuous functions in the interval $[0,l]$ defined by

$$W = \{ v \in Lip([0,l]) \colon v(0) = 0, v(l) = A;\ 0 \le v'(x) \le \rho(x), \forall \text{ a.e. } x \in [0,l] \}. \tag{4.3}$$

Let $\rho = \rho(x)$ be a nonnegative, bounded and measurable function in $[0,l]$. We assume that the total amount of food present in the habitat is sufficient to the animal; i.e.

$$\int_0^{+\infty} \rho(x)\, dx > A \tag{4.4}$$

for some fixed $A \in \mathbb{R}$. This condition guarantees that $T(v,l)$ in (4.2) is finite for some (v,l). Finally we define the set I of possible values of l

$$I = \left\{ l \in \mathbb{R} \colon \int_0^l \rho(x)\, dx \ge A \right\} . \tag{4.5}$$

Theorem *The integral T in (4.2) has a minimizer $(u,l) \in W \times I$ and there exists a real positive number λ such that*

$$u'(x) = \begin{cases} 0 & \text{if } \rho(x) \le \lambda \\ \rho(x) - \lambda & \text{if } \rho(x) > \lambda \end{cases} \qquad \text{a.e. } x \in [0,l] . \tag{4.6}$$

We call λ *the threshold*; it is the *critic density of food*: if $\rho(x)$ is below λ, then the animal does not eat at the point x; vice-versa, if $\rho(x)$ is greater than λ, then the animal eats with *density of acquisition of food* $u'(x) = \rho(x) - \lambda$.

The proof is divided into 3 steps.

Step 1 Let us prove that, for every $l \in I$, there exists a unique $\lambda \geq 0$ such that

$$\int_0^l \varphi_\lambda(x)\, dx = A\,, \tag{4.7}$$

where

$$\varphi_\lambda(x) = \max\{\rho(x), \lambda\} - \lambda\,.$$

To this aim let us first prove that $\varphi_\lambda(x)$ is Lipschitz continuous with respect to λ, consequence of the inequality

$$0 \leq \varphi_{\lambda_1}(x) - \varphi_{\lambda_2}(x) \leq \lambda_2 - \lambda_1\,, \qquad \forall\, \lambda_2 \geq \lambda_1 \geq 0\,. \tag{4.8}$$

In fact, if $\rho(x) \leq \lambda_1 \leq \lambda_2$, then $\varphi_{\lambda_1}(x) = \varphi_{\lambda_2}(x) = 0$ and (4.8) holds. If $\lambda_1 < \rho(x) \leq \lambda_2$, then $\varphi_{\lambda_1}(x) = \rho(x) - \lambda_1$ and $\varphi_{\lambda_2}(x) = 0$; thus $\varphi_{\lambda_1}(x) - \varphi_{\lambda_2}(x) \geq 0$ and

$$\varphi_{\lambda_1}(x) - \varphi_{\lambda_2}(x) = \rho(x) - \lambda_1 \leq \lambda_2 - \lambda_1\,;$$

again (4.8) holds. Finally, if $\lambda_1 \leq \lambda_2 < \rho(x)$, then

$$\varphi_{\lambda_1}(x) - \varphi_{\lambda_2}(x) = \rho(x) - \lambda_1 - (\rho(x) - \lambda_2) = \lambda_2 - \lambda_1\,, \tag{4.9}$$

which gives the proof of (4.8).

It follows that the integral in (4.7) is Lipschitz continuous with respect to λ; therefore our conclusion (4.7) is consequence of the fact that

$$\int_0^l \varphi_0(x)\, dx = \int_0^l \rho(x)\, dx \geq A$$

and that $\varphi_\lambda(x) = 0$ when $\lambda \geq \|\rho\|_\infty$.

Step 2 Let us prove the analogous of the statement of the theorem with $l \in I$ fixed. To this aim, let $\lambda = \lambda(l)$ be the solution of the equation (4.7). Let $u = u(x)$ be the function of $W^{1,\infty}([0,l])$, such that

$$u(0) = 0 \qquad \text{and} \qquad u'(x) \;=\; \max\{\rho(x), \lambda\} \;-\; \lambda\;; \tag{4.10}$$

we will prove that u is the minimizer of the functional $v \to T(v,l)$ in the class W.

First we observe that u satisfies the boundary conditions $u(0) = 0$,

$$u(l) = \int_0^l u'(x)\,dx \;=\; \int_0^l \left[\max\{\rho(x), \lambda\} \;-\; \lambda\right] dx \;=\; A$$

and the constraints $0 \le u'(x) \le \rho(x)$. Secondly, let us show that

$$T(v,l) \;\ge\; T(u,l)\,, \qquad \forall\, v \in W\,. \tag{4.11}$$

In fact $G(\xi)$, by (4.1), is a convex function; we get

$$\begin{aligned} \int_0^l G(\rho - v')\,dx \;&\ge\; \int_0^l \{G(\rho - u') + G'(\rho - u')\cdot(u' - v')\}\,dx \;= \\ &= \int_0^l G(\rho - u')\,dx \;+\; \int_0^l G'(\lambda)\cdot(u' - v')\,dx \;+ \\ &\quad + \int_0^l \{G'(\rho - u') - G'(\lambda)\}\cdot(u' - v')\,dx\,. \end{aligned} \tag{4.12}$$

Moreover, since the difference $u - v$ is zero at the boundary, we have

$$\int_0^l G'(\lambda)\cdot(u' - v')\,dx \;=\; G'(\lambda)\cdot\int_0^l (u' - v')\,dx \;=\; 0\,. \tag{4.13}$$

Finally, the quantity $G'(\rho - u') - G'(\lambda)$ is different from zero only if $u' = 0$ (otherwise $\rho(x) - u'(x) = \lambda$) and in this case

$$\{G'(\rho - u') - G'(\lambda)\}\cdot(u' - v') \;=\; \{G'(\rho(x)) - G'(\lambda)\}\cdot(u' - v') \;\ge\; 0 \tag{4.14}$$

for all x; in fact, being $\rho(x) \le \lambda$, then $G'(\rho(x)) - G'(\lambda) \le 0$, since G' is increasing and $u' - v' = \;-v' \le 0$.

By (4.12), (4.13), (4.14) we obtain (4.11) and u is a minimizer of the time T for l fixed. We also obtain that u is the unique minimizer for l fixed, by the strict convexity of the function $G(\xi)$, consequence of the assumption that $\Phi(\xi)$ is strictly increasing.

Step 3 We fist minimize with respect to u (as in the previous step) and then with respect to $l \in I$. Fixed l, the minimizer u of Step 2 satisfies the identity

$$\rho(x) - u'(x) = \min\{\lambda, \rho(x)\}\,. \tag{4.15}$$

Therefore we are let to minimize, for $l \in I$, the function $g(l)$ defined by

$$g(l) = \int_0^l \left\{\frac{1}{V_0(x)} + G\big(\min\{\lambda, \rho(x)\}\big) - G(\rho(x))\right\} dx\,. \tag{4.16}$$

It is easy to verify (see for details Focardi and Marcellini, 1995; note that the function g depends on l not only through the right endpoint of the interval $[0, l]$ used in the integration, but also through $\lambda = \lambda(l)$) that g is continuous with respect to $l \in I$ and that it diverges to $+\infty$ if $l \to +\infty$. Thus $g(l)$ achieves its minimum in the closed interval I.

References

Arditi, R. and Dacorogna, B. (1985). Optimal foraging in nonpatchy habitats. 1. Bounded one-dimensional resource. *Mathem. Bioscen.*, **76**, 127-145.

Arditi, R. and Dacorogna, B. (1987). Optimal foraging in nonpatchy habitats. 2. Unbounded one dimensional habitat. *SIAM Journal of Appl. Math.*, **47** (4), 800-821.

Arditi, R. and Dacorogna, B. (1988). Optimal foraging on arbitrary food distributions and the definition of habitat patches. *American Naturalist*, **131**, 837-846.

Astrom, M., Lundberg, P. and Danell, K. (1990). Partial prey consumption by browsers: tree as patches. *Journal of Animal Ecology*, **59**, 287-300.

Botteron, B. and Dacorogna B. (1990). Existence of solutions for a variational problem associated to models in optimal foraging theory, *Journal of Math. Analysis Appl.*, **147**, 263-276.

Charnov, E.L. (1976). Optimal foraging, the marginal value theorem. *Theoretical Population Biology*, **9**, 129-136.

Demmet, M.W. and Van Soest, P.J. (1985). A nutritional explanation for body-size patterns of ruminant and nonruminant herbivores. *American Naturalist*, **125**, 641-672.

Focardi, S. and Marcellini, P. (1995). A mathematical framework for optimal foraging of herbivores. *Journal of Mathematical Biology*, **33**, 365-387.

Focardi, S., Marcellini, P. and Montanaro, P. (in prep.). Do ungulates exhibit a food density threshold? Optimal foraging in non-patchy habitats.

Hofmann, R.R. (1989). Evolutionary steps of ecophysiological adaptation and diversification of ruminants: a comparative view of thier digestive system. *Oecologia*, **78**, 443-457.

Jiang, Z. and Hudson, R.J. (1993). Optimal grazing of wapiti (Cervus elaphus) on grassland: patch and feeding station departure rules. *Evolutionary Ecology*, **7**, 488-498.

Myers, J.P. (1983). Commentary: optimal foraging. *Perspectives in ornithology* (A. H. Brush and G.A. Clark eds), Cambridge Un. Press. Cambridge.

Owen Smith, N. (1993). Assessing the constraints for optimal diet models. *Evolutionary Ecology*, **7**, 530-531.

Provenza, F.D. and Balph, D.F. (1990). Applicability of five diet selection models to various foraging challenges ruminant encounter. *Behavioural mechanisms of food selection* (R.N. Hughes ed.), Nato ASI Series G **20**, 423-459. Springer-Verlag Berlin.

Real, L.A. (1979). Ecological determinants of functional response. *Ecology* **60**, 481-485.

Saether, B.E. (1990). The impact of different growth pattern on the utilization of tree species by a generalist herbivore, the moose Alces alces: implications of optimal foraging theory. (in Behavioural mechanisms of food selection, R.N. Hughes ed.) Nato ASI Series G **20**, 323-340. Springer-Verlag Berlin.

Short, J. (1985). The functional response of Kangaroos, sheep and rabbits in an arid grazing system. *Journal of Appl. Ecol.*, **22**, 435-447.

Stephens, D.W. and Krebs, J.R. (1986). *Foraging Theory*, Princeton Univ. Press. Princeton, N.J.

Ward, D. (1992). The role of satisfacing in foraging theory. *Oikos*, **63**, 312-317.

15. A Limit Model of a Soft Thin Joint

Giuseppe Geymonat LMT-FNS de Cachan, Cachan, France

Françoise Krasucki Laboratoire de Modélisation en Mécanique, Université Pierre et Marie Curie, Paris, France

1 INTRODUCTION

Let $\mathbb{E}^3$ be the Euclidian space referred to the orthonormal frame $(0, \vec{e}_1, \vec{e}_2, \vec{e}_3)$ and let Ω_- and Ω_+ be open, disjoint, connected domains with boundaries $\partial\Omega_-$ and $\partial\Omega_+$ piecewise of class C^2. Ω_- and Ω_+ are bonded together on a surface $S = \partial\Omega_- \cap \partial\Omega_+$. We assume that S is a set of positive 2−measure, is described by a global piecewise C^2 chart, given by $x_3 = \varphi(x_1, x_2)$. We also set $\bar{\Omega} = \bar{\Omega}_+ \cup \bar{\Omega}_-$. In fact, we describe the bonding of the two solids occupying Ω_- and Ω_+ by a thin adhesive layer, whose thermomechanical coefficients are small with respect of those of the adherents. More precisely, for a fixed $\varepsilon > 0$ we define the domain Ω_0^ε occupied by the thin layer :

$$\bar{\Omega}_0^\varepsilon = \{\vec{x}_S + \varepsilon z \vec{e}_3, \quad 0 \leq z \leq 1, \quad \vec{x}_S \in S\}$$

Let also set

$$S^\varepsilon = \{\vec{x}_S + \varepsilon \vec{e}_3, \vec{x}_S \in S\} \;\; ; \;\; \Omega_+^\varepsilon = \left\{\vec{x} + \varepsilon \vec{e}_3, \vec{x} \in \Omega^+\right\} .$$

and $\bar{\Omega}^\varepsilon = \bar{\Omega}_- \cup \bar{\Omega}_0^\varepsilon \cup \bar{\Omega}_+^\varepsilon$.

Let us point out that $\varepsilon > 0$ is the thickness characteristic parameter.

As for thermomechanical model of the adherents and of the adhesive, we consider two typical situations : time-independent heat conduction and linear elasticity.

a) Heat conduction

In this case the local equations are :

$$\begin{aligned} -\left(a_{ij}^{+}\, u_{,j}^{\varepsilon}\right)_{,i} &= f \quad \text{in } \Omega_+^\varepsilon \\ -\varepsilon\left(a_{ij}^{0}\, u_{,j}^{\varepsilon}\right)_{,i} &= f \quad \text{in } \Omega_0^\varepsilon \\ -\left(a_{ij}^{-}\, u_{,j}^{\varepsilon}\right)_{,i} &= f \quad \text{in } \Omega_- \end{aligned}$$

with the usual transmission conditions (continuity of u^ε and of the heat flux) on the surfaces S and S^ε. In order that the problem has a unique solution for all $f \in L^2(\Omega)$, we assume $u^\varepsilon = 0$ on $\Gamma^\varepsilon_u \subset \partial\Omega^\varepsilon$ with $2-meas(\Gamma^\varepsilon_u) > 0$ while the heat flux g is given on $\Gamma^\varepsilon_g = \partial\Omega^\varepsilon \backslash \Gamma^\varepsilon_u$. Here $a^{\pm}_{ij}$ are the thermal conductivity coefficients of the adherents and εa^0_{ij} those of the adhesive. All the matrices $\mathbf{a}^+ = (a^+_{ij})$, $\mathbf{a}^- = (a^i_{ij})$, $\mathbf{a}^0 = (a^0_{ij})$ satisfy the usual smoothness, symmetry and ellipticity conditions and do not depend on ε. For later use, we define the bilinear symmetric forms

$$a^\varepsilon_+(u,v) = \int_{\Omega^\varepsilon_+} a^+_{ij}\, u_{,j}\, v_{,i}\, dV$$
$$a^\varepsilon_0(u,v) = \int_{\Omega^\varepsilon_0} a^0_{ij}\, u_{,j}\, v_{,i}\, dV$$
$$a_-(u,v) = \int_{\Omega_-} a^-_{ij}\, u_{,j}\, v_{,i}\, dV$$

and the linear form :

$$L^\varepsilon(v) = \int_{\Omega^\varepsilon} f.v\; dV + \int_{\Gamma^\varepsilon_g} g.v\; dS.$$

At last we introduce the space :

$$\mathcal{V}^\varepsilon = \{u \in H^1(\Omega^\varepsilon)\; ; \quad u = 0 \quad \text{on } \Gamma^\varepsilon_u\}$$

so the variational formulation for $\varepsilon > 0$ fixed is :

$$(P^\varepsilon)\begin{cases} \text{Find } u^\varepsilon \in \mathcal{V}^\varepsilon \text{ such that for all } v \in \mathcal{V}^\varepsilon \\ a^\varepsilon_+(u^\varepsilon, v) + \varepsilon a^\varepsilon_0(u^\varepsilon, v) + a_-(u^\varepsilon, v) = L^\varepsilon(v) \end{cases}$$

Thanks to the assumptions, (P^ε) has a unique solution.

b) Linear elasticity

Since the statement of the problem is essentially the same, we directly define the bilinear symmetric forms :

$$a^\varepsilon_+(\vec{u},\vec{v}) = \int_{\Omega^\varepsilon_+} A^+_{ijk\ell}\gamma_{k\ell}(\vec{u})\gamma_{ij}(\vec{v})dV$$
$$a^\varepsilon_0(\vec{u},\vec{v}) = \int_{\Omega^\varepsilon_0} A^0_{ijk\ell}\gamma_{k\ell}(\vec{u})\gamma_{ij}(\vec{v})dV$$
$$a_-(\vec{u},\vec{v}) = \int_{\Omega_-} A^-_{ijk\ell}\gamma_{k\ell}(\vec{u})\gamma_{ij}(\vec{v})dV$$

where $\gamma_{ij}(\vec{u}) = 1/2(\nabla\vec{u} + (\nabla\vec{u})^T)_{ij}$ is the linearized deformation tensor and $\mathbf{A}^+$, $\mathbf{A}^-$, $\varepsilon\mathbf{A}^0$ are the anisotropic elasticity tensors. They satisfy the usual smoothness, symmetry and positivity properties. We also define :

$$L^\varepsilon(v) = \int_{\Omega^\varepsilon} \vec{f}.\vec{v}\; dV + \int_{\Gamma^\varepsilon_g} \vec{g}.\vec{v}\; dS$$

$$\mathcal{V}^\varepsilon = \left\{ v \in \left(H^1(\Omega^\varepsilon)\right)^3 \quad ; \quad \vec{v} = 0 \quad \text{on } \Gamma_u^\varepsilon \right\}$$

With these notations, the variational formulation of the problem, denoted by (Q^ε), is the following

$$(Q^\varepsilon) \begin{cases} \text{Find } \vec{u}^\varepsilon \in \mathcal{V}^\varepsilon \text{ such that for all } \vec{v} \in \mathcal{V}^\varepsilon \\ a_+^\varepsilon(\vec{u}^\varepsilon, v) + \varepsilon\ a_0^\varepsilon(\vec{u}^\varepsilon, \vec{v}) + a_-(\vec{u}^\varepsilon, \vec{v}) = L^\varepsilon(\vec{v}) \end{cases}$$

Thanks to the assumptions this problem has a unique solution.

Let us explicitely remark that the "small" parameter $\varepsilon > 0$ is not only a geometric parameter but also a thermomechanical one.

Our aim is to study with two different methods the asymptotic behavior of the solutions $u^\varepsilon \in \mathcal{V}^\varepsilon$ as $\varepsilon \to 0$. At first, we obtain the model of adhesion using a formal method of asymptotic expansions, see e.g. Ciarlet (1990) and Sanchez-Palencia (1980), and then we prove the convergence in the framework of the De Giorgi $\Gamma-$convergence.

This problem has been investigated under different forms by many authors. Without being exhaustive let us quote at first, Brezis et al. (1980) and Suquet (1988), who give convergence proofs in a sligthly different functional framework. Our use of $\Gamma-$convergence follows mainly the works of Acerbi and Buttazo (1986a), (1986b) on boundary reinforcement. As far as it concerns formal asymptotic expansions see e.g. Klarbring (1991), Gilibert and Rigolot (1979). When the thermomechanical coefficients of the thin layer are "not small" with respect of those of the adherents, the limit problem behaves in a very different way, see e.g. Caillerie (1980), Lemrabet (1987). For other models of the thermomechanical behavior of the adhesive see Suquet (1988), Licht (1993)...

2 ASYMPTOTIC DEVELOPMENT

As is usual, we transform the problem into a problem posed over fixed domains. For that we apply to $\bar{\Omega}_0^\varepsilon$ a dilatation of amount $1/\varepsilon$ in the direction $\vec{e}_3$, obtaining

$$\bar{\Omega}_0 = \{\vec{x}_S + z\vec{e}_3, \quad 0 \le z \le 1, \quad \vec{x}_S \in S\}.$$

We also apply a rigid translation of Ω_+^ε into $\Omega_+^\varepsilon + (1-\varepsilon)\vec{e}_3$ and we leave fixed Ω_- (N.B. $\Omega_+^\varepsilon + (1-\varepsilon)\vec{e}_3 = \Omega_+ + \vec{e}_3$). The field $u(\varepsilon)(x) : \Omega_0 \to \mathbb{R}^3$ is defined by $u(\varepsilon)(x) = u^\varepsilon(x^\varepsilon)$ (i.e. no scaling). We choose the same correspondence (i.e. no scaling) between the applied forces fields and between the material coefficients.

The space $\mathcal{V}^\varepsilon$ is naturally transformed in $\mathcal{V}$ and P^ε becomes :

$$\begin{cases} \text{Find } u(\varepsilon) \in \mathcal{V} \text{ such that for all } v \in \mathcal{V} \\ a_+(u(\varepsilon), v) + a_-(u(\varepsilon), v) + a_n^0(u(\varepsilon), v) + \\ +\varepsilon\ a_{nt}^0(u(\varepsilon), v) + \varepsilon^2 a_t^0(u(\varepsilon), v) = L_0(v) + \varepsilon L_1(v) \end{cases}$$

In the case of heat conduction the forms are the following :

$$\begin{aligned} a_\pm(u, v) &= \int_{\Omega_\pm} a_{ij}^\pm u_{,i} v_{,j} dV \\ a_n^0(u, v) &= \int_{\Omega_0} a_{ij}^0 n_i n_j\ u_{,3} v_{,3} dV, \end{aligned}$$

where $\vec{n}$ denotes the unit normal to S and $\vec{t_1}$ and $\vec{t_2}$ are the tangent vectors to S, $\left(\vec{n} = \frac{\vec{t_1} \times \vec{t_2}}{\|\vec{t_1} \times \vec{t_2}\|}\right)$,

$$a^0_{nt}(u,v) = \int_{\Omega_0} a^0_{\alpha i} n_i \left[\frac{\partial u}{\partial \vec{t_\alpha}} V_{,3} + u_{,3} \frac{\partial v}{\partial \vec{t_\alpha}} \right] dV,$$

$$a^0_t(u,v) = \int_{\Omega_0} a^0_{\alpha\beta} \frac{\partial u}{\partial \vec{t_\alpha}} \frac{\partial v}{\partial \vec{t_\beta}} dV,$$

$$L_0(v) = \int_{\Omega_+ \cup \Omega_-} f.v \, dV + \int_{\Gamma_g^+ \cup \Gamma_g^-} g.v \, dS,$$

$$L_1(v) = \int_{\Omega_0} f.v \, dV + \int_{\Gamma_g^0} g.v \, dS,$$

where $\Gamma_g = \Gamma_g^0 \cup \Gamma_g^+ \cup \Gamma_g^-$ and $\Gamma_g^0 = \Gamma_g \cap (\partial\Omega_0 \backslash (S \cup S + \vec{e}_3))$

$$\Gamma_g^{\pm} = \Gamma_g \cap (\partial\Omega_{\pm} \backslash S)$$

We write a priori $u(\varepsilon)$ as a formal expansion :

$$u(\varepsilon) = u^0 + \varepsilon u^1 + ... \tag{2.1}$$

The leading term u^0 is solution of the problem :

$$(P_0) \begin{cases} \text{Find } u^0 \in \mathcal{V} \text{ such that for all } v \in \mathcal{V} \\ a_+(u^0, v) + a_-(u^0, v) + a^0_n(u^0, v) = L_0(v) \end{cases}$$

We can easily see that u^0 is, in Ω_0, affine with respect to x_3.
Therefore, we have :

$$a^0_n(u^0, v) = a_S(u^0, v)$$

where we have set :

$$a_S(u,v) = \int_S \frac{1}{< h^{-1} >} (u_S^+ - u_S^-)(v_S^+ - v_S^-) dS$$

$$h = \vec{n}.\mathbf{a}^0.\vec{n} \quad ; \quad < h^{-1} > = \int_0^1 h^{-1}(\vec{x}_S + \varepsilon t \vec{e}_3) dt = \frac{1}{\varepsilon} \int_0^\varepsilon h^{-1}(x_S + r\vec{e}_3) dr$$

$$u_S^+ = u|_{S+\vec{e}_3} \quad ; \quad u_S^- = u|_S$$

At last, we introduce the following space :

$$\mathcal{W} = \{ u \in L^2(\Omega) \; ; \; u^+ = u|_{\Omega_+} \in H^1(\Omega_+) \, , \; u^- = u|_{\Omega_-} \in H^1(\Omega) \; ; \; u = 0 \text{ on } \Gamma_u \}$$

and we have that u^0 is completely determined by $u \in W$, unique solution of the problem (P) :

$$(P)\begin{cases} \text{Find } u \in \mathcal{W} \text{ such that for all } v \in \mathcal{W} \\ a_+(u^+, v^+) + a_-(u^-, v^-) + a_S(u, v) = L_0(v) \end{cases}$$

This problem is our modelization of the bonding at S, of the two adherents Ω_+ and Ω_-, for the heat conduction.
For the elasticity, the corresponding (Q) is obtained in a similar way. The surface bonding integral is now :

$$\int_S (< \mathbf{H}^{-1} >)^{-1} \quad (\vec{u}_s^+ - \vec{u}_s^-)(\vec{v}_s^+ - \vec{v}_s^-) dS$$

where $\mathbf{H} = \vec{n}.\mathbf{A}^0.\vec{n}$ and $\mathbf{H}^{-1}$ denotes its inverse matrix.

3 APPLICATION OF THE Γ-CONVERGENCE

As is well known, the Γ–convergence of a sequence of functionals is strictly related to the convergence of their minimum points and minimum values. When X is a metric space, $(F_\varepsilon)_{\varepsilon>0}$ a family of mappings from X to $\bar{\mathbb{R}}$, we set for $x \in X$

$$F^-(x) = inf \left\{ \underset{\varepsilon \to 0}{liminf}\ F_\varepsilon(x_\varepsilon) \ ; \ x_\varepsilon \to x \text{ in } x \right\}$$

$$F^+(x) = inf \left\{ \underset{\varepsilon \to 0}{limsup}\ F_\varepsilon(x_\varepsilon) \ ; \ x_\varepsilon \to x \text{ in } x \right\}$$

if $F^-(x) = F^+(x)$ one says that $F_\varepsilon(x)$ $\Gamma(X)$–converges to $F(x)$ as $\varepsilon \to 0$, with

$$F(x) = F^-(x) = F^+(x).$$

We only sketch the proof for the heat conduction since the case of elasticity is essentially the same. In order to work with a fixed metric space, we remark that for every $\varepsilon > 0$ fixed, the solution u^ε of (P^ε) is the minimum of the functional $I^\varepsilon(\tilde{w})$ defined on $X = L^2(\mathbb{R}^3)$ by :

$$\begin{cases} I^\varepsilon(\tilde{w}) = \frac{1}{2}a_+^\varepsilon(w,w) + \frac{1}{2}a_-(w,w) + \frac{\varepsilon}{2}a_0^\varepsilon(w,w) - L^\varepsilon(w) & \text{if } w \in \mathcal{V}^\varepsilon \\ I^\varepsilon(\tilde{w}) = +\infty & \text{elsewhere.} \end{cases}$$

As a rule, we denote the restrictions of $\tilde{w}$ by w. In the same way, the solution u of (P) is the minimum of the functional $I(\tilde{w})$ defined on $L^2(\mathbb{R}^3)$ by

$$\begin{cases} I(\tilde{w}) = \frac{1}{2}a_+(w^+,w^+) + \frac{1}{2}a_-(w^-,w^-) + \frac{1}{2}a_S(w,w) - L_0(w) & \text{if } w \in \mathcal{W} \\ I^\varepsilon(\tilde{w}) = +\infty & \text{elsewhere.} \end{cases}$$

In order to prove that for all $w \in \mathcal{W}$, $I^\varepsilon(w)$ $\Gamma(L^2(\mathbb{R}^3))-$ converges to $I(w)$ one has to verify the following statements.

(i) For all $w \in \mathcal{W}$ there exists a sequence $w_\varepsilon \in \mathcal{V}^\varepsilon$ such that $\tilde{w}_\varepsilon \to \tilde{w}$ in $L^2(\mathbb{R}^3)$, where $\tilde{w} = w\chi_\Omega$ and χ_Ω is the characteristic function of Ω, and such that

$$\limsup_{\varepsilon \to 0} I^\varepsilon(\tilde{w}_\varepsilon) \leq I(\tilde{w})$$

(ii) For all sequences $w_\varepsilon \in \mathcal{V}_\varepsilon$ such that $\tilde{w}_\varepsilon \to \tilde{w} = w\chi_\Omega$ in $L^2(\mathbb{R}^3)$ one has :

$$I(\tilde{w}) \leq \liminf_{\varepsilon \to 0} I^\varepsilon(\tilde{w}_\varepsilon)$$

Proof of (i) :

Let us define at first $\hat{w}^+ \in H^1(\mathbb{R}^3)$ an extension of $w^+ \in H^1(\Omega_+)$ and let $\hat{w}_\varepsilon^+ \in H^1(\mathbb{R}^3)$ defined by $\hat{w}_\varepsilon^+(x) = \hat{w}^+(\vec{x} - \varepsilon\vec{e}_3)$. Let also be $\hat{w}^- \in H^1(\mathbb{R}^3)$ an extension of $w^- \in H^1(\Omega^-)$. We then define :

$$w_\varepsilon(\vec{x}) = \begin{cases} \hat{w}_\varepsilon^+(\vec{x}) \quad \text{for } x \in \Omega_+^\varepsilon \\ \dfrac{\int_0^z h^{-1}(\vec{x}_S + \varepsilon t\vec{e}_3)dt}{< h^{-1} >} \hat{w}_\varepsilon^+(\vec{x}) + \dfrac{\int_z^1 h^{-1}(\vec{x}_S + \varepsilon t\vec{e}_3)dt}{< h^{-1} >} \hat{w}^-(\vec{x}) \\ \qquad \text{for } \vec{x} = \vec{x}_S + \varepsilon z\vec{e}_3 \in \Omega_0^\varepsilon \\ \hat{w}^-(\vec{x}) \quad \text{for } x \in \Omega_- \end{cases}$$

and at last we set $\tilde{w}_\varepsilon \in H^1(\mathbb{R}^3)$ an extension of $w_\varepsilon \in \mathcal{V}^\varepsilon$.

We have that $\tilde{w}_\varepsilon \to w\ \chi_\Omega$ in $L^2(\mathbb{R}^3)$ and obviously $\hat{w}_\varepsilon^+ \to \hat{w}^+$ in $H^1(\mathbb{R}^3)$. Thanks to the definition of w_ε we have the following estimate :

$$\left\| \nabla w_\varepsilon - \frac{h^{-1}(\vec{x})}{< h^{-1} >} \left[\hat{w}_\varepsilon^+ - \hat{w}^-\right] \frac{1}{\varepsilon}\, \vec{n} \right\|_{L^2(\Omega_0^\varepsilon)} \leq C \tag{3.2}$$

Since $a_0^\varepsilon(u,u)$ is convex and positive with respect to ∇u we have immediately for every $0 < t < 1$

$$\begin{aligned} I^\varepsilon(\tilde{w}_\varepsilon) = {} & \frac{1}{2}a_+^\varepsilon(\hat{w}_\varepsilon^+, \hat{w}_\varepsilon^+) + \frac{1}{2}a_-(\hat{w}^-, \hat{w}^-) + \frac{\varepsilon}{2}\int_{\Omega_0^\varepsilon} \mathbf{a}^0 \nabla w_\varepsilon \nabla w_\varepsilon \, dx - L^\varepsilon(w_\varepsilon) \\ \leq {} & \frac{1}{2}a_+(w^+, w^+) + \frac{1}{2}a_-(w^-, w^-) - L_0(w) + \\ & \frac{1}{2\varepsilon t}\int_{\Omega_0^\varepsilon} \frac{h^{-1}(\vec{x})}{< h^{-1} >^2}(\hat{w}_\varepsilon^+ - \hat{w}^-)^2 dV + \frac{c_1\varepsilon}{(1-t)} - \int_{\Omega_0^\varepsilon} f w_\varepsilon dV - \int_{\Gamma_g^{0\varepsilon}} g w_\varepsilon . dS \end{aligned}$$

A simple adaptation of lemma [III.1] in Acerbi-Buttazzo (1986b) implies that

$$\lim_{\varepsilon \to 0} \frac{1}{\varepsilon}\int_{\Omega_0^\varepsilon} \frac{h^{-1}(\vec{x})}{< h^{-1} >^2}(\hat{w}_\varepsilon^+ - \hat{w}^-)^2 dV = \int_S \frac{1}{< h^{-1} >}(w_s^+ - w_s^-)^2 dS = a_S(w, w)$$

It then follows that for $0 < t < 1$

$$limsup\ I^{\varepsilon}(\tilde{w}_{\varepsilon}) \leq \frac{1}{2}a_{+}(w^{+},w^{+}) + \frac{1}{2}a_{-}(w^{-},w^{-}) - L_{0}(w) + \frac{1}{2t}a_{S}(w,w)$$

and $t \to 1$ yields the result.

Proof of (ii) : Without loss of generality, we may suppose that $liminf I^{\varepsilon}(\tilde{w}_{\varepsilon}) = lim\ I^{\varepsilon}(\tilde{w}_{\varepsilon}) < +\infty$. Since $\tilde{w}_{\varepsilon} \to w\chi_{\Omega}$ in $L^2(\mathbb{R}^3)$ it then follows from the assumptions on $\mathbf{a}^{+}$, $\mathbf{a}^{-}$ and $\mathbf{a}^{0}$ that :

$$\int_{\Omega_-} |\nabla w_{\varepsilon}|^2\, dV + \int_{\Omega_+^{\varepsilon}} |\nabla w_{\varepsilon}|^2\, dV + \varepsilon \int_{\Omega_0^{\varepsilon}} |\nabla w_{\varepsilon}|^2\, dV \leq C$$

and so, $w_{\varepsilon}|_{\Omega_-} \to w|_{\Omega_-} = w^-$ weakly in $H^1(\Omega_-)$.
Let us define $\tilde{w}_{\varepsilon}^{+}(\vec{x}) = \tilde{w}_{\varepsilon}(\vec{x} + \varepsilon\vec{e}_3)$, then $\tilde{w}_{\varepsilon}^{+}|_{\Omega_+} \to w|_{\Omega_+} = w^+$ weakly in $H^1(\Omega_+)$.
Therefore $w \in \mathcal{W}$. The convexity of the quadratic forms implies :

$$\begin{aligned} &\liminf_{\varepsilon \to 0} \left\{\frac{1}{2}a_{+}^{\varepsilon}(w_{\varepsilon},w_{\varepsilon}) + \frac{1}{2}a_{-}(w_{\varepsilon},w_{\varepsilon}) - L(w_{\varepsilon})\right\} \\ &\geq \frac{1}{2}a_{+}^{\varepsilon}(w^{+},w^{+}) + \frac{1}{2}a_{-}(w^{-},w^{-}) - L_{0}(w) \end{aligned}$$

Let now z_{ε} be defined in Ω_0^{ε} from w^+, w^- as in (3.1). The convexity of a_0^{ε} implies

$$\begin{aligned} \frac{\varepsilon}{2}a_0^{\varepsilon}(w_{\varepsilon},w_{\varepsilon}) = & \frac{\varepsilon}{2}\int_{\Omega_0^{\varepsilon}} a_{ij}^{0} w_{\varepsilon,j} w_{\varepsilon,i}\, dV \\ \geq & \frac{1}{2\varepsilon}\int_{\Omega_0^{\varepsilon}} \frac{h^{-1}(\vec{x})}{< h^{-1} >^2}(\hat{w}_{\varepsilon}^{+} - \hat{w}^{-})^2\, dV \\ & + \int_{\Omega_0^{\varepsilon}} a_{ij}^{0}(w_{\varepsilon} - z_{\varepsilon})_{,j}\frac{h^{-1}(\vec{x})}{< h^{-1} >}(w^{+} - w^{-})n_i\, dV \\ & + \int_{\Omega_0^{\varepsilon}} a_{ij}^{0}\left((z_{\varepsilon})_{,j} - \frac{h^{-1}(\vec{x})}{< h^{-1} >}(w^{+} - w^{-})\frac{1}{\varepsilon}n_j\right)\frac{h^{-1}(\vec{x})}{< h^{-1} >}(w^{+} - w^{-})n_i\, dV \end{aligned}$$

From the quoted lemma of Acerbi-Buttazzo and (3.2), we need only to prove that the second integral vanishes as $\varepsilon \to 0$, and this follows integrating by parts.

4 CONCLUDING REMARKS

4.1. We write the system of local equations corresponding to (P)

$$\begin{cases} -div(\mathbf{a}^{+}\nabla u^{+}) = f & \text{in } \Omega_+ \\ -div(\mathbf{a}^{-}\nabla u^{-}) = f & \text{in } \Omega_- \\ \left.\begin{aligned} \vec{n}^{+}.\mathbf{a}^{+}\nabla u^{+} &= -\vec{n}^{-}.\mathbf{a}^{-}.\nabla u^{-} \\ \vec{n}^{+}.\mathbf{a}^{+}\nabla u^{+} &= < h^{-1} >^{-1}(u^{-} - u^{+}) \end{aligned}\right\} & \text{on } S \\ u = 0 & \text{on } \Gamma_u \\ \vec{n}.\mathbf{a}.\nabla u = g & \text{on } \Gamma_g \end{cases}$$

where $\vec{n}^+$ (resp. $\vec{n}^-$) denotes the outer unit normal to Ω_+ (resp. Ω_-).
The transmission conditions can be written

$$\begin{cases} \vec{n}^+.\mathbf{a}^+.\nabla u^+ = -\vec{n}^-.\mathbf{a}^-.\nabla u^- \\ \vec{n}^+.\mathbf{a}^+.\nabla u^+ + 2 < h^{-1} >^{-1} u^+ = \vec{n}^-.\mathbf{a}^-.\nabla u^- + 2 < h^{-1} >^{-1} u^- \end{cases}$$

The second equation is of Fourier-Robin type. It suggest the use of a decomposition domain method to compute the solution of problem (P) see Geymonat, Krasucki, Marini (in preparation). For a numerical study with a more classical F.E.M. see Morante (1996).

4.2. The results concerning the regularity of the classical transmission problem (e.g. Lemrabet (1977)) cannot be immediately applied to the Fourier-Robin condition. Since the determination of the singularity seems interesting for the delamination (see for example, Davet-Destuynder (1985), Leguillon (1994)), one should try to apply some ideas of Dauge et al (1990) to this situation. From a pure numerical point of view, following a suggestion of F. Brezzi, A. Morante is developping some numerical experiments when the thickness ε of Ω_0^ε is of the same order of the F.E.M. used near the singular point.

4.3. The result of the $\Gamma-$convergence suggests to look for the equations satisfied by the first order term u^1 in the asymptotic development (2.1). As it has been pointed out by Klarbring (1991) this can lead to a boundary layer, see also e.g. Davet-Destuynder (1985), Leguillon (1994).

Aknowledgements - This research was partly done as the authors were visiting the Mathematics Department of the University of Pavia and the "Istituto di Analisi Numerica" del CNR at Pavia. The authors warmly thank the professors F. Brezzi, G. Gilardi and D. Marini for some very useful discussions.

References

Acerbi E., Buttazzo G. (1986a), Reinforcement problems in the calculus of variations, *Ann. Inst. Henri Poincaré, Analyse non linéaire*, 3, pp. 273-284.

Acerbi E., Buttazzo G. (1986b), Limit problems for plates surrounded by soft material, *Arch. Rational Mech. Anal.*, 92, pp. 355-370.

Brezis H., Caffarelli L., Friedman A. (1980), Reinforcement problems for elliptic equations and variational inequalities, *Ann. Mat. Pura Appl.*, 4, 123, pp. 219-246.

Caillerie D. (1980), The effect of a thin inclusion of high rigidity in an elastic body, *Math. Meth. Appl. Sci.*, 2, pp. 251-270.

Ciarlet P. (1990), *Plates and junctions in elastic multi-structures. An asymptotic analysis*, Masson, Paris, Springer-Verlag, Berlin.

Dauge M., Nicaise S., Bourlard M., Lubumba J.M.S. (1990), Coefficients des singularités pour des problèmes aux limites elliptiques sur un domaine à points coniques, I and II, *M2.A.N.*, 24, pp. 27-52 and pp. 343-367.

Davet J.L., Destuynder P. (1985), Singularités logarithmiques dans les effets de bord d'une plaque en matériaux composites, *J. de Mécanique Théo. et Appl.*, 4, pp. 357-373.

Gilibert Y., Rigolot A. (1979), Analyse asymptotique des assemblages collés à double recouvrement sollicités au cisaillement en traction, *J. Mec. Appl.*, 3, pp. 341-37 .

Klarbring A. (1991), Derivation of a model of adhesively bonded joints by the asymptotic expansion method, *Int. J. Engng. Sci.*, 29, pp. 493-512.

Leguillon D. (1994), Un exemple d'interaction singularité-couche limite pour la modélisation de la fracture dans les composites, *C. R. Acad. Sci. Paris*, 319, série II, pp. 161-166.

Lemrabet K., (1977), Régularité de la solution d'un problème de transmission, *J. Math. Pures et Appl.*, 56, pp. 1-38.

Lemrabet K. (1987), Problème de Ventcel pour le système de l'élasticité dans un domaine de $\mathbb{R}^3$., *C. R. Acad. Sci. Paris*, 304, série I, pp. 151-154.

Licht C. (1993), Comportement asymptotique d'une bande dissipative mince de faible rigidité, *C. R. acad. Sci. Paris*, 317, série I, pp. 429-433.

Morante A. (1996), Thesis University Paris VI.

Sanchez-Palencia (1980), *Non-homogeneous media and vibration theory*, Springer-Verlag, Berlin.

Suquet P.M. (1988), Discontinuities and plasticity, *Non-smooth mechanics and applications*, CISM. Courses and Lectures. (J.J. Moreau and P.D. Panagiotopoulos ed.), Springer-Verlag, Wien, pp. 280-340.

16. Projective Invariants of Complete Intersections

Francesco Gherardelli Dipartimento di Matematica, Università di Firenze, Florence, Italy

Summary. We define some projective invariants, related to the second fundamental form, of an algebraic manifold M that is a complete intersection in $\mathbb{CP}^N$ and has $\operatorname{codim} M \geq 2$. If M is canonical ($\mathcal{O}_M(1) \simeq K_M$), our remarks show that the field of rational functions of M contains a birationally invariant subfield. In this Note we consider the cases $\operatorname{codim} M = 2$ and $N = 6, 7$.

1. In this short Note we define some projective invariants of algebraic manifolds M, complete intersections in $\mathbb{CP}^N$ ($\operatorname{codim} M \geq 2$). We follow a method very similar to that of [1], where we have defined biholomorphic invariants of real submanifold of $\mathbb{C}^N$. The same arguments may be used to abtain orthogonal invariants of submanifold of differentiable manifolds.

If M is algebraic and canonical or pluricanonical ($\mathcal{O}_M(1) \simeq rK_M$, $r \geq 1$), our remarks show that the field $K(M)$ of rational functions contains a birationally invariant subfield.

We consider here only the cases $\operatorname{codim} M = 2$ and $N = 6, 7$.

2. Let M be a submanifold of a differentiable manifold V of codimension $p \geq 2$. The second fundamental form of M is a symmetric bilinear map of $T(M) \times T(M)$ into the normal bundle $T(M)^{\perp}$. Let $\xi^1, \dots, \xi^m$ be a local basis of $T(M)$ ($m = \dim M$) and let $v_1, \dots, v_p$ be p local fields of unit normal vectors, orthogonal at each point. With these choices the second fundamental form can be written

$$\sum_{i,j=1}^{m} \sum_{\alpha=1}^{p} h_{ij}^{\alpha} \xi^i \xi^j v_\alpha .$$

If $\lambda_1, \dots, \lambda_p$ are indeterminates,

$$\det \left| \sum_{\alpha=1}^{p} h_{ij}^{\alpha} \lambda_\alpha \right|$$

is a homogeneous polynomial in $\lambda_1, \dots, \lambda_p$ of degree equal to $\dim M$. Its orthogonal invariants are local orthogonal invariants of M. I hope to return on this subject in a forthcoming paper.

3. The previous remarks apply to a non-singular algebraic manifold M, complete intersection in a complex projective space $\mathbb{P}^N$. In this case we shall obtain global projective invariants.

The classical invariant theory is mainly developed for binary forms; for this reason we shall consider essentially the codim-two case: M is a complete intersection of two hypersurfaces

$$F(\xi_0,\dots,\xi_N) = 0 = G(\xi_0,\dots,\xi_N)$$

of degrees m and n respectively ($m \leq n$) and in general position.

If $U = \{\xi_0 \neq 0\}$, $V = \{\xi_1 \neq 0\}$ are affine open in $\mathbb{P}^N$, we put

$$x_1 = \xi_1/\xi_0 \,, \dots, x_N = \xi_N/\xi_0 \,,$$

$$y_1 = \xi_0/\xi_1 \,, y_2 = \xi_2/\xi_1 \,, \dots, y_N = \xi_N/\xi_1 \,,$$

$$F(\xi_0,\dots,\xi_N) := \xi_0^m f_U(x_1,\dots,x_N) = \xi_1^m f_V(y_1,\dots,y_N) \,,$$

$$G(\xi_0,\dots,\xi_N) := \xi_0^n g_U(x_1,\dots,x_N) = \xi_1^n g_V(y_1,\dots,y_N) \,,$$

so that in $U \cap V$

$$f_U(x) = x_1^m f_V(y) \quad , \quad g_U(x) = x_1^n g_V(y) \,.$$

Consider in $U \cap M$

$$P_U(\lambda,\mu) := \left| \begin{matrix} 0 & 0 & f_{U,1} & \cdots & f_{U,N} \\ 0 & 0 & g_{U,1} & \cdots & g_{U,N} \\ f_{U,1} & g_{U,1} & & & \\ \vdots & \vdots & & \lambda f_{U,ij} + \mu g_{U,ij} & \\ f_{U,N} & g_{U,N} & & & \end{matrix} \right|_{f_U = g_U = 0} \,,$$

where $f_{U,i} = \partial f_U/\partial x_i$, $g_{U,i} = \partial g_U/\partial x_i$, $f_{U,ij} = \partial^2 f_U/\partial x_i \partial x_j$, $g_{U,ij} = \partial^2 g_U/\partial x_i \partial x_j$ $(i,j = 1,\dots,N)$.

$P_U(\lambda,\mu)$ is a homogeneous polynomial of degree $N-2$ in the indeterminates λ,μ. We shall write

$$P_U(\lambda,\mu) = \sum_{p=0}^{N-2} \binom{N-2}{p} a_p \,\lambda^{N-p-2} \mu^p \,.$$

In a more intrinsic way, $P_U(\lambda,\mu)$ is the coefficient of the double form on M:

$$(df_U)^{\otimes 2} \wedge (dg_U)^{\otimes 2} \wedge \Big[\sum_{i,j} \big(\lambda f_{U,ij} + \mu g_{U,ij}\big) dx^i \otimes dx^j\Big]^{\wedge (N-2)},$$

where the tensor products are the symmetric ones. We are interested in the transformation laws of $P_U(\lambda,\mu)$ and its linear invariants under changes of coordinates, e.g., in $U \bigcap V$. Moreover we have to see how $P_U(\lambda,\mu)$ depends on the representation of M as complete intersection.

The main result is given by the following

Proposition. *The invariants — with respect to $SL_2(\mathbb{C})$ — of $P_U(\lambda,\mu)$ are projectively determined global sections of suitable line-bundles of M and the absolute invariants are projectively invariant rational functions on M.*

4. We shall prove this proposition when $N = 6$: this case is sufficient to explain the general one.

$P_U(\lambda,\mu)$ is now a homogeneous polynomial of degree four:

$$P_U(\lambda,\mu) := a_0\lambda^4 + 4a_1\lambda^3\mu + 6a_2\lambda^2\mu^2 + 4a_3\lambda\mu^3 + a_4\mu^4 .$$

The algebra of its invariants is generated by

$$i_U := a_0a_4 - 4a_1a_3 + 3a_2^2 ,$$

$$j_U := \begin{vmatrix} a_0 & a_1 & a_2 \\ a_1 & a_2 & a_3 \\ a_2 & a_3 & a_4 \end{vmatrix} .$$

Let

$$P_V(\lambda,\mu) := a_0'\lambda^4 + 4a_1'\lambda^3\mu + 6a_2'\lambda^2\mu^2 + 4a_3'\lambda\mu^3 + a_4'\mu^4 .$$

be the analogous polynomial in the affine set $V \bigcap M$, and let i_V, j_V be its invariants.

It is easy to verify that

$$P_U(\lambda,\mu) = x_1^{2(m+n-3)}\big\{x_1^{4(m-2)}a_0'\lambda^4 + 4x_1^{3(m-2)+n-2}a_1'\lambda^3\mu + 6x_1^{2(m-2)+2(n-2)}a_2'\lambda^2\mu^2 + 4x_1^{m-2+3(n-2)}a_3'\lambda\mu^3 + x_1^{4(n-2)}a_4'\mu^4\big\}$$

and then

(1) $$i_U = x_1^{8(m+n)-28}\, i_V \quad , \quad j_U = x_1^{12(m+n)-42}\, j_V .$$

Of course (f_U, g_U) is not the unique basis for the ideal of M in V. If $m = n$ we may substitute $af_U + bg_U, cf_U + dg_U$ to f_U, g_U $(a,b,c,d \in \mathbb{C}, ad-bc \neq 0)$: i_U and j_U change by a power of the modulus $ad-bc$, but i_U^3/j_U^2 does not change. If $m < n$, up to constants, another basis of the ideal of M is given by $(f_U, g_U + h_Uf_U)$, where h_U is a polynomial of degree $n-m$. An easy computation shows that this amounts to pass from (λ,μ) to $\begin{pmatrix}\lambda+\mu h_U\\ \mu\end{pmatrix}=\begin{pmatrix}1 & h_U\\ 0 & 1\end{pmatrix}\begin{pmatrix}\lambda\\ \mu\end{pmatrix}$, and the unimodular transformation $\begin{pmatrix}1 & h_U\\ 0 & 1\end{pmatrix}$ does not affect i_U and j_U.

Relations (1) say that $\{i_U\}$ is a global section of $\mathcal{O}_M(8(m+n)-28)$ and $\{j_U\}$ is a global section of $\mathcal{O}_M(12(m+n)-42)$. The divisors $i_U=0, j_U=0$ are projectively invariant and so is the rational function

$$i_U^3/j_U^2 = i_V^3/j_V^2 \,.$$

If $m+n=8$, M is a canonical manifold ($\mathcal{O}_M(1)\simeq K_M$) and the field defined by i^3/j^2 is birationally invariant in the field of all rational functions on M. In other words, there exists a canonical projection of $M\backslash\{i_U=j_U=0\}$ onto $\mathbb{P}^1$.

5. If $N=7$, the computations are only longer. M is now of dim 5 and the polynomial $P(\lambda,\mu)$ has degree 5. The algebra of the invariants of $P(\lambda,\mu)$ is well known ([3] p.284, [4] p.61): it is generated by four invariants $\varphi_1,\varphi_2,\varphi_3,\varphi_4$, homogeneous and isobaric in the coefficents of $P(\lambda,\mu)$, of degree respectively $4,8,12,18$; moreover $\varphi_1,\varphi_2,\varphi_3$ are algebraically independent over $\mathbb{C}$ and $\varphi_4^2\in\mathbb{C}[\varphi_1,\varphi_2,\varphi_3]$. $\varphi_1,\varphi_2,\varphi_3,\varphi_4$ are global sections of line bundles over M, multiples of that of the hyperplane sections. If $m+n=9$, these sections define a birationally invariant subfield (of trascendence degree 3) of the field of all rational functions on M.

Remarks. a) If $N=4,5$, the polynomial $P(\lambda,\mu)$ has degree 2 and 3, respectively. There is only one invariant: its determinant.

b) The definition of the polynomial P extends obviously to algebraic manifolds, complete intersections in $\mathbb{P}^N$ of $r\leq N-3$ hypersurfaces in general position. For instance, if $N=7$ and $r=3$, P is a homogeneous polynomial of degree four in three variables. Some of its invariants are well known and part of the previous observations can be repeated.

Bibliography

[1] F.Gherardelli, Biholomorphic invariants of real submanifolds of $\mathbb{C}^n$. Pages 189-194 in *Geometry and complex variables,* Lecture notes in Pure and Appl. Math. 132, S.Coen editor, Dekker, 1991.

[2] D.Hilbert, *Theory of algebraic invariants* (translation of a series of lectures), Cambridge University Press, 1993.

[3] G.Salmon, *Vorlesungen über die Algebra der linearen Transformationen,* Teubner, 1877.

[4] T.A.Springer, *Invariant theory,* Lecture Notes in Math. 585, Springer-Verlag, 1977.

17. *M*-Hyperbolicity, Evenness, and Normality

G. Gigante Dipartimento di Matematica, Università di Parma, Parma, Italy

Giuseppe Tomassini Dipartimento di Matematicà, Scuola Normale Superiore, Pisa, Italy

0. Introduction

Many papers concerning Kobayashi's theory on hyperbolic manifolds M pointed out how the existence of an invariant distance on M and its completeness are equivalent to the evenness and the normality of the family of holomorphic maps from the unit disk to M ([1], [7], [8]).
The definition of M-*hyperbolicity* for real analytic submanifolds V of complex manifolds M which has been introduced in [3], leads to the construction of a distance $d_{V,M}$ on V which is invariant by the group of holomorphic transformations of M leaving V invariant.
In this paper we want to carry further this study by exploring the relationship between M-hyperbolicity and properties of evenness and normality of the family of holomorphic maps from the unit disk to M that send the real interval into V.
We show that analogous results of complex theory of Kobayashi's hyperbolicity can be formulated in the setting of M-hyperbolicity and among these the classical theorem of Cartan ([6]). Furthermore, in section 2, we sketch how some application of tautness of complex manifolds in the theory of iteration of holomorphic maps can be carried out also in our context.

1. Evenness and tautness

1. We denote by Δ_r the disk $\{z \in \mathbb{C} : |z| < r\}$ and by I_r the real interval $\Delta_r \cap \mathbb{R}$. In particular, we set $\Delta = \Delta_1$ and $I = I_1$.

The authors were supported by the project 40% M.U.R.S.T. "Geometria reale e complessa".

Let $M^\star, M$ be complex manifolds and $V^\star \subset M^\star, V \subset M$ generic CR-analytic submanifolds.
We denote by $H(M^\star, M)$ the space of all holomorphic maps $f : M^\star \to M$ and by $H((M^\star, V^\star), (M, V))$ the space $\{f \in H(M^\star, M) : f(V^\star) \subset V\}$. In particular, we set $H(M, V) = H((M, V), (M, V))$.
$d_{V,M}, F_{V,M}$ denote respectively the invariant distance and the pseudometric on V as defined in [3]: namely, for every $p \in V$ and every $\xi \in T_pV$, $F_{V,M}(p, \xi)$ is the greatest lower bound of the positive real a for which there exists an $f \in H((\Delta, I), (M, V))$ such that $f(0) = p$ and $f'(0) = a^{-1}\xi$; $d_{V,M}$ is the integrated form of $F_{V,M}$.
Let us give the definitions of "evenness" and "tautness" of M along V.
M is said to be *even* along V if, for any pair as above $(M^\star, V^\star)$ the family $H((M^\star, V^\star), (M, V))$ has the following property: given $x \in V^\star$, $y \in V$ and a neighbourhood U of y in M, then there exist a neighbourhood W of y in V and a neighbourhood $U^\star$ of x in $M^\star$ such that $f(x) \in W$ implies $f(U^\star) \subset U$ for any $f \in H((M^\star, V^\star), (M, V))$.
M is said to be *taut* along V if for any sequence $\{f_n\}$ in $H((M^\star, V^\star), (M, V))$ the following holds true: either there is a subsequence which is uniformly convergent on any compact subset of a neighbourhood of $V^\star$ to a holomorphic map f, or there is a subsequence $\{f_{n(k)}\}$ which is *compactly divergent* i.e. for all compact subsets $K \subset V^\star$ and $K' \subset V$ there is an integer k' such that $f_{n(k)}(K) \cap K' = \emptyset$ for $k > k'$.
It is an easy matter to verify that evenness implies tautness. We also observe that the above definitions can be given if V is only CR. For this it is enough to replace neighbourhoods of a point $x \in V$ by neighbourhoods of x in a complexification of V in M ([9]).

Remark 1.1 In the most part of examples given in [3] of M-hyperbolic submanifolds, M is even along V. Moreover, in the situation explained in Proposition 3.11 of [3], using the same notations we get the following: if $V_{\mathbb{C}}$ is hyperbolic (resp. taut) then $\mathbb{C}^n$ is even along V (resp. taut along V).

Theorem 1.1 *If $H((\Delta, I), (M, V))$ is an even family, then M is even along V.*

Proof. We refer to [5], p.237 for the definition of even family.
Let us start with a remark. Given x in $V^\star$ there is a coordinate polydisc neighbourhood of x in $M^\star$ where $(z_1, \ldots, z_m), |z_j| < 1$ are coordinates, such that $x = (0, \ldots, 0)$ and $\{Im\, z_1 = \ldots = Im\, z_m = 0\} \subset V^\star$ ([9]). Thus for any $p = (a_1, \ldots, a_m)$ close to x we construct $\varphi \in H((\Delta, I), (M^\star, V^\star))$, such that $p \in \varphi(\Delta)$ in the following way: either is such that $Im\, a_j = 0$ for all j and then we set $\varphi(z) = (st_1 z, \ldots, st_m z)$ where $t_j = Re\, a_j$, $s = 1/t$, $t = \max\{|t_1|, \ldots, |t_m|\}$; or $\mu = \max\{|Im\, a_1|, \ldots, |Im\, a_m|\} \neq 0$; in this case assuming $\mu = |Im\, a_1|$ we define $\varphi(z) = (r_1 z + a_1 - r_1 a_1, \ldots, r_m z + a_m - r_m a_1)$ where $r_j = Im\, a_j / Im\, a_1$.
Let us prove the theorem. Suppose that M is not even along V; then there exist

$x \in V^\star, y \in V$ and a neighbourhood U of y (in M) such that for any choice of neighbourhoods W of y in V and $U^\star$ of x (in $M^\star$) there is a map $f \in H((M^\star, V^\star), (M, V))$ such that $f(x) \in W$ but $f(U^\star) \not\subset U$. Therefore, given x, y, U as above we can find $y_n \to y$ in V, $x_n \to x$ in $V^\star$ and maps f_n so that $f_n(x) = y_n$ and $f_n(x_n) \notin U$. Now, owing to the above remark we get $\varphi_n \in H((\Delta, I), (M^\star, V^\star))$ such that $\varphi_n(0) = Re\, x_n$ and $\varphi_n(b_n) = x_n$ where $b_n \to 0$; consequently $\psi_n = f_n \circ \varphi_n \in H((\Delta, I), (M, V))$ and $\psi_n(0) \in V$, $\psi_n(b_n) \notin U$. Let us prove that $\psi_n(0) \to y$ and for this assume that $M^\star$ is a polydisc and $V^\star$ is given by $Im\, z_1 = \ldots = Im\, z_m = 0$. Now using the decreasing property of the distances defined in $V^\star$ and V ([3]) we obtain

$$d_{V,M}(f_n(Re\, x_n), f_n(x_n)) \leq d_{V,M}(Re\, x_n, x).$$

This is in contradiction with the evenness of $H((\Delta, I), (M, V))$.

Theorem 1.2 *If M is even along V, then V is M-hyperbolic.*

Proof. Let $x \in V$ and let D be a polydisc around x in M. The evenness of M along V implies that there is a neighbourhood W, $\bar{W} \subset D$, of x and a disk Δ_μ with $|\mu| < 1$, such that, for all $\psi \in H((\Delta, I), (M, V))$, $\psi(0) \in W$ implies $\psi(\Delta_\mu) \subset D$. Now consider the pseudometric $F_{V,M}$ defined in [3]:

$$F_{V,M}(y, \xi) = inf\, \{1/R : \exists \psi \in H_R, \psi(0) = y, \psi'(0) = \xi\}$$

where $H_R = H((\Delta_R, I_R), (M, V))$.
If F_D denotes the pseudometric of Royden([8]), then $\mu F_D(y, \xi) \leq F_{V,M}(y, \xi)$. Thus in order to end the proof it is enough to recall that $d_{V,M}$ coincides with the integrated form of $F_{V,M}(y, \xi)$.

The following result is the analogous in this setting of the classical theorem of Cartan ([6]).

Theorem 1.3 *Let M be even along V, $f \in H(M, V)$, $f(0) = 0$ with $0 \in V$ and $df(0) = id$. Then the restriction of f to V is the identity map.*

Proof. Let us consider the family $\mathcal{H} = \{g \in H(M, V) : g(0) = 0\}$.
In view of theorem 1.1, for any neighbourhood U of 0 there exist two neighbourhoods W and $U^\star$ of 0 such that $h \in H(M, V)$ and $h(0) \in W$ imply $h(U^\star) \subset U$.
Choose U of type $\Delta \times \ldots \times \Delta$ and $U^\star$ of type $\Delta_{r_1} \times \ldots \Delta_{r_m}$, where $r_j < 1$, $j = 1, \ldots, m$ and consider the family $\mathcal{H}_{rest}$ of those $g \in H(U^\star, U)$ which are restriction to $U^\star$ of some $g \in \mathcal{H}$. Since Δ is taut, $\mathcal{H}_{rest}$ is a compact family for the compact open topology. Moreover, as the iterates of f belong to $\mathcal{H}$ there is a subsequence $f^{n(k)}$ which is uniformly convergent on all compact subsets to a map $f^\star \in H(U^\star, U)$. Now to conclude the proof we proceed as in theorem 3.3 of [6].

The following examples show why evenness along V is necessary:
(a) $(\mathbb{C}^2, \mathbb{R}^2)$ and $(z,w) \to (z, z+w^2)$,
(b) (C^2, V), $V = (t,t^2), t \in \mathbb{R}$ and
$(z,w) \to (z+\psi(z,w), w+\psi(z,w)^2+2z\psi(z,w))$, $\psi \in H(\mathbb{C}^2, \mathbb{C}), \psi(t,t') \in \mathbb{R}$, $\psi(0,0) = \psi_z(0,0) = \psi_z(0,0) = 0$.

2. We want to explore now tautness of M along V.

Theorem 1.4 *If $H((\Delta, I),(M,V))$ is a normal family, then M is taut along V.*

Proof. Consider $\{f_n\} \subset H((M^\star, V^\star),(M,V))$, $x \in V^\star$ and suppose first that $S(x) = \cup_n\{f_n(x)\}$ is relatively compact in V.
Since $H((\Delta,I),(M,V))$ is even, in view of theorem 1.1 we have the following: let $\{f_{n(k)}\}$ be a subsequence converging to y; then, given a taut neighbourhood U of y in M there is a neighbourhood W of y in V and a neighbourhood N of x in $M^\star$ such that $f(x) \in W$ implies $f(N) \subset U$. The restriction of $\{f_{n(k)}\}$ to N has a subsequence which is uniformly convergent on compact subsets of N to a map $f \in H(N,V)$, thus using the diagonal process (with respect to a countable covering of V), we find a subsequence which is uniformly convergent on compact susbsets all over a neighbourhood of V to a holomorphic map f.
Suppose now that $S(x)$ is not relatively compact; then there is a subsequence $\{f_{n(k)}\}$ of $\{f_n\}$ such that for each compact subset K of V one has $f_{n(k)}(x) \notin K$ for $k \gg 0$.
Consider the subset $A = \{x \in V^\star : S(x) \subset\subset V\}$ and the subset B of $V^\star$ so defined: $x \in B$ iff there a subsequence $\{f_{n(k)}\}$ such that for all compact subsets K in V one has $f_{n(k)} \notin K$ for $k \gg 0$. Then $A \cap B = \emptyset$, $A \cup B = V^\star$ and we show that A and B are both open in $V^\star$.
In order to prove that A is open consider the compact subset $\bar{S}(x)$, the closure of $S(x)$ and for each $y \in T(x)$ choose a compact neighbourhood U_y. Let N and W be such that $f_n(x) \in W$ implies $f_n(N) \subset U_y$. Then by standard arguments, it is easy to construct a neighbourhood N' of x contained in A.
B is open. If this is not true we find $x_\nu \to x$, $x_\nu \in A$ and $x \in B$. Now, for $\nu > \nu^\star$, we find $\varphi_\nu \in H((\Delta,I),(M^\star,V^\star))$ such that $\varphi_\nu(1/2) = x_\nu, \varphi_\nu(0) = x$ for $\nu > \nu^\star$. Take a subsequence $\{f_{n(k)}\}$ as indicated in the definition of B; thus there is no subsequence of $\{f_{n(k)} \circ \varphi_\nu\}$ which converge at 0. Observe that $\{f_{n(k)}\}$ itself is such that $\{f_{n(k)} \circ \varphi_\nu\}$ is compactly divergent for all $\nu > \nu^\star$ otherwise we should have, for some ν a subsequence $\{f_{n(k')}\}$ of $\{f_{n(k)}\}$ such that $\{f_{n(k')} \circ \varphi_\nu(1/2)\}$ is convergent, while $\{f_{n(k')} \circ \varphi_\nu(0)\}$ has no convergent subsequence, which contradicts the normality of $H((\Delta,I),(M,V))$. Thus $x_\nu \in B$ for $\nu > \nu^\star$. Since $V^\star$ is connected, then either $A = V^\star$ or $B = V^\star$.
Let us suppose that $B = V^\star$ and keep in mind that we have shown the following: for each $x \in V^\star$ there is a neighbourhood U of x in $V^\star$ and a subsequence $\{f_{n(k)}\}$ such that $\{f_{n(k)}(y)\}$ is compactly divergent for all y in U and consequently one subsequence exists (we call it again $\{f_{n(k)}\}$) which is compactly divergent at each point

of $V^\star$. Let $K^\star \subset\subset V^\star$ and $K \subset\subset V$; we are going to show that there exists k' such that $f_{n(k)}(K^\star) \cap K = \emptyset$ for all $k > k'$
If this is not true, then we can easily realize the following situation: there are sequences $x_\nu \to x$ in $K^\star$ and $\{g_n\} \subset H((M^\star, V^\star), (M, V))$ such that $\{g_n(y)\}$ is compactly divergent at $y \in V$ and $g_n(x_n) \to y$ in K. Take ϵ so small that $B(y, \epsilon)$, the ϵ-ball with respect to $d_{V,M}$ centered at y is relatively compact in V; then since

$$\begin{aligned} d_{V,M}(g_n(x), y) &\leq d_{V,M}(g_n(x), g_n(x_n)) + d_{V,M}(g_n(x_n), y) \leq \\ &\leq d_{V^\star,M^\star}(x, x_n) + d_{V,M}(g_n(x_n), y), \end{aligned}$$

(if $V^\star$ is not $M^\star$-hyperbolic, we can reduce the problem to the case where $M^\star$ is a ball in $\mathbb{C}^n$ and $V^\star = B \cap \mathbb{R}^n$ ([3], Prop.4.1)) we obtain the following: there is n' such that $g_n(x) \in B(y, \epsilon)$ for every $n > n'$ which is a contradiction as $\{g_n\}$ is compactly divergent at x.

In the same vein we have the following

Theorem 1.5 *If M is even along V and $d_{V,M}$ is complete (in the sense that Cauchy sequences are convergent), then M is taut along V.*

Proof. The proof runs as above; we need only to change it at the point where it is shown that B is open.
Let $x_\nu \to x, x_\nu \in A, x \in B$ and let $\{g_k\}$, $g_k = f_{n(k)}$ be compactly divergent at x. Let $\epsilon > 0$, ν' be such that $d_{V^\star,M^\star}(x, x_{\nu'}) < \epsilon$, and take a subsequence $\{g_{k(s)}\}$ of $\{g_k\}$ such that $\{g_{k(s)}(x_{\nu'})\}$ is convergent. We find $s^\star$ such that $d_{V,M}(g_{k(s)}(x_{\nu'}), g_{k(s')}(x_{\nu'})) < \epsilon$ for all $s, s' > s^\star$. Now we get,

$$\begin{aligned} d_{V,M}(g_{k(s)}(x), g_{k(s')}(x)) &\leq d_{V,M}(g_{k(s)}(x), g_{k(s)}(x_{\nu'})) + \\ &+ d_{V,M}(g_{k(s)}(x_{\nu'}), g_{k(s')}(x_{\nu'})) + \\ &+ d_{V,M}(g_{k(s')}(x_{\nu'}), g_{k(s)}(x)) < 3\epsilon. \end{aligned}$$

So $\{g_{k(s)}(x)\}$ is a Cauchy sequence, which is absurd.

2. Applications

This section is devoted to some applications and remarks.
Let

$$Aut(M, V) = \{f \in Aut(M) : f(V) \subset V\}.$$

Theorem 2.1 *Let M be even along V, $\{f_n\} \subset Aut(M, V)$ be such that $f_n \to f$ uniformly on compact subsets. Then f is an analytic automorphism on V.*

Proof. Let $x_o \in V$ and $g_n = f_n^{-1}$ so that $g_n(f_n(x_o)) = x_o$ and $f_n(x_o) \to f(x_o)$. In view of evenness of M along V, for any neighbourhood U of x_o there is a neighbourhood $U^\star$ of $f(x_o)$ such that $g_n(U^\star) \subset U$ for $n \gg 0$. Thus it is clear that $\cup_n\{g_n(f(x_o))\}$ is relatively compact in V; with the notations of Theorem 1.4, we get $V = A$ and we can find a subsequence $\{g_{n(k)}\}$ which converges on all compact subsets of a neighbourhood N of V to a holomorphic map g such that $f \circ g = id$.

In the theory of iteration of holomorphic maps on a complex manifold M a central rule is played by the construction of limit manifolds and limit retractions (see, for example [1], p.143). In this context, we state an analogous construction.

Theorem 2.2 *Let M be taut along V and $f \in H(M, V)$. Assume that the sequence $\{f^n\}$ of iterates of f is not compactly divergent on V. Then there exists a real analytic submanifold W of V and an analytic retraction $\rho : V \to W$ such that every limit point h of $\{f^n\}$ in a neighbourhood of V is of the form $h = \gamma \circ \rho$ where γ is analytic automorphism of V. Moreover, the retraction ρ is a limit point of $\{f^n\}$, f is an automorphism of W and the set of fixed point of f on V is contained in W.*

Proof. We need only to follow step by step the proof of theorem 2.1.29 of [1], keeping in mind that we get convergence of subsequences of $\{f^n\}$ only on neighbourhoods of V and using the fact that Lemma 2.1.28 of [1] holds also in the real analytic case ([2]).

Some other results, quoted in [1], as theorems 2.4, 2.4.2, 2.4.3 can be easily restated in this setting.
The following result, which is the samespirit of a theorem of Kaup ([4]), holds for any real analytic submanifold V of M provided V is M-hyperbolic.

Theorem 2.3 *Let V be a compact M-hyperbolic submanifold of M and $f \in H(M, M)$ be such that $f(V) = V$. Then f is injective on V.*

Proof. The family of restrictions on V of $\{f^n\}$ is equicontinuous with respect to $d_{V,M}$. By the Ascoli-Arzela's theorem, there exist a subsequence $\{f^{n(k)}\}$ which converges uniformly on V to a continuous map $g : V \to V$. Suppose that $n(k+1)-n(k) \to +\infty$. Then, the sequence $\{f^{n(k+1)-n(k)}\}$ has a subsequence which converges on V to a continuous map $h : V \to V$; we get $h \circ g = g$ on V. Moreover, since $f(V) = V$, we have $g(V) = V$, so $h = id$ on V.

Remark 2.1 The condition of M-hyperbolicity cannot be removed from theorem 2.3. as it is shown by the following example: $(M, V) = (\mathbb{C}, S^1)$ and f given by $z \to z^2$.

Bibliography

[1] M.ABATE,*Iteration of holomorphic maps on taut manifolds*, Mediterranean Press, Rende, Cosenza, 1990.

[2] H.CARTAN, *Sur les rétractions d'une variété*, C.R. Acad. Sc. Paris Sefie 1 Math. 303 (1986), p.715-716.

[3] G.GIGANTE-G.TOMASSINI-S.VENTURINI, *M-hyperbolic real subsets of complex spaces* , (to appear in Pac. Journ. of Math.).

[4] W.KAUP, *Hyperbolische komplexe Räume*, Ann. Inst. Fourier 18 (1968), p.303-330.

[5] J.L.KELLEY, *General Topology*, Van Nostrand, Princeton 1955.

[6] S.KOBAYASHI, *Hyperbolic manifolds and holomorphic mappings* , Dekker, New York, 1970.

[7] S.KOBAYASHI, *Intrinsic distances, measures and geometric function theory* , Bull. A. M. S. 82 (1976), p.375-416.

[8] H.ROYDEN, *Remarks on the Kobayashi metric* , Several Complex Variables II, Lec. Notes in Math.189, Springer, Berlin, 1971, p.125-137.

[9] G.TOMASSINI, *Tracce delle funzioni olomorfe sulle sottovarietà analitiche reali d'una varietà complessa*, Ann. Scuola Norm. Sup. Pisa, (3) 20, 1 (1966), p.31-43.

18. An Extended Variational Principle

Richard Jordan Center for Nonlinear Analysis, Carnegie Mellon University, Pittsburgh, Pennsylvania

David Kinderlehrer Department of Mathematics, Carnegie Mellon University, Pittsburgh, Pennsylvania

1 INTRODUCTION

The paradigm of the calculus of variations is the functional

$$I(v) = \int_\Omega W(\nabla v)\, dx. \tag{1.1}$$

In the absence of quasiconvexity of W, I may fail to assume its minimum in a given admissible class, even when this class is very reasonable. In this situation, we may seek to represent the solution in terms of the oscillatory statistics developed by a minimizing sequence, or its Young Measure, [37]. For definiteness, suppose that

$$c(|\lambda|^p - 1)^+ \le W(\lambda) \le C(1 + |\lambda|^p), \quad \lambda \in \mathbb{M}, 0 < c < C, 1 < p < \infty. \tag{1.2}$$

where $\mathbb{M}$ denotes $m \times n$ matrices and that $\Omega \subset \mathbb{R}^n$ is a domain with smooth boundary. Given $u_0 \in H^{1,p}(\Omega, \mathbb{R}^m)$, consider the variational principle

$$\inf_V I(v) \qquad \text{where } V = u_0 + H_0^{1,p}(\Omega, \mathbb{R}^m). \tag{1.3}$$

In recent years we have learned much about minimizing sequences (u^k) of (1.1) and their Young Measures. For example, (u^k) may be chosen so that in addition to

$$u^k \rightharpoonup u \text{ in } V,$$

we have that there is a $g \in L^1(\Omega)$ for which

This research supported by the Air Force Office of Scientific Research, the Army Research Office, and the National Science Foundation

$$|\nabla u^k|^p \rightharpoonup g \text{ in } L^1(\Omega) \tag{1.4}$$

and the Young Measure representation is valid for $\psi \in C(\mathbb{M})$ satisfying

$$|\psi(\lambda)| \leq C(1 + |\lambda|^p) \tag{1.5}$$

This means that there is a family $\nu = (\nu_x)_{x \in \Omega}$ of probability measures on $\mathbb{M}$ such that when ψ satisfies (1.5)

$$\psi(\nabla u^k) \rightharpoonup \bar{\psi} \text{ in } L^1(\Omega) \quad \text{where} \quad \bar{\psi}(x) = \int_{\mathbb{M}} \psi(\lambda)\, d\nu_x(\lambda) \text{ in } \Omega \text{ a.e.} \tag{1.6}$$

In particular, the limit gradient ∇u is recaptured as

$$\nabla u(x) = \int_{\mathbb{M}} \lambda\, d\nu_x(\lambda) \text{ in } \Omega \text{ a.e. and} \tag{1.7}$$

$$\int_{\Omega\times\mathbb{M}} W(\lambda)\, d\nu_x(\lambda)dx = \lim_{k\to\infty} \int_{\Omega} W(\nabla u^k)\, dx = \inf_{V} I(v). \tag{1.8}$$

This suggests introducing a class A of measures μ on $\Omega \times \mathbb{M}$, for example, the Young Measures μ generated by a sequences of gradients (∇v^k), $v^k \in V$, and considering, in place of (1.4), the variational principle, with $d\mu(x,\lambda) = d\mu_x(\lambda)dx$,

$$\inf_{A} I_W(\mu) \quad \text{with} \quad I_W(\mu) = \langle T,\mu \rangle = \int_{\Omega\times\mathbb{M}} W(\lambda)\, d\mu(x,\lambda). \tag{1.9}$$

The functional $\langle T,\mu \rangle$ is a linear function on A. Thus every variational principle is linear.

To what extent can we formulate principles in the form (1.9) so they become meaningful? In this note we shall discuss, somewhat informally, this isssue. We shall also address its natural successor, namely, the analysis of questions which depend in some more complicated fashion on the measure and are not linear functionals. To offer a short and readable account, we shall be guided by a precept of Carlo Pucci's old friend Hans Lewy: emphasize the obvious and eschew the difficult.

2 FRAMEWORK AND BASIC PROPERTIES

We introduce a few notations and review briefly the general features of Young Measures. Let

$$E^p = \{ \psi \in C(\mathbb{R}^N): \lim_{|\lambda|\to\infty} \frac{\psi(\lambda)}{1 + |\lambda|^p} \text{ exists } \}, \; 1 \leq p < \infty. \tag{2.1}$$

E^p is a separable Banach Space whose dual we denote by $E^{p\prime}$. For technical reasons, it has an advantage over the inseparable space of functions X^p defined by the inequality (1.5). Given a domain $\Omega \subset \mathbb{R}^n$, note that, [10],

$$L^1(\Omega;E^p)' = L^\infty(\Omega; E^{p'}).$$

A measure $\mu \in L^\infty(\Omega; E^{p'})$ is a parametrized measure or Young Measure if and only if there is a sequence $f^k \in L^p(\Omega;\mathbb{R}^N)$ and $f \in L^p(\Omega;\mathbb{R}^N)$ and $g \in L^1(\Omega)$ such that

$$f^k \rightharpoonup f \text{ in } L^p, \quad |f^k|^p \rightharpoonup g \text{ in } L^1, \text{ and}$$
(2.2)
$$\psi(f^k) \rightharpoonup \bar{\psi} \text{ in } L^1(\Omega) \quad \text{where} \quad \bar{\psi}(x) = \int_{\mathbb{M}} \psi(\lambda)\, d\nu_x(\lambda) \text{ in } \Omega \text{ a.e. for } \psi \in E^p.$$

The formula in (2.2) is called the "Young Measure representation" and it permits us to represent the statistics of the oscillations developed by the sequence (f^k) by means of the formula

$$\bar{\psi}(a) = \lim_{\rho\to 0} \lim_{k\to\infty} \frac{1}{|B_\rho|} \int_{B_\rho(a)} \psi(f^k)\, dx, \quad a \in \Omega \text{ a.e.} \tag{2.3}$$

A situation of special interest is when $f^k = \nabla u^k$, that is, when (f^k) is a sequence of gradients. In this case we call μ an $H^{1,p}$ - Young measure, or simply a gradient Young Measure, when the value of p is not of concern.

An important feature of Young Measures as they appear in a variational context is their duality with lower semicontinuous functionals exhibited by means of Jensen's Inequality [24,25,26]. The characterizing property of parametrized measures in $L^\infty(\Omega; E^{p'})$ is Jensen's Inequality for convex functions, cf. (2.11). Gradient Young Measures are dual to quasiconvex functions: a parametrized measure $\nu \in L^\infty(\Omega; E^{p'})$ is a gradient Young Measure if and only if

(i) $\varphi(F(a)) \leq \int_{\mathbb{M}} \varphi(\lambda)\, d\nu_a(\lambda)$ where $F(a) = \int_{\mathbb{M}} \lambda\, d\nu_a(\lambda)$ in Ω a.e. for every quasiconvex $\varphi \in E^p$,

(2.4)

(ii) there is a $u \in H^{1,p}(\Omega,\mathbb{R}^m)$ with $F(x) = \nabla u(x)$, and (2.5)

(iii) $\Psi(x) = \int_{\mathbb{M}} |\lambda|^p\, d\nu_x(\lambda) \in L^1(\Omega).$

(2.6)

Condition (i) is akin to a local condition, [25,26],

(i)' ν_a is a homogeneous gradient Young Measure for $a \in \Omega$ a.e.
(2.4)'

There is an extensive literature on this subject beginning with L. C. Young's own interpretation in control theory [37]. Unfortunately, we lack space to adequately cite all the important recent work. These methods were enhanced and generalized by Tartar [34,35], who studied conservation laws and compensated compactness. The past ten years have witnessed further extensions. The tool of the Young Measure has become fundamental to the study of microstructure in solids, where the weak continuity properties of the minors of (∇u^k) have made it possible to establish kinematical restrictions on minimum energy configurations. This has led to new understanding of structural phase transformations, [4,5,6,7,8,9,17,18,19]. The interpretation of coherent structures in turbulence by means of statistical equilibrium theory has led to maximum entropy principles and a Young Measure description of most probable states via the theory of large deviations, [20,21,31,32,36]. The Young Measure is an extremely useful device for understanding nonlinear processes across widely disparate length scales.

Although the presentation here may be viewed in some ways as a synthesis of the two areas discussed above, our motivation was an attempt to investigate the nature of metastability in some physical systems. We treat this in a separate paper. Details and extensions of our discussion here are in [22,23].

The first issue at hand is what happens, in general, to a bounded sequence of measures in $L^\infty(\Omega; E^{p'})$. A sequence $\mu^k \in L^\infty(\Omega; E^{p'})$ of parametrized measures with

$$\|\mu^k\| = \int_{\Omega\times\mathbb{R}^N} (1 + |\lambda|^p)\, d\mu^k(x,\lambda) \leq M < \infty \tag{2.7}$$

admits a subsequence, not relabeled, such that

$$\mu^k \overset{*}{\rightharpoonup} \tau \in L^\infty(\Omega; E^{p'}), \tag{2.8}$$

where τ need not be a probability measure on $\mathbb{R}^N$, as we know well. Nonetheless, we may isolate from τ a probability measure μ, and it is this we wish to clarify. For simplicity, assume that the μ^k are homogeneous, i.e., independent of $x \in \Omega$, so we may regard $\mu^k \in E^{p'}$. It follows from elementary methods that there is a measure $\mu \in C_0(\mathbb{R}^N)'$ such that

$$\mu^k \overset{*}{\rightharpoonup} \mu \text{ in } C_0(\mathbb{R}^N)'. \tag{2.9}$$

We claim

(a) $\mu \in E^{p'}$ and

(b) $$\int_{\mathbb{R}^N} \psi\, d\mu = \lim_{k\to\infty} \int_{\mathbb{R}^N} \psi\, d\mu^k \text{ whenever } \psi \in E^p \text{ with}$$

$$\lim_{|\lambda|\to\infty} \frac{\psi(\lambda)}{1 + |\lambda|^p} = 0.$$

Part (a) follows from the Monotone Convergence Theorem. Part (b) follows from (a) and the bound (2.7). In particular

$$\int_{\mathbb{R}^N} d\mu = \lim_{k\to\infty} \int_{\mathbb{R}^N} d\mu^k = 1,$$

so μ is a probability measure.

Returning now to the characterization of $\tau \in E^{p'}$ with

$$\mu^k \overset{*}{\rightharpoonup} \tau \quad \text{in } E^{p'},$$

it is easy to check that $\tau \geq 0$ and may be expressed (as the measure on $\mathbb{R}^N \cup \{\infty\}$)

$$\tau = c\,\delta_\infty + \mu, \quad c \geq 0,$$

where $\mu \in E^{p'}$ satisfies (a) and (b),

$$\langle \delta_\infty, \psi \rangle = \lim_{|\lambda| \to \infty} \frac{\psi(\lambda)}{1 + |\lambda|^p}, \quad \text{and} \quad c = \lim_{r \to \infty} \langle T, \chi_{\{|\lambda| \geq r\}} |\lambda|^p \rangle.$$

Because of this, we are justified in referring to μ as the *probability measure determined by the* (μ^k). Similarly, for $\tau \in L^\infty(\Omega; E^{p'})$ satisfying (2.8), there is a parametrized family of probability measures $\mu = (\mu_x)_{x \in \Omega} \in L^\infty(\Omega; E^{p'})$ and $\gamma \in L^1(\Omega)$ such that

$$\tau = \gamma\,\delta_\infty + \mu, \quad \gamma \geq 0, \quad \text{and}$$

$$\int_{\Omega \times \mathbb{R}^N} \psi \, d\mu = \lim_{k \to \infty} \int_{\Omega \times \mathbb{R}^N} \psi \, d\mu^k \quad \text{whenever } \psi \in L^\infty(\Omega; E^p) \text{ with} \tag{2.10}$$

$$\lim_{|\lambda| \to \infty} \frac{\psi(x,\lambda)}{1 + |\lambda|^p} = 0 \text{ a.e. in } \Omega$$

The characterization theorem for Young Measures ensures that the conditions (2.2) are satisfied: there is a sequence $f^k \in L^p(\Omega; \mathbb{R}^N)$ which generates μ.

It is easy to verify Jensen's Inequality directly in this situation from (a) and (b). Let $\varphi \in E^p$, $p > 1$, be convex and let

$$g(\lambda) = \varphi(\lambda_o) + L \cdot (\lambda - \lambda_o) \leq \varphi(\lambda), \quad \lambda_o = \int_{\mathbb{R}^N} \lambda \, d\mu_{x_o}(\lambda), \quad x_o \in \Omega.$$

Then from (b),

$$\varphi(\lambda_o) = \lim_{k \to \infty} \int_{\mathbb{R}^N} g(\lambda) \, d\mu^k_{x_o}(\lambda) = \int_{\mathbb{R}^N} g(\lambda) \, d\mu_{x_o}(\lambda) \leq \int_{\mathbb{R}^N} \varphi(\lambda) \, d\mu_{x_o}(\lambda), \text{ in } \Omega \text{ a.e..} \tag{2.11}$$

This property does not seem so obvious for other constraints, in particular, if μ^k are gradient Young Measures, is μ a gradient Young Measure?

Proposition 2.1 *Let* $\mu^k \in L^\infty(\Omega; E^{p'})$ *be a weak* convergent sequence of gradient Young Measures and let* μ *be the parametrized measure determined by the* (μ^k). *Then* $\mu \in L^\infty(\Omega; E^{p'})$ *is a gradient Young Measure.*

The localization property (2.4)' makes it possible to assume that the (μ^k) are homogeneous. Each μ^k is generated by a sequence $v^{k,j} \in H^{1,p}(\Omega, \mathbb{R}^m)$ with

$$\| v^{k,j} \|_{H^{1,p}(\Omega)} \leq C.$$

Choosing a diagonal sequence w^k of the $v^{k,j}$ we obtain a sequence which generates μ as an $H^{1,q}$ Young Measure for $q < p$ and as a biting Young Measure, cf. [26]. From Theorem1.1 of [26], we are assured that μ is an $H^{1,p}$ Young Measure. Note that the sequence which generates μ as as an $H^{1,p}$ Young Measure is not in general the (w^k).

The localization feature makes it possible to construct variations of the Proposition. For example, we may specify a subdomain $\Omega' \subset \Omega$ such that $\mu^k|_{\Omega'}$ are gradient Young Measures. Then $\mu|_{\Omega'}$ is also a gradient Young Measure. In the next section we outline a different method.

The variational principle (1.9) may now be placed into a rigorous context. We choose A to be the gradient Young Measures generated by sequences in V, or what is the same, the gradient Young Measures for which $u \in V$ in (2.5).

Proposition 2.2 *Suppose that* $v^k \in A \subset L^\infty(\Omega;E^{p'})$ *with*

$$I_W(v^k) \to \inf_A I_W(\mu) . \tag{2.12}$$

Then there is a subsequence, not relabeled, of the v^k *and a* $v \in A$ *(a gradient Young Measure) such that*

$$v^k \overset{*}{\rightharpoonup} v \quad \textit{and} \quad I_W(v) = \inf_A I_W(\mu). \tag{2.13}$$

From (1.2) we have that (v^k) are bounded in $L^\infty(\Omega;E^{p'})$ and hence, after extraction of a subsequence, have a weak limit of the form

$$\tau = \gamma\delta_\infty + v \quad \text{with } \gamma \geq 0 \text{ and } v \in A . \tag{2.14}$$

It is easy to check from this that $\gamma = 0$ so that $(2.13)_1$ holds.

Clearly, in this first try at the "every variational principle is linear" idea, the lower bound in (1.2) is crucial, cf. [12,13,28].

3 Local Constraints

A local constraint on a parametrized measure might be one determined by some quasiconvex functions rather than all of them. For example, write $\mathbb{R}^N = \mathbb{M}\times\mathbb{R}^d$, where $\mathbb{M}$ denotes $m \times n$ matrices, and write $\lambda = (A,\alpha)$, $A \in \mathbb{M}$ and $\alpha \in \mathbb{R}^d$. Suppose that v satisfies Jensen's Inequality for functions φ satisfying, $|\Omega| = 1$,

$$\varphi(F,p) \leq \int_\Omega \varphi(F + \nabla\zeta, p + q)\, dx \quad \text{for } \zeta \in H_0^{1,\infty}(\Omega,\mathbb{R}^m) \text{ and}$$
$$q \in L^\infty(\Omega;\mathbb{R}^d) \text{ with } \int_\Omega q\, dx = 0. \tag{3.1}$$

We would expect ν to be generated by a sequence of the form $(\nabla u^k, p^k)$, and this is in fact the case, [14]. Here we wish to give a different result in a similar direction.

Given $0 \le \varphi \in E^p$ satisfying (3.1), determine the set $K_\varphi \subset L^\infty(\Omega; E^{p'})$ of parametrized measures ν for which

$$\psi(F,p) \le \int_{\mathbb{M}\times\mathbb{R}^d} \psi(A,\alpha)\, d\nu_x(A,\alpha) \quad \text{where} \quad (F,p) = \int_{\mathbb{M}\times\mathbb{R}^d} (A,\alpha)\, d\nu_x(A,\alpha)\,,\ x \in \Omega \text{ a.e.},$$

whenever $\psi \in E^p$ satisifes (3.1) and $\psi \le \varphi$. (3.2)

Theorem 3.1 *Let $\mu^k \in K_\varphi$ be a weak* convergent sequence and let μ be the parametrized measure determined by (μ^k). Then $\mu \in K_\varphi$.*

To prove this, we really only have to show that if $\varphi \in E^p$ is nonnegative and satisfies (3.1), then there is a sequence, which we shall call $(\psi^s)^\#$ such that $(\psi^s)^\# \in E^{sp}$, $0 < s < 1$, and $(\psi^s)^\# \to \varphi$, pointwise as $s \to 1^-$, $(\psi^s)^\# \le \varphi$. For this, introduce

$$\psi^s(A,\alpha) = \begin{cases} \varphi(A,\alpha) & \text{in } \{\varphi(A,\alpha) \le 1\} \\ \varphi(A,\alpha)^s & \text{in } \{\varphi(A,\alpha) > 1\} \end{cases}$$

(3.3)

and its "relaxation"

$$(\psi^s)^\#(A,\alpha) = \inf \int_\Omega \psi^s(A + \nabla\zeta, \alpha + q)\, dx \in E^{sp} \tag{3.4}$$

for $\zeta \in H_0^{1,\infty}(\Omega, \mathbb{R}^m)$ and $q \in L^\infty(\Omega; \mathbb{R}^d)$ with $\int_\Omega q\, dx = 0$. Note that we may choose $\Omega = Q$, a unit cube, and replace the boundary condition on ζ by one of periodicity. To show that $(\psi^s)^\#$ converges to φ, we avail ourselves of the argument of Marcellini and Sbordone [29] to establish that a minimizing sequence for (3.4) may be chosen equi-integrable in L^p. This argument combines the Ekeland Distance Lemma [11] with the Meyers - Elcrat [30] form of the reverse Hölder Inequality.

Obviously, the theorem above implies the other statements we have been discussing.

4 EXTENDED VARIATIONAL PRINCIPLES

We now seek to enlarge the scope of questions amenable to Young Measure methods with a new paradigm. Let

$$\begin{aligned} &\psi \in E^p \text{ satisfy } \psi \ge 0 \text{ and} \\ &\varphi \in C(\mathbb{R}^+) \text{ satisfy } \rho\varphi(\rho) \text{ convex and increasing for large } \rho \text{ and} \\ &\int_\Omega \varphi\left(\frac{d\nu}{d\mu}\right) d\nu \ge 0 \text{ for all probability measures } \nu \in E^{p'}. \end{aligned} \tag{4.1}$$

For a fixed probablility measure $\mu^o \in E^{p'}$, with $d\mu = d\mu^o dx$ and $\sigma > 0$, consider the functional

$$I_\sigma(\nu) = \int_{\Omega\times\mathbb{R}^N} \psi \, d\nu + \sigma \int_{\Omega\times\mathbb{R}^N} \varphi(\frac{d\nu}{d\mu}) \, d\nu, \quad \nu \in L^\infty(\Omega; E^{p'}), \tag{4.2}$$

with the convention that $I_\sigma(\nu) = +\infty$ if ν fails to be absolutely continuous with respect to μ. It will be clear in what follows that μ^o need not be a probability measure. The most common choice of φ is $\varphi(\rho) = \log \rho$ and in this case the Kullback Entropy is minus the second integral. (4.1) is satisfied in this case. Also in this case, the minimizer of (4.2) is given by

$$d\nu^\sigma = \frac{1}{Z(\sigma)} e^{-\frac{\psi}{\sigma}} d\mu, \text{ where } Z(\sigma) = \int_{\mathbb{R}^N} e^{-\frac{\psi}{\sigma}} d\mu^o \tag{4.3}$$

is the well known "partition function," when the competition is among all parametrized measures and $Z(\sigma)$ is finite. We wish to make prominent some extremely elementary properties of (4.2).

Theorem 4.1

Suppose that $I_\sigma(\nu^*) < +\infty$ *for some* $\nu^* \in K$ *and* $\sigma = \sigma_0$. *Then*

$$\inf_K \int_{\Omega\times\mathbb{R}^N} \psi \, d\nu = \lim_{\sigma\to 0} \inf_K I_\sigma(\nu) \tag{4.4}$$

If $\sigma < \sigma'$, then $I_\sigma(\nu) \le I_{\sigma'}(\nu)$, whence $f(\sigma) = \inf_K I_\sigma(\nu)$ is decreasing as $\sigma \to 0$. Let

$$a = \inf_K \int_{\Omega\times\mathbb{R}^N} \psi \, d\nu$$

and given $\varepsilon > 0$, choose ν such that

$$\int_{\Omega\times\mathbb{R}^N} \psi \, d\nu \le a + \varepsilon.$$

Then

$$\int_{\Omega\times\mathbb{R}^N} \psi \, d\nu + \sigma \int_{\Omega\times\mathbb{R}^N} \varphi(\frac{d\nu}{d\mu}) \, d\nu \le a + \varepsilon + \sigma \int_{\Omega\times\mathbb{R}^N} \varphi(\frac{d\nu}{d\mu}) \, d\nu$$

and

$$\inf_K I_\sigma(\nu) \le a + \varepsilon + \sigma \int_{\Omega\times\mathbb{R}^N} \varphi(\frac{d\nu}{d\mu}) \, d\nu ,$$

whence

$$\lim_{\sigma\to 0} \inf_K I_\sigma(\nu) \le a + \varepsilon.$$

Suppose that the set where ψ assumes its minimum is compact and

$$\inf_K \int_{\Omega\times\mathbb{R}^N} \psi \, d\nu = |\Omega| \min \psi.$$

Assume that we have in hand ν^σ such that $I_\sigma(\nu^\sigma) = \inf_K I_\sigma(\nu)$. Then the (ν^σ) are bounded in $L^\infty(\Omega;E^{p'})$ and thus, according to our discussion, we may select a subsequence (not relabeled) which determines a parametrized measure $\nu^o \in L^\infty(\Omega;E^{p'})$. (Note that ν^o is not generally absolutely continuous with respect to μ.) From the Monotone Convergence Theorem (viz. the argument we used to prove (a) in §2),

$$\int_{\Omega\times\mathbb{R}^N} \psi \, d\nu^o \leq \liminf_{\sigma\to 0} \int_{\Omega\times\mathbb{R}^N} \psi \, d\nu^\sigma \leq \lim_{\sigma\to 0} I_\sigma(\nu^\sigma) = |\Omega| \min \psi.$$

It follows that supp $\nu^o \subset \{\psi = \min \psi\}$ is compact and $\nu^\sigma \overset{*}{\rightharpoonup} \nu^o$ in $L^\infty(\Omega;E^{p'})$ and, of course, ν^o realizes the minimum for $\sigma = 0$. Facts like these are well known for the particular case of (4.3) but are proved by direct computation. Passing to the limit as $\sigma \to 0$ gives the stationary distribution of a random variable and is related to the simulated annealing algorithm. In practical situations, it is quite important to choose the sequence of σ carefully, [16].

We now address the existence question.

Theorem 4.2 *Let $K \subset L^\infty(\Omega;E^{p'})$ be a set of parametrized measures which enjoys the closure property*

if $\mu^k \in K$ and $\mu^k \overset{*}{\rightharpoonup} \tau$ in $L^\infty(\Omega;E^{p'})$, then the parametrized measure μ determined by (μ^k) satisfies $\mu \in K$. (4.5)

Assume that (4.1) *is satisfied and that $\varphi(\rho) \to \infty$ as $\rho \to \infty$. Then there exists a parametrized measure minimizer $\nu^\sigma \in K$ of I_σ. Of course $\nu^\sigma << \mu$.*

For a minimizing sequence ν^k, write $d\nu^k = \rho^k d\mu$, $\rho^k \in L^1(\Omega\times\mathbb{R}^N,\mu)$. Then

$$\sup \int_{\Omega\times\mathbb{R}^N} \varphi(\rho^k)\rho^k \, d\mu < \infty \quad \text{with} \quad \lim_{|\rho|\to\infty} \frac{\varphi(\rho)\rho}{\rho} = \infty.$$

This is a well known condition for weak relative compactness in L^1 and it follows from the de la Vallée Poussin Criterion that there is a subsequence of the (ρ^k) weakly convergent in L^1. A standard technique shows that the second term of I_σ is lower semi-continuous with respect to weak convergence in L^1 from which it is easy to show that the weakly convergent subsequence converges to a minimizer.

Theorem 4.3 *Assume that* (4.1) *is satisfied and set $g(\rho) = \varphi(\rho)\rho$. Suppose that*

$$\frac{d^2}{d\rho^2} g > 0 \textit{ for } \rho > 0, \quad \text{range } (g')^{-1} \subset (0,\infty),$$

and that there is a constant α such that

$$\int_{\Omega\times\mathbb{R}^N} \rho^*(\lambda) \, d\mu^o(\lambda) = 1 \quad \textit{where} \quad \rho^*(\lambda) = (g')^{-1}\left(\frac{\alpha - \psi(\lambda)}{\sigma}\right).$$

Then $\nu^* = \rho^*\mu$ *is an absolute minimizer of* I_σ.

Minimization falls into this framework. Let $K \subset \mathbb{R}^N$ be compact and let μ^o denote normalized Lebesgue measure on K. Consider a function $\psi \in C^2$, for example, and the functional

$$I_\sigma(\nu) = \int_K \psi \, d\nu + \sigma \int_K \log(\frac{d\nu}{d\mu}) \, d\nu, \quad \nu \in EP', \ \sigma > 0.$$

As mentioned above, the minimizer is given by (4.3). Assuming that ψ has M isolated global minimizers at $\lambda_1, \ldots, \lambda_M$ in K, our theorem asserts that

$$\nu^\sigma \overset{*}{\rightharpoonup} \sum c_i \delta_{\lambda_i} \text{ as } \sigma \to 0 \text{ where } c_i \geq 0 \text{ and } \sum c_i = 1.$$

This can also be established by calculating the Taylor expansion of ψ, which reveals that the c_i are related to $\nabla^2\psi(\lambda_i)$. These considerations form the basis of the simulated annealing algorithm for global minimization. More generally, there is an intimate connection between the type of variational principle we are discussing and the stationary Fokker-Planck Equation, [1,15,16].

5 SOME EXAMPLES

5.1 The Langevin function and constrained theory

Consider an ensemble of N identical particles governed in equilibrium by thermal motion in, say, $-1 \leq \xi \leq 1$, each of which tends to orient under a field f with strength $\eta\xi$. The probability distribution $d\nu^\sigma = \rho^\sigma d\xi$ which describes the state of a given particle at "temperature" σ is the extremal of

$$I_\sigma(\nu) = -\int_K f\eta\xi\rho \, d\xi + \sigma \int_K \rho \log \rho \, d\xi, \quad K = [-1,1], \tag{5.1}$$

and is given by the formula (4.3). Note that since K is compact, we may arrange that (4.1) is satisfied. The expected value of the "strength", the state of a particle, is

$$\eta\langle \xi \rangle = \eta \int_K \xi\rho^\sigma \, d\xi = \eta \, L(\frac{f\eta}{\sigma}), \text{ where } L(x) = \coth x - \frac{1}{x} \tag{5.2}$$

is the Langevin Function. The expected strength of the ensemble is

$$S = N\eta \, L(\frac{f\eta}{\sigma}). \tag{5.3}$$

Langevin used this analysis successfully to explain paramagnetism [27]. For small values of x, $L(x) = \frac{1}{3}x + O(x^3)$, and leads to the notion of the susceptibility

$$\chi = \frac{S}{f} = \frac{1}{3}\frac{N\eta^2}{\sigma}. \tag{5.4}$$

Note the interesting scaling properties here. If the N particles are grouped into clusters of size M where each cluster responds as a unit, then the expected strength of a cluster is $M\eta\langle M\xi\rangle$. The expected strength of the ensemble and the susceptibility are

$$S_M = \frac{N}{M}M\eta L(\frac{fM\eta}{\sigma}) \quad \text{and} \quad \chi_M = \frac{S_M}{f} = \frac{1}{3}\frac{MN\eta^2}{\sigma}. \tag{5.5}$$

Thus the effective susceptibility is enhanced when the particles can act as clusters. This is among the mechanisms considered in magnetic nanocomposites, a subject of current research, [33]. The linear approximation to $L(x)$ breaks down if M is large, which leads to an optimization problem for M.

A second view is given by what we call the "constrained theory," [3]. Here a nonlinear elastic body, for example, is assumed to reside in a collection of potential wells Σ even when subjected to a modest constant field T. This gives rise to the functional

$$\inf_{K} - \int_{\Omega\times\Sigma} T\cdot\lambda \, d\nu \quad \text{where } K = \text{gradient Young Measures with support in } \Sigma. \tag{5.6}$$

The solution of (5.6) may be approximated by extemals of

$$I_\sigma(\nu) = - \int_{\Omega\times\Sigma} T\cdot\lambda \, d\nu + \sigma \int_{\Omega\times\Sigma} \varphi(\frac{d\nu}{d\mu}) \, d\nu,$$

where μ is a fixed reference measure, e.g., a Gaussian. A generalized Langevin Function is given by

$$L(\frac{T}{\sigma}) = \frac{1}{\sigma}\int_{\Sigma} T\cdot\lambda \, d\nu^\sigma(\lambda). \tag{5.7}$$

Thus, in view of Theorem 4.1, the constrained theory may be realized as the zero temperature limit of a system governed by thermal motion confined to a given collection of potential wells.

5.2 Coherent structures identified by maximum entropy principles

We present an example of the analysis related to a problem that arises in modeling coherent structures in 2D microtearing turbulence and in 2D magnetohydrodynamic turbulence. Consider the functional

$$F_\sigma(\nu) = \frac{1}{2}\int_{\Omega\times\mathbb{R}^N} (|A|^2 + |\alpha|^2)d\nu + \sigma \int_{\Omega\times\mathbb{R}^N} \log(\frac{d\nu}{d\mu}) \, d\nu \quad (A \in \mathbb{M}^{n\times m}, \alpha \in \mathbb{R}^n) \tag{5.8}$$

$$d\mu = d\mu^o(A,\alpha)dx, \quad d\mu^o(A,\alpha) = \frac{1}{(2\pi)^{N/2}} e^{-\frac{1}{2}(|A|^2 + |\alpha|^2)} \, dAd\alpha. \tag{5.9}$$

a standard Gaussian on $\mathbb{R}^N$, $N = nm + n$. We shall minimize F_σ over the set K defined as the set of parametrized measures $\nu \in L^\infty(\Omega;E^{2'})$ such that ν is an $H^1(\Omega) \times L^2(\Omega)$ Young Measure satisfying

$$J_1(\nu) = \frac{1}{2}\int_\Omega |\nu|^2 dx = J_1^o \text{ and } J_2(\nu) = \int_\Omega \nu \cdot q\, dx = J_2^o, \text{ where} \tag{5.10}$$

$$(\nabla v(x), q(x)) = \int_{\mathbb{R}^N} (A,\alpha)\, d\nu_x(A,\alpha) \text{ with } (v,q) \in H_0^1(\Omega;\mathbb{R}^m) \times L^2(\Omega;\mathbb{R}^n). \tag{5.11}$$

The statement that ν is an $H^1(\Omega) \times L^2(\Omega)$ Young Measure means that $\nu \in K_\varphi$ whenever φ satisfies the conditions of (3.1)-(3.2). It is easy to check that K has the closure property (4.5) by using Theorem 3.1 and the Rellich Compactness Theorem. Hence by Theorem 4.3, we are assured the existence of

$$\nu^\sigma \in K: F_\sigma(\nu^\sigma) = \min_K F_\sigma(\nu) \tag{5.12}$$

with first moment $(\nabla u^\sigma, p^\sigma)$ and from Theorem 4.1 we know that

$$\inf_K \int_{\Omega\times\mathbb{R}^N} \psi\, d\nu = \lim_{\sigma\to 0} \inf_K I_\sigma(\nu), \text{ where } \psi(A,\alpha) = \frac{1}{2}(|A|^2 + |\alpha|^2).$$

Let

$$E(\nu) = \frac{1}{2}\int_{\Omega\times\mathbb{R}^N} (|A|^2 + |\alpha|^2) d\nu \text{ and } E^\#(v,q) = \frac{1}{2}\int_\Omega (|\nabla v|^2 + |q|^2) dx. \tag{5.13}$$

By Jensen's Inequality, $E^\#(v,q) \le E(\nu)$ when the relation (5.11) holds with equality when $\nu_x = \delta_{(\nabla v(x),q(x))}$, so it follows that, with A the set of (v,q) which satisfy (5.10) and $(5.11)_2$,

$$\inf_A E^\#(v,q) \le \inf_K E(\nu) = \mathit{inf}.$$

From this we see that *inf* is attained at $\nu^o \in K$ with $\nu_x^o = \delta_{(\nabla u(x),p(x))}$, where $(u,p) \in A$ minimizes $E^\#$. (Note that the constants J_i^o must be given consistently so that the class K is not empty.) There are multipliers a and b such that

$$u \in H_0^1(\Omega;\mathbb{R}^m): \quad -\Delta u + au + bp = 0 \text{ and } bu + p = 0 \text{ in } \Omega, \tag{5.14}$$

and so it turns out that u is an eigenfunction corresponding to the smallest eigenvalue of $-\Delta$ with homogeneous boundary conditions.

Although we have invented this problem to illustrate the method, it has several interesting features. The solution ν^σ $(\sigma > 0)$ is for each x a Gaussian with mean $(\nabla u^\sigma(x), p^\sigma(x))$, each component having variance $\sigma/(1+\sigma)$, and with independent components, i.e., zero covariance. Let μ^σ denote the Young Measure which is for each x Gaussian with mean $(\nabla u(x), p(x))$, with the same variance $\sigma/(1+\sigma)$, and with independent components. Then

$$F_\sigma(\nu^\sigma) = (\sigma + 1)\, E^\#(u^\sigma,p^\sigma) + \sigma\,|\,\Omega\,|\log\left(\left(\frac{1+\sigma}{\sigma}\right)^{N/2}\right) \text{ and}$$

$$F_\sigma(\mu^\sigma) = (\sigma + 1)\, E^\#(u,p) + \sigma\,|\,\Omega\,|\log\left(\left(\frac{1+\sigma}{\sigma}\right)^{N/2}\right). \tag{5.15}$$

Thus $(u^\sigma,p^\sigma) = (u,p)$ and the Young Measure delivers the limit deformation at each "temperature".

Finally, ν^σ minimizes F_σ over the larger set K' of $L^2(\mathbb{R}^N)$ Young measures defined analogously to K with the boundary condition applied to the curl-free part of the appropriate portion of the first moment, as determined by the Helmholtz decomposition.

ACKNOWLEDGEMENTS

We would like to thank Yury Grabovsky and Pablo Pedregal for helpful conversations.

REFERENCES

[1] Aluffi-Pentini, F., Parisi, V., and Zirilli, F. 1985 Global optimization and stochastic differential equations, J. Opt. Theory and Appl., 47, 1-16

[2] Ball, J. M. 1989 A version of the fundamental theorem for Young measures, *PDE's and continuum models of phase transitions,* Lecture Notes in Physics, 344,(Rascle, M., Serre, D., and Slemrod, M., eds.) Springer, 207-215

[3] Ball, J. M. and James, R. 1987 Fine phase mixtures as minimizers of energy, Arch. Rat. Mech. Anal.,100, 15-52

[4] Ball, J. M. and James, R. to appear

[5] Ball, J.M. and James, R. 1991 Proposed experimental tests of a theory of fine microstructure and the two well problem, Phil. Trans. Roy. Soc. Lond. A338, 389-450

[6] Battacharya, K. 1991 Wedge-like microstructure in martensite, Acta Metal., 39, 2431-2444

[7] Battacharya, K. 1992 Self accomodation in martensite, Arch. Rat. Mech. Anal., 120, 201-244

[8] Chipot, M. and Kinderlehrer, D. 1988 Equilibrium configurations of crystals, Arch. Rat. Mech. Anal., 103, 237-277

[9] DeSimone, A. 1993 Energy minimizers for large ferromagnetic bodies, Arch. Rat. Mech. Anal., 125, 99-144

[10] Edwards, R.E. 1965 *Functional Analysis,* Holt, Rinehart, and Winston

[11] Ekeland, I. 1979 Nonconvex minimiztion problems, Bull. AMS, 1.3, 443-474

[12] Fonseca, I. and Maly, J. Relaxation of multiple integrals below the growth exponent, Anal. Nonlineare (to appear)

[13] Fonseca, I. and Marcellini, P. Relaxation of multiple integrals in subcritical Sobolev spaces, J. Geom. Anal.. (to appear)

[14] Fonseca, I., Kinderlehrer, D., and Pedregal, P. 1994 Energy functionals depending on elastic strain and chemical composition, Calc. Var., 2, 283-313

[15] Gardiner, C.W. 1990 *Handbook of Stochastic Methods,* Second Edition, Springer-Verlag

[16] Geman, S. and Hwang, C.-R. 1986 Diffusions for global optimization, SIAM J. Control, 24.5, 1031-1045

[17] James, R. D. and Kinderlehrer, D. Laminate structures in martensite (to appear)

[18] James, R. D. and Kinderlehrer, D. 1989 Theory of diffusionless phase transitions, *PDE's and continuum models of phase transitions,* Lecture Notes in Physics, 344,(Rascle, M., Serre, D., and Slemrod, M., eds.) Springer, 51-84

[19] James, R.D. and Kinderlehrer, D. 1993 A theory of magnetostriction with application to TbDyFe2, Phil. Mag. B, 68.2, 237-274

[20] Jordan, R. 1995 A statistical equilibrium model of coherent structures in magnetohydrodynamics, Nonlinearity, 8, 585-613

[21] Jordan, R. Maximum entropy states and coherent structures in two-dimensional microtearing turbulence (to appear)

[22] Jordan, R. and Kinderlehrer, D. in preparation

[23] Jordan, R., Kinderlehrer, D., and Pedregal, P. in preparation

[24] Kinderlehrer, D. and Pedregal, P. 1991 Charactérisation des mesures de Young associées à un gradient, C.R.A.S. Paris, 313, 765-770

[25] Kinderlehrer, D. and Pedregal, P. 1991 Characterizations of Young measures generated by gradients, Arch. Rat. Mech. Anal., 115, 329-365

[26] Kinderlehrer, D. and Pedregal, P. 1994 Gradient Young measures generated by sequences in Sobolev spaces, J. Geom. Anal., 4, 59-90

[27] Langevin, P. 1905 J. Phys., 4, 678

[28] Marcellini, P. 1991 Regularity and existence of solutions of elliptic equations with p,q growth conditions, J. Diff. Eqns., 90, 1-30

[29] Marcellini, P. and Sbordone, C. 1983 On the existence of minima of multiple integrals in the calculus of variations, J. Math. Pures. Appl., 62, 1-9

[30] Meyers, N. and Elcrat, A. 1975 Some results on regularity for solutions of non linear elliptic systems and quasiregular functions, Duke Math. J., 42 (1), 121-136

[31] Robert, R. 1991 A maximum entropy principle for two-dimensional perfect fluid dynamics, J. Stat. Phys., 65, 531-553

[32] Robert, R. and Sommeria, J. 1991 Statistical equilibrium states for two-dimensional flows, J. Fluid Mech., 229, 291-310

[33] Shull, R. 1995 Magnetic applications of nanocomposite materials, lecture at Carnegie Mellon University

[34] Tartar, L. 1979 Compensated compactness and applications to partial differential equations, *Nonlinear analysis and mechanics: Heriot Watt Symposium, Vol I V*(Knops, R., ed.) Pitman Res. Notes in Math. 39, 136-212

[35] Tartar, L. 1983 The compensated compactness method applied to systems of conservation laws, *Systems of nonlinear partial differential equations* (Ball, J. M., ed) Riedel

[36] Turkington, B. and Jordan, R. Turbulent relaxation of a megnetofluid: a statistical equilibrium model, to appear

[37] Young, L. C. 1969 *Lectures on calculus of variations and optimal control theory*, W.B. Saunders

19. Instability Criteria for Solutions of Second Order Elliptic Quasilinear Differential Equations

B. Kawohl Department of Mathematics, University of Cologne, Cologne, Germany

1 INTRODUCTION

Let $\Omega \subset \mathbb{R}^n$ be a bounded domain with smooth boundary $\partial\Omega$ and suppose that $v : \Omega \to \mathbb{R}$ satisfies an elliptic differential equation, e.g.

$$-\Delta v = f(x, v) \quad \text{in } \Omega \tag{1}$$

and suitable boundary conditions, e.g.

$$v = 0 \qquad \text{on } \partial\Omega. \tag{2}$$

One can interpret $v(x)$ as a stationary solution of a corresponding parabolic problem

$$u_t - \Delta u = f(x, u) \quad \text{in } (0, \infty) \times \Omega \tag{3}$$

under boundary conditions, say

$$u(t, x) = 0 \qquad \text{on } (0, \infty) \times \partial\Omega \tag{4}$$

and under initial data

$$u(0, x) = u_0(x) \qquad \text{in } \Omega \tag{5}$$

Suppose that the initial data $u_0(x)$ do not deviate too much from a stationary state $v(x)$. Does the solution of (3) (4) (5) return to $v(x)$ as $t \to \infty$? Under structural assumptions on f the answer is known to be negative, and $v(x)$ is then called an unstable solution of (1) (2), see [2,4]. In the present note this result is extended to quasilinear elliptic equations of divergence type and to more general boundary conditions, see Theorems A and B in section 2.

A decisive tool is the second variation of the underlying Ljapunov-functional. Other results, for instance on symmetry of solutions, can be obtained by means of this tool as well, see Theorems C and D in section 4. In Theorem D structural assumptions on the nonlinearities are given up in exchange for assumptions on the geometry of the underlying domain.

2 STATEMENT OF THE PROBLEM AND RESULT

We consider the following class of problems

$$-\mathrm{div}(P_q(v,|\nabla v|^2)\nabla v) = f(x,v) - \frac{1}{2}P_v(v,|\nabla v|^2) \quad \text{in } \Omega, \tag{6}$$

$$Bv(x) = 0 \quad \text{on } \partial\Omega, \tag{7}$$

where $P = P(v,q)$ is of class $C^2(\mathbb{R}\times\mathbb{R}^+)$, P_v and P_q denote partial derivatives and

$$Bv(x) = v(x) \tag{7a}$$

or

$$Bv(x) = P_q(v,|\nabla v|^2)\frac{\partial v}{\partial n} + g(x,v). \tag{7b}$$

Here $\frac{\partial v}{\partial n}$ denotes the outer normal derivative. We want (6) to be an elliptic equation, therefore we shall assume

$$P_q + 2qP_{qq} > 0 \quad \text{in } \mathbb{R}^+\times\mathbb{R}^+\,. \tag{8}$$

Associated with (6) and (7) is the parabolic problem

$$u_t - \mathrm{div}(P_q(u,|\nabla u|^2)\nabla u) = f(x,u) - \frac{1}{2}P_v(u,|\nabla u|^2) \quad \text{in } (0,\infty)\times\Omega, \tag{9}$$

$$Bu(t,x) = 0 \quad \text{in } (0,\infty)\times\partial\Omega, \tag{10}$$

$$u(0,x) = u_0(x) \quad \text{in } \Omega. \tag{11}$$

We call a solution v of (6) (7) *stable* iff for every $\varepsilon > 0$ there exists a $\delta > 0$ such that

$$\|u_0(x) - v(x)\|_{L^\infty(\Omega)} < \delta \quad \text{implies} \quad \|u(t,x) - v(x)\|_{L^\infty((0,\infty)\times\Omega)} \le \varepsilon$$

for the solution of (9) (10) (11). Otherwise v is said to be *unstable.*

The result will state that under structural assumptions on P, f and g most solutions are unstable, and only constant solutions are stable. In case of the Dirichlet condition

(7a) this means that nontrivial solutions must be unstable. Let me therefore list the structural assumptions:

$$f(x,0) \le 0 \text{ in } \Omega, \tag{12}$$

$$\left.\begin{array}{c} f(x,u) \text{ is convex in } u \text{ for all } x \in \Omega \text{ , i.e.} \\ (t-s)f_u(x,s) \le f(x,t) - f(x,s) \text{ for all real } s,t \text{ ; } x \in \Omega \text{ , } s \ne t. \end{array}\right\} \tag{13}$$

$$\left.\begin{array}{l} R(u,q) \le 0 \text{ in } \mathbb{R}^+ \times \mathbb{R}^+ \text{ ,} \\ \text{where } R(u,q) := 2q^2 P_{qq} + 2uqP_{uq} + \frac{1}{2}u^2 P_{uu} - \frac{1}{2}uP_u, \end{array}\right\} \tag{14}$$

$$-g(x,0) \le 0 \text{ in } \partial\Omega \tag{15}$$

$$\left.\begin{array}{c} -g(x,u) \text{ is convex in } u \text{ for all } x \in \partial\Omega \text{ , i.e.} \\ -(t-s)g_u(x,s) \le -g(x,t) + g(x,s) \text{ for all real } s,t \text{ , } x \in \Omega \text{ , } s \ne t. \end{array}\right\} \tag{16}$$

Now we can state the results, first for the case of Dirichlet boundary condition (7a).

Theorem A:
Let $u \in C^2(\Omega) \cap C^1(\overline{\Omega})$ be a nonconstant, nonnegative solution of (6) (7a). *Let f satisfy* (12) *and* (13) *and P satisfy* (8) *and* (14).
If at least one of the inequalities (12), (13) *or* (14) *is strict, then u is unstable.*

For the Neumann type boundary condition (7b) the result is as follows.

Theorem B:
Let $u \in C^2(\Omega) \cap C^1(\overline{\Omega})$ be a nonconstant, nonnegative solution of (6) (7b). *Let f satisfy* (12) *and* (13), *P satisfy* (8) *and* (14) *and g satisfy* (15) *and* (16).
If at least one of the inequalities (12) *through* (16) *is strict, then u is unstable.*

3 PROOFS

To prove Theorem A set $F(x,v) := \int_0^v f(x,s)ds$ and

$$E(v) := \int_\Omega \frac{1}{2}P(v, |\nabla v|^2) - F(x,v)\, dx$$

for $v \in W_0^{1,\infty}(\Omega)$. The functional E is a Ljapunov functional for the parabolic problem (9) (10) (11), since a simple calculation shows that E decreases along solutions of (9) (10) (11):

$$\begin{aligned} \frac{d}{dt}E(u) &= \int_\Omega \frac{1}{2}P_v u_t + P_q \nabla u \nabla u \nabla u_t - f u_t\, dx \\ &= -\int_\Omega u_t \left[\operatorname{div}(P_q \nabla u) + f - \frac{1}{2}P\right] dx + \int_{\partial\Omega} u_t P_q \frac{\partial u}{\partial n}\, ds \\ &= -\int_\Omega u_t^2\, dx \le 0. \end{aligned} \tag{17}$$

Note that the boundary integral vanishes, since $u = u_t = 0$ on $\partial\Omega$. Moreover, stationary solutions of (9) (10) (11), or solutions of (6) (7a) are critical points of E, i.e. their first variation vanishes:

$$\delta E(v,w) = \int_\Omega \frac{1}{2}P_v(v,|\nabla v|^2)w + P_q(v,|\nabla v|^2)(\nabla v \nabla w) - f(x,v)w \; dx = \\ \int_\Omega -w\left[\operatorname{div}(P_q\nabla v) + f - \frac{1}{2}P_v(v,|\nabla v|^2)\right] dx = 0 \quad \text{for } w \in W_0^{1,\infty}(\Omega). \tag{18}$$

In other words, (6) (7a) is the Euler-Lagrange equation associated to E. Let us now calculate the second variation of E at v in direction w:

$$\delta^2 E(v,w) = \int_\Omega \{\frac{1}{2}P_{vv}w^2 + P_{vq}w(\nabla v\nabla w) + P_{qv}w(\nabla v \nabla w) + 2P_{qq}(\nabla v\nabla w)^2 \\ + P_q|\nabla w|^2 - f_v(x,v)w^2\} \; dx \tag{19}$$

If there is a function w for which the second variation is strictly negative, then v is unstable. Let us therefore choose $w = v$ in (19). We obtain

$$\delta^2 E(v,v) = \int_\Omega \frac{1}{2}v^2 P_{vv} + 2v|\nabla v|^2 P_{vq} + 2P_{qq}|\nabla v|^4 \\ - \frac{1}{2}vP_v + v[f - vf_v] \; dx \\ + \int_\Omega \frac{1}{2}vP_v + |\nabla v|^2 P_q - vf \; dx, \tag{20}$$

and note that the second integral vanishes because of (18).

The integrand $v[f - vf_v]$ is nonpositive, because (12) and (13) imply

$$f(x,v) - vf_v(x,v) \leq f(x,0) \leq 0 \quad .$$

Finally we see that $\delta^2 E(v,v)$ is nonpositive from (14). This completes the proof of Theorem A. □

For the proof of Theorem B we modify the Ljapunov-functional E. We set

$$G(x,v) = \int_0^v g(x,s)ds \qquad \text{and}$$

$$E(v) := \int_\Omega \frac{1}{2}P(v,|\nabla v|^2) - F(x,v) \; dx + \int_{\partial\Omega} G(x,v) \; ds \; .$$

Proceeding as before we arrive at a modification of (20) which contains the additional term

$$+\int_{\partial\Omega} v[-g + vg_v]ds \quad .$$

But now $-g(x,v) + vg_v(x,v) \leq -g(x,0) \leq 0$ due to (15) and (16) and the statement follows as before. □

4 APPLICATIONS AND COMMENTS, FURTHER RESULTS

The Theorems A and B apply in particular to the following special cases:

Example 1: Prescribed mean curvature equations

$$-\operatorname{div}\left(\frac{\nabla v}{\sqrt{1+|\nabla v|^2}}\right) = f(x,v) \quad \text{in } \Omega \tag{21}$$

Example 2: p-Laplace equations with $p \in (1,2]$

$$-\operatorname{div}(|\nabla v|^{p-2}\nabla v) = f(x,v) \quad \text{in } \Omega \tag{22}$$

Example 3: Equations involving porous medium type operators

$$-\operatorname{div}(ku^{m-1}\nabla u) = f(x,u) - \frac{m-1}{2}ku^{m-2}|\nabla u|^2 \quad \text{in } \Omega \tag{23}$$

where $k > 0$ and $m \in [-1,1]$.

Remark 1: On Symmetry

The particular choice of $w = v$ in (19) led to Theorem A. Other choices of testfunctions can lead to other results. Here is one of them:

Let Ω be a ball or a radially symmetric domain with center 0 and $w = v_\theta$ an angular derivative of v, and let v be a solution of (6) (7a). For simplicity let us suppose that P does not depend on v and f does not depend on θ. Then (6) reads

$$-\operatorname{div}(P'(|\nabla v|^2)\nabla v) = f(r,v)\ . \tag{24}$$

Now the second variation $\delta^2 E(v, v_\theta)$ boils down to

$$\delta^2 E(v, v_\theta = \int_\Omega \{2P''(\nabla v, \nabla v_\theta)^2 + P'|\nabla v_\theta|^2 - f_v(v)v_\theta^2\}\, dx. \tag{25}$$

The term $f_v v_\theta^2$ can also be generated if we differentiate (6) with respect to v_θ and multiply the resulting expression with v_θ. Thus we obtain:

$$-\operatorname{div}[P'(|\nabla v|^2)\nabla v_\theta + P''(|\nabla v|^2)2(\nabla v \nabla v_\theta)\nabla v] = f_v v_\theta\ ,$$

and after integration by parts

$$\int_\Omega P'(|\nabla v|^2)|\nabla v_\theta|^2 + P''(|\nabla v|^2)2(\nabla v \nabla v_\theta)^2\, dx = \int_\Omega f_v v_\theta^2\, dx.$$

We plug this into (25) to arrive at $\delta^2 E(v, v_\theta) = 0$. Thus one is led to the following result:

Theorem C:
Let $v \in C^2(\Omega) \cap C^0(\overline{\Omega})$ be a strictly stable solution of (24) (7a), i.e. suppose that $\delta^2 E(v, \varphi) > 0$ for any nonzero φ.
Then v is a radially symmetric function.

For the special case that $P(q) = q$ and $f = f(v)$, this result can be found in [1]. Notice that no structure conditions like (12) – (14) are used in the proof of Theorem C.

Remark 2: On Convex Domains
There is a variant of Theorem C for homogeneous Neumann conditions, which holds not only for radially symmetric but for convex domains. Consider the problem

$$-\text{div}(P'(|\nabla v|^2)\nabla v) = f(v) \text{ in } \Omega\ , \tag{26}$$

$$P'(|\nabla v|^2)\frac{\partial v}{\partial n} = 0 \text{ on } \partial\Omega\ . \tag{27}$$

Theorem D:
Let $v \in C^3(\overline{\Omega})$ be a strictly stable solution of (26) (27) and suppose that Ω is convex and $P' \geq 0$.
Then v is constant.

For the proof one chooses $w = \frac{\partial v}{\partial x_i}$ as variations in (19) and obtains

$$\delta^2 E(v, v_{x_i}) = \int_\Omega \{2P''(\nabla v\ \nabla v_{x_i})^2 + P'|\nabla v_{x_i}|^2 - f_v(v)v_{x_i}^2\}\ dx\ . \tag{28}$$

As in the proof of Theorem C the last term $f_v(v)v_{x_i}^2$ in (28) can also be generated by differentiating (26) and multiplying with v_{x_i}. Then partial integration leads to

$$\begin{aligned}\int_\Omega f_v|\nabla v|^2\ dx = &\int_\Omega 2P''(\nabla v|^2)(\nabla v, \nabla v_{x_i})^2 + P'(|\nabla v|^2)|\nabla v_{x_i}|^2\ dx \\ &- \int_{\partial\Omega} 2P''(|\nabla v|^2)(\nabla v, \nabla v_{x_i})\frac{\partial v}{\partial n}v_{x_i}\ ds \\ &- \int_{\partial\Omega} P'(|\nabla v|^2)\frac{\partial v_{x_i}}{\partial n}v_{x_i}\ ds\ ,\end{aligned} \tag{29}$$

and a combination of (28), (29) and (27) gives

$$\delta^2 E(v, v_{x_i}) \leq + \int_{\partial\Omega} P'(|\nabla v|^2)\frac{\partial}{\partial n}(|\nabla v|^2)\ ds\ . \tag{30}$$

But $\frac{\partial}{\partial n}(|\nabla v|^2) \leq 0$, since Ω is convex, see [5, Lemma 5.3]; and so $\nabla v \equiv 0$ and the proof is complete.□

Theorem D was previously known in the case $P(q) = q$. In fact all Theorems A through D are now extended from semilinear to quasilinear equations.

REFERENCES

[1] N.D. Alikakos & P.W. Bates, On the singular limit in a phase field model of phase transitions, Analyse Nonlineaire, Ann. Inst. Henry Poinc. **5** (1988) 141-178.
[2] K.J. Brown & Shivaji, Instability of nonnegative solutions for a class of semipositone problems. Proc. Amer. Math. Soc. **112** (1991) 121-124.
[3] R.G. Casten & C.J. Holland, Instability results for reaction-diffusion equations with Neumann boundary conditions. J. Differ. Equations **27** (1978) 266-273.
[4] B. Kawohl, A short note on stability. Expositiones Mathem. **11** (1993), 141-144.
[5] H. Matano, Asymptotic behaviour and stability of solutions of semilinear diffusion equations. Publ. Res. Inst. Math. Sci. **15** (1979) 401-451.

20. Maximum Principles for Difference Operators

Hung-Ju Kuo Department of Applied Mathematics, National Chung-Hsing University, Taichung, Taiwan

Neil S. Trudinger Center for Mathematicas and Its Applications, Australian National University, Canberra, A.C.T., Australia

Let L be a second order elliptic differential operator of the form,

(1) $$Lu = a^{ij}D_{ij}u \quad ,$$

acting on functions $u \in C^2(\Omega)$, where Ω is a domain in Euclidean n–space. Utilizing geometric ideas of Aleksandrov, Bakelman [4] established the estimate,

(2) $$\sup_{\Omega} u \leq \sup_{\partial\Omega} u + \frac{\operatorname{diam}\Omega}{n\omega_n^{1/n}} \left\| \frac{(Lu)^-}{\mathcal{D}^*} \right\|_{L^n(\Omega)}$$

where C is a constant depending on n, $\mathcal{D}^* = \mathcal{D}^{1/n}$ where $\mathcal{D} = \det[a^{ij}]$ (> 0 in Ω). The estimate (1) was subsequently extended to general elliptic operators of the form,

(3) $$Lu = a^{ij}D_{ij}u + b^iD_iu + cu,$$

by Aleksandrov [1,2] and Pucci [13], using different methods and the precision of the corresponding estimates was also carefully

Research supported by Australian Research Council and National Science Council of Taiwan

investigated by Aleksandrov [3]. For further information the reader is referred to the books [5], [6].

In our papers [8], [12], we have treated analogous results for difference operators, with the purpose of deriving local estimates and eventually results for nonlinear schemes as in [9]. In this paper, we present a general discussion of the corresponding maximum principle for difference operators of the form

$$Lu = \sum_y a(x,y)u(y) \tag{4}$$

acting on mesh function u, which are defined on a mesh E which is an arbitrary discrete set in $\mathbb{R}^n$. The coefficients a are defined on E×E, with compact support in y for each x value. The operator L is called *monotone* if

$$a(x,y) \geq 0 \quad , \text{for all } x\,,\, y \in E\,, \tag{5}$$

and *positive* if, in addition,

$$c(x) \equiv \sum_y a(x,y) \;\leq\; 0 \qquad \text{for all } x \in E\;. \tag{6}$$

Furthermore, we shall call L *balanced* if

$$b(x) \equiv \sum_y a(x,y)(x-y) = 0 \qquad \text{for all } x \in E\;. \tag{7}$$

Let D be a subset of E. The interior of D, with respect to L, is defined by

$$D^o = \left\{ x \in D \mid a(x,y) = 0\,,\, \forall\, y \notin D \right\}\,, \tag{8}$$

and the *boundary* of D, with respect to L, is defined by

$$(9) \qquad D^b = D - D^o .$$

We shall establish a maximum principle, analogous to the estimate (2). Again, the fundamental geometric ideas of Aleksandrov are crucial. For a mesh function u, we define the normal mapping $\chi = \chi_u$ over a domain D in E by

$$(10) \qquad \chi(x) = \{p \in \mathbb{R}^n \mid u(y) \le u(x) + p \cdot (y - x),\ \forall\ y \in D\} ,$$

that is $\chi(x)$ consists of the slopes of the support hyperplanes at x lying above the graph of u over D. For a fixed point $x \in D$, let us denote by $k = k_x$ the mesh function given by

$$(11) \qquad k_x(y) = 0 \quad \text{if } x \ne y$$

$$k_x(x) = 1$$

and set

$$(12) \qquad D^*(x) = \chi_k(x) = \chi_k(D)$$

It follows that $D^*(x)$ is a convex polyhedron in $\mathbb{R}^n$, the Lebesgue measure of which we denote by μ, that is

$$(13) \qquad \mu = \mu(D,x) = |D^*(x)| .$$

To get an idea of the meaning of μ, we let $\hat{D}$ denote the convex hull of D in $\mathbb{R}^n$ and suppose that

$$(14) \qquad B_\rho(x) \subset \hat{D} \subset B_R(x) ,$$

where $B_R(x)$ denotes the ball of radius R and center x in $\mathbb{R}^n$. Then we have the inequality

(15) $$\omega_n R^{-n} \le \mu(D,x) \le \omega_n \rho^{-n}$$

Let us now assume that the difference operator L is positive and balanced, and that D is a bounded subset of E. For any point $x \in D^o$, we define a discrete set $Y \subset \mathbb{R}^n$ by

(16) $$Y = Y_x = \left\{ x + a(x,y)(y - x) \mid y \in E \right\}$$

The condition that L is balanced implies that Y_x is centered at x. We then have the following discrete analogue of the Aleksandrov, Bakelman and Pucci estimates for the operator (1).

THEOREM 1. *Let* u *be a mesh function satisfying the difference inequality*

(17) $$Lu + f \ge 0$$

in the interior D^o *of a bounded set* D *in* E, *with* $u \le 0$ *on the boundary* D^b. *Then at any point* $x \in D^o$, *we have the estimate*

(18) $$u(x) \le \left\{ \frac{1}{\mu(D,x)} \sum_{z \in D^o} \mu(Y_z, z) \, |f(z)|^n \right\}^{1/n}$$

Note that by defining

(19) $$\rho(x) = \sup \left\{ \tau \mid B_\tau(x) \subset \hat{Y}_x \right\} ,$$

that is $\rho(x)$ is the internal radius of $\hat{Y}_x$, we infer from (18), using (15), the cruder estimate

(20) $$u \le \operatorname{diam} D \left\{ \sum_{D^o} \left(|f|/\rho \right)^n \right\}^{1/n} ,$$

which implies Theorem 1.1 of [12]. More generally, if $\hat{Y}_z \supset \tilde{Y}_z$ where, for each $z \in D^o$, $\tilde{Y}_z$ is a convex domain in $\mathbb{R}^n$, centered at z, we can estimate in (18), (see [2], [3], [5]) ,

$$(21)\qquad \frac{\mu(Y_z,z)}{\mu(D,x)} \leq C(n)\frac{|\hat{D}|}{|\tilde{Y}_z|} ,$$

where C(n) is a constant depending on n, and hence obtain from (18)

$$(22)\qquad u \leq \left\{C(n)|\hat{D}|\sum_{D^o}\frac{|f|^n}{|\tilde{Y}|}\right\}^{1/n} .$$

We also remark that in the estimates (18), (20) and (22) the summation over D^o may be replaced by summation over the upper contact set $\Gamma^+ = \Gamma_u^+$ defined by

$$\Gamma^+ = \left\{x \in D^o \mid \chi_u(x) \neq \emptyset \right\} .$$

The proof of Theorem 1 follows that of Theorem 2.1 in [12]. The main idea is to fix a point $x \in \Gamma^+$ with $u(x) > 0$ and a vector $p \in \chi_u(x)$, so that writing

$$(23)\qquad w(z) = u(z) - p\cdot(z - x) ,$$

we have $w(y) \leq w(x)$ for all $y \in D$. Using the difference inequality (17) and our hypotheses on L, we then have

$$(24)\qquad \begin{aligned} &\sum a(x,y)(w(x) - w(y)) \\ &= \sum a(x,y)(u(x) - u(y)) \\ &= - Lu(x) + c(x)u(x) \end{aligned}$$

$$\leq f(x)$$

Let $Y = Y_x$ be the set given by (16) and let $z_1, \dots, z_r$ denote the extreme points of Y so that each z_i , $i = 1, \dots, r$, is given by

$$z_i = x + a(x,y_i)(y_i - x) \tag{25}$$

for points $y_1, \dots, y_r \in D^o$. Defining a new function $\tilde{w}$ on $\{x, z_1, \dots, z_r\}$ by

$$\begin{aligned} \tilde{w}(x) &= w(x), \\ \tilde{w}(z_i) &= w(x) + a(x,y_i)(w(y_i) - w(x)), \qquad i = 1, \dots, r, \end{aligned} \tag{26}$$

it follows that

$$\chi_w(x) \subset \chi_{\tilde{w}}(x) \tag{27}$$

and by (23),

$$\sum_{i=1}^{r} (\tilde{w}(x) - \tilde{w}(z_i)) \leq f(x) \tag{28}$$

Consequently

$$\chi_{\tilde{w}}(x) \subset \chi_{f(x)k}(x) = f(x)\chi_k(x) \tag{29}$$

where k is the function (11) for D = Y, and hence, by (24) and (27),

$$|\chi_u(x)| \leq \mu(Y_x,x)\, |f(x)|^n . \tag{30}$$

The remainder of the proof of Theorem 1 is now standard, as replacing u by its positive part u^+, we have for any $x \in D^o$,

$$u(x)\, \chi_k(D) \subset \chi_u(D) = \chi_u(\Gamma^+) \tag{31}$$

and hence (18) follows.

The extension of Theorem 1 to unbalanced operators follows as in [12] and, as above, the considerations of Aleksandrov [2,3] can be used to refine the dependence on the domain D. In particular, we obtain in place of (18), an estimate of the form

$$(32) \qquad u(x) \le \nu(x,D)\, C(n,\|b\|)\, \|f\|$$

where, for a mesh function g, we have defined

$$(33) \qquad \|g\| = \|g\|_{n;D^\circ} = \left\{ \sum_{D^\circ} \mu(Y_z,z)|g(z)|^n \right\}^{1/n}$$

and ν and C are positive quantities depending respectively on x, D and n, $\|b\|$. In particular,

$$(34) \qquad \nu(x,D) \le C(n)\,\mathrm{diam}\, D \ ,$$

but, as with (20), we may replace the diameter of the domain D by the mean breadth of its convex hull $\hat{D}$, [2].

In our papers [8], [12] we proceed from the global maximum principle to the derivation of local estimates, including local maximum principles, Hölder estimates and Harnack inequalities. The original paper [8] treats the special case of a cubic mesh and the Hölder estimate there is applied to the proof of stability for nonlinear schemes in [9]. In the remainder of this article we describe the Harnack inequality and Liouville theorem which result from Theorem 1. For these results we need to impose non-degeneracy results on the operator L, analogous to uniform ellipticity, with the effect that, unlike Theorem 1, the level of generality is already present in [12]. Accordingly, the reader is referred to [12] for further details.

In order to simplify our presentation we shall restrict attention to the homogeneous equation

$$(35) \qquad Lu \equiv \sum a(x,y)u(y) = 0$$

with L positive and balanced on the whole mesh E. Moreover let us assume $c \equiv 0$, so that (35) may be written in the form

$$(36) \qquad u(x) = \sum_{y \neq x} p(x,y)u(y) \quad ,$$

by writing

$$(37) \qquad p(x,y) = \frac{a(x,y)}{\sum_{y \neq x} a(x,y)}$$

Any further conditions on L may then be expressed in terms of the coefficients p. Harnack and Liouville theorems relate to properties of non-negative solutions. Let us assume that u is non-negative in a domain D and there is a point $z \in D^0$ where u is positive so that without loss of generality we may assume $u(z) = 1$. In order to guarantee propagation of the positivity of u, we need to assume that points in D^0 are effectively linked through the operator L. That is, we assume for any two points $x, y \in D^0$, there exists points $x_0 = x, x_1, x_2, \ldots, x_k = y$ in D^0 such that

$$(38) \qquad p(x_i, x_{i+1}) \geq \lambda , \qquad i = 0, 1, \ldots, k-1 ,$$

for some positive constant λ. Then, by iteration of equation (36), we have for any $x \in D^0$,

$$(39) \qquad u(x) \geq \lambda^k$$

where $k = k(x,z)$. The estimate (39) shows immediately that u cannot vanish in D, but it is unsatisfactory in the sense that there is no control on the number k which should be expected to grow as the mesh is refined. To describe the size of the mesh E, we introduce quantities

$$(40) \qquad h(x) = \min_{y \neq x} |x - y| \quad , \quad \bar{h}(x) = \max_{p(x,y) \neq 0} |x - y|$$

and suppose there exist constants h_0, $\bar{h}_0$ such that

$$(41) \qquad h_0 \leq h \leq \bar{h} \leq \bar{h}_0$$

for all $x \in E$. It is reasonable to assume that the number k satisfies

$$(42) \qquad k(x,y) \leq \frac{k_0 |x-y|}{h(x)}$$

for some constant k_0, that is the number of steps to link two points is controlled by the ratio of the distance between them to the minimum mesh length. Clearly the condition yields no control on k nor therefore on the positivity of u in (39) as $h \to 0$. The latter is achieved by invoking Theorem1 and the measure-theoretic argument of Krylov-Safonov [6], [7]. For this, we need the further assumption,

$$(43) \qquad \rho \geq \rho_0 \quad ,$$

where ρ_0 is a positive constant and ρ is defined by (19), with a = p. As in [12], we shall express the Harnack inequality in terms of mesh balls. Namely, if R > 0 and $z \in \mathbb{R}^n$, we define the mesh ball $E_R(z)$ of radius R and center z by

(44) $$E_R(z) = \{x \in E \mid |x - z| < R\} \quad .$$

THEOREM 2. *Let* u *be a non-negative solution of equation* (35) *in* D *with* u(z) = 1 *at some point* $z \in D^o$. *Then if* $E_{2R}(z) \subset D$, $\bar{h}_o < R$, *we have*

(45) $$\min_{E_R(z)} u \geq \gamma$$

where γ *is a positive constant depending only on* n, k_o, λ, $\bar{h}_o/\rho_o$, $\bar{h}_o/h_o$.

From Theorem 2, we conclude immediately the following Liouville theorem.

THEOREM 3. *A non-negative solution of equation* (35) *in the whole mesh* E *must be a constant.*

Theorems 2 and 3 are slightly more general than the corresponding results, Theorem 1.2 and Corollary 1.3 in [12]. They follow from [12] by observing that the quantities $\lambda\rho$, $\lambda\rho^2$ in the estimates (1.13), (1.28) in [12] correspond to ρ, ρ^2 in our setup here.

REFERENCES

[1] A.D. Aleksandrov : *Certain estimates for the Dirichlet problem,* Dokl. Akad. Nauk. SSSR **134**(1960), 1001-1004, [Russian], English translation in Soviet Math. Dokl. **1**(1960),1151-1154 .

[2] A.D. Aleksandrov : *Uniqueness conditions and estimates for the solution of the Dirichlet problem,* Vestnik Leningrad Univ. **18**, 5-29, [Russian], English translation in Amer. Math. Soc. Transl. (2) **68**(1968), 89-119.

[3] A.D. Aleksandrov : *Majorization of solutions of second-order linear equations.* Vestnik Leningrad Univ. **21**(1966), 5-25 [Russian]. English translation in Amer. Math. Soc. Transl. (2) **68**(1968), 120-143.

[4] I. Ya. Bakel'man : *Theory of quasilinear elliptic equations.* Sibirsk. Mat. Zh. **2**(1961), 179-186. [Russian].

[5] I. J. Bakelman : *Convex analysis and nonlinear geometric elliptic equations,* Springer-Verlag, 1994.

[6] D. Gilbarg & N.S. Trudinger : *Elliptic partial differential equations of second order,* 2nd edition, Springer–Verlag,1983.

[7] N.Y. Krylov & M.V. Safonov : *Certain properties of solutions of parabolic equations with measurable coefficients,* Izv. Akad. Nauk SSSR **44**(1980) 161-175, [Russian], English translation in Math. USSR-Izv. **16**(1981),151-164 .

[8] H. J. Kuo and N. S. Trudinger : *Linear elliptic difference inequalities with random coefficients,* Math. Comp., **55**(1990), 37-53.

[9] H. J. Kuo and N. S. Trudinger : *Discrete methods for fully nonlinear elliptic equations,* SIAM J. Numer. Anal., **29**(1992), 123-135.

[10] H. J. Kuo and N. S. Trudinger : *On the discrete maximum principle for parabolic difference operators,* RAIRO Modél. Math. Anal. Numér., **27**(1993), 719-737.

[11] H. J. Kuo and N. S. Trudinger : *Local estimates for parabolic difference operators,* J. Differential Equations, to appear.

[12] H. J. Kuo and N. S. Trudinger : *Positive difference operators on general meshes,* to appear; (see also Australian National Univ., Centre for Math. Appl. Research Report MRR031-95)

[13] C. Pucci : *Limitazioni per soluzioni di equazioni ellittiche,* Ann. Mat. Pura Appl (4) **74**(1966), 15-30.

21. A Generic Uniqueness Result for the Stokes System and Its Control Theoretical Consequences

Jacques-Louis Lions Department of Mathematics, Collège de France, Paris, France

Enrique Zuazua Departamento de Matemática Aplicada, Universidad Complutense, Madrid, Spain

Abstract We consider the Stokes system in a three-dimensional cylinder $\Omega = G \times (0, L)$ of $\mathbb{R}^3$, G being a bounded smooth domain of $\mathbb{R}^2$. We study the following uniqueness property: If u is a solution of the Stokes system in $\Omega \times (0, T)$ with Dirichlet boundary conditions, T being a positive time, and its third component vanishes, i.e. $u_3 \equiv 0$, then can we ensure that $u \equiv 0$? We prove that this property does hold for "almost every" cross-section G. By using the Fourier expansion of solutions the problem is reduced to show that, generically with respect to the cross-section G, there is no eigenfunction of the Stokes system with third component identically zero. We also show how this uniqueness result can be applied to obtain approximate controllability properties of the Stokes system with scalar controls oriented in the direction (0,0,1) of $\mathbb{R}^3$. Actually, it was while working on the approximate controllability problem that we were led to the problem of generic uniqueness studied in the present paper. We also prove that the results above fail when G is a ball of $\mathbb{R}^2$.

1. INTRODUCTION AND MAIN RESULTS

Let G be a smooth, bounded domain of $\mathbb{R}^2$ and $L > 0$. We consider the three-dimensional cylinder $\Omega = G \times (0, L) \subset \mathbb{R}^3$. Let $T > 0$ be a positive number and let us consider the evolution Stokes system:

$$\begin{cases} u_t - \Delta u = -\nabla p & \text{in } \Omega \times (0,T) \\ \text{div } u = 0 & \text{in } \Omega \times (0,T) \\ u = 0 & \text{in } \partial\Omega \times (0,T) \\ u(x,0) = u^0(x) & \text{in } \Omega. \end{cases} \tag{1.1}$$

Let $\omega \subset \Omega$ be an open and non-empty subset of Ω.

This paper is devoted to study the following uniqueness property:

$$\textit{Does } u_3 = 0 \textit{ in } \omega \times (0,T), \textit{ imply that } u \equiv 0 \textit{ in } \Omega \times (0,T)? \tag{1.2}$$

The following main result shows that this holds for "almost every" smooth cross section G:

Theorem 1

Generically with respect to the smooth, bounded cross-section G, the unique-continuation property (1.2) holds for all $L > 0$ and $T > 0$.

More precisely, given any bounded domain G of $\mathbb{R}^2$ of class C^k, with $k \geq 3$, we can find another domain $\tilde{G}$ arbitrarily close to G in the C^k topology such that (1.2) holds in $\tilde{G}$ for any $T > 0$ and $L > 0$.

Remarks 1.1

a) The method of proof of Theorem 1 we shall use shows that, in fact, Theorem 1 holds if all the eigenvalues of the Laplacian in $H_0^1(G)$ are simple. This is well known to be a generic property of smooth domains (see J. H. Albert [A], A. M. Micheletti [M] and K. Uhlenbeck [U]).

b) In Section 3 we shall show that the uniqueness property (1.2) fails when G is a ball of $\mathbb{R}^2$.

c) If we impose periodic or Neumann type boundary conditions at $x_3 = 0, L$ for $x \in G$ and Dirichlet ones on $\partial G \times (0,L)$ it is easier to construct counter-examples. In fact, in those cases there is no cross-section G such that the analogue of (1.2) holds. To see this it is sufficient to take a solution (u,p) of the two-dimensional Stokes problem in $G \times (0,T)$, with $u = (u_1, u_2)$, u_j and p being functions of x_1, x_2 and t, and to observe that $(\tilde{u}, p)$ with $\tilde{u} = (u_1, u_2, 0)$ solves the Stokes problem in $\Omega \times (0,T)$ for those boundary conditions. This type of counter-example is developed in detail by Díaz and Fursikov [DF].

As a consequence of Theorem 1, applying Hahn-Banach Theorem we can deduce immediately an approximate controllability result for the Stokes system with controls supported in ω and oriented in the direction $e_3 = (0,0,1)$.

Indeed, let us denote by $v = v(x,t)$ the scalar control function and consider the controlled Stokes system:

$$\begin{cases} y_t - \Delta y = -\nabla p + v\chi_\omega e_3 & \text{in } \Omega \times (0,T) \\ \text{div } y = 0 & \text{in } \Omega \times (0,T) \\ y = 0 & \text{in } \partial\Omega \times (0,T) \\ y(x,0) = y^0(x) & \text{in } \Omega \end{cases} \tag{1.3}$$

where χ_ω denotes the characteristic function of ω.

Assume that

$$y^0 \in L^2_{\text{div}}(\Omega) = \left\{y^0 \in \left(L^2(\Omega)\right)^3 : \ \text{div } y^0 = 0\right\}$$

and let us define the reachable set

$$R(y^0;T) = \left\{y(x,T) : \ v \in L^2\left(\omega \times (0,T)\right)\right\}. \tag{1.4}$$

The system is said to be approximately controllable if $R(y^0,T)$ is dense in $L^2_{\text{div}}(\Omega)$ for any $y^0 \in L^2_{\text{div}}(\Omega)$.

We have the following result:

Theorem 2

If Ω, ω and T are such that the uniqueness property (1.2) holds, the system (1.3) is approximately controllable.

Remarks 1.2

Several open problems naturally arise. We list below some of them.

a) We consider the system given by

$$y_t - \Delta y + b(x,t)\nabla y = -\nabla p + v\chi_\omega e_3 \tag{1.5}$$

where $b(x,t)$ is a given (possibly smooth) vector function such that $\text{div} b(x,t) = 0$, with all other conditions in (1.3) unchanged.

Do we still have the same result than in Theorem 2, i. e. the generic uniqueness property? Or, in other words, *do we have the generic approximate controllability*?

b) Do we have generic approximate controllability when Ω is arbitrary in $\mathbb{R}^3$ and not of the particular cylindrical structure above?

c) Let us consider now the system given by

$$\begin{cases} y_t - \Delta y = -\nabla p & \text{in } \Omega \times (0,T) \\ \text{div } y = v & \text{in } \Omega \times (0,T) \\ y = 0 & \text{in } \partial\Omega \times (0,T) \\ y(x,0) = y^0(x) & \text{in } \Omega \end{cases}$$

with a *boundary control* v of the form

$$v = (0,0,v_3); \quad \int_\Gamma v_3 n_3 d\Gamma = 0.$$

Do we still have generic approximate controllability?

d) Let us now consider the "full" Navier-Stokes system

$$\begin{cases} y_t - \Delta y + y\nabla y = -\nabla p + v\chi_\omega e_3 & \text{in } \Omega \times (0,T) \\ \text{div } y = 0 & \text{in } \Omega \times (0,T) \\ y = 0 & \text{in } \partial\Omega \times (0,T) \\ y(x,0) = y^0(x) & \text{in } \Omega. \end{cases}$$

Do we still have generic approximate controllability? It has been conjectured (J. L. Lions [L3]) that the non linear terms in the Navier Stokes system actually *help* for the controllability (because, among other things, of the "mixing properties" they bring to the system). An interesting result by J. M. Coron [C1,2] for the $2D$ Euler equations goes in this direction.

The plan of the paper is as follows. We present the proof of Theorem 1 in Section 2 below. We show in Section 3 that there exist domains such that Theorem 1 is false making the "generic statement" a necessity. The proof of Theorem 2 is given in Section 4 and it is followed by further comments on (approximate) controllability questions.

2. PROOF OF THE UNIQUENESS RESULT

We proceed in several steps.

Step 1

Fist of all we observe that if (u,p) solve (1.1), then $p(x,t)$ is harmonic in Ω with respect to x for all $t \in (0,T)$. Indeed, applying the divergence operator to the first equation in (1.1) and taking into account that div $u = 0$ we deduce that

$$\Delta p = 0 \quad \text{in} \quad \Omega \times (0,T)$$

On the other hand, from the fact that

$$u_3 = 0 \quad \text{in} \quad \omega \times (0,T)$$

we deduce immediately that

$$\frac{\partial p}{\partial x_3} = 0 \quad \text{in} \quad \omega \times (0,T).$$

Therefore, by elliptic unique-continuation we deduce that

$$\frac{\partial p}{\partial x_3} \equiv 0 \quad \text{in} \quad \Omega \times (0,T) \tag{2.1}$$

i.e.

$$p = p(x_1, x_2, t) \quad \text{in} \quad \Omega \times (0,T) \quad \text{is independent of } x_3.$$

But then u_3 satisfies the heat equation

$$u_{3,t} - \Delta u_3 = 0 \quad \text{in} \quad \Omega \times (0,T).$$

Since $u_3 = 0$ in $\omega \times (0,T)$, by the parabolic unique continuation property

$$u_3 \equiv 0 \quad \text{in} \quad \Omega \times (0,T). \tag{2.2}$$

Notice that we have applied the unique continuation theorem to equations with constant coefficients. Thus, Holmgren's theorem suffices.

The semigroup generated by the Stokes system in $L^2_{\text{div}}(\Omega)$ is analytic. Therefore, the function

$$t \in (0,\infty) \longrightarrow u_3(t) \in L^2(\Omega)$$

is analytic. This fact combined with (2.2) implies that

$$u_3 \equiv 0 \quad \text{in} \quad \Omega \times (0,\infty). \tag{2.3}$$

We have now to show that (1.2) holds true when $u_3 \equiv 0$ in $\Omega \times (0,T)$ (and not only in $\omega \times (0,T)$).

Step 2

We now use a representation of the solution of (1.1) using the eigenvalues and eigenfunctions of the Stokes system (so that this method of proof does not apply to the situation described in Remarks 1.2, a).)

Let $\{\lambda_j\}$ be the distinct eigenvalues of the Stokes system (1.4). Let $m_j \geq 1$ be the multiplicity of λ_j. We denote by $\{\omega_j^k, \pi_j^k\}_{k=1}^{m_j}$ the eigenfunctions and eigenpressures associated to each eigenvalue λ_j. Le us recall that $\{\omega_j^k\}, j \in \mathbb{N}, k = 1, \ldots, m_j$ can be chosen to constitute an orthonormal basis of $L^2_{\text{div}}(\Omega)$.

Expanding the solution of (1.1) in Fourier series we obtain that

$$u = \sum_{j=1}^{\infty} \sum_{k=1}^{m_j} a_j^k e^{-\lambda_j t} \omega_j^k(x)$$

with

$$a_j^k = \int_\Omega u_0(x) \cdot \omega_j^k(x)\, dx$$

(by $\cdot$ we denote the scalar product in $\mathbb{R}^3$).

In particular, if $\omega_j^k = (\omega_{j,1}^k\,,\,\omega_{j,2}^k\,,\,\omega_{j,3}^k)$, by (2.3) we have:

$$u_3(x,t) = \sum_{j=1}^{\infty}\sum_{k=1}^{m_j} a_j^k\, e^{-\lambda_j t}\,\omega_{j,3}^k(x) = 0 \quad \text{in} \quad \Omega\times(0,\infty). \tag{2.4}$$

Multiplying in (2.4) by $e^{\lambda_1 t}$ and letting $t\to\infty$ we obtain that

$$\sum_{k=1}^{m_1} a_1^k\,\omega_{1,3}^k(x) = 0 \quad \text{in} \quad \Omega.$$

By induction over j, repeating this argument we deduce that

$$\sum_{k=1}^{m_j} a_j^k\,\omega_{j,3}^k(x) = 0 \quad \text{in} \quad \Omega\,, \quad \forall j\geq 1. \tag{2.5}$$

We define

$$z = \sum_{k=1}^{m_j} a_j^k\,\omega_j^k; \quad \sigma = \sum_{k=1}^{m_j} a_j^k\,\pi_j^k. \tag{2.6}$$

Then

$$\begin{cases} -\Delta z = \lambda_j z - \nabla\sigma & \text{in} \quad \Omega \\ z = 0 & \text{on} \quad \Gamma \\ \operatorname{div} z = 0 & \text{in} \quad \Omega \end{cases} \tag{2.7}$$

and by virtue of (2.5)

$$z_3 = 0 \quad \text{in } \Omega. \tag{2.8}$$

We shall have proven Theorem 1 if we verify the following Lemma:

Lemma 1

Generically with respect to the bounded, smooth cross-section G, for any $L>0$, there is no solution of (2.7)–(2.8) except $z=0$, $\nabla z = 0$.

Step 3

We write λ instead of λ_j in (2.7). Because of (2.8)

$$\frac{\partial\sigma}{\partial x_3} = 0$$

so that

$$\begin{cases} -\Delta z_1 = \lambda z_1 - \partial_1\sigma \quad \text{in} \quad \Omega \;\; ; \;\; z_1 = 0 \quad \text{on} \quad \partial\Omega \\ -\Delta z_2 = \lambda z_2 - \partial_2\sigma \quad \text{in} \quad \Omega \;\; ; \;\; z_2 = 0 \quad \text{on} \quad \partial\Omega \\ \sigma = \sigma(x_1,x_2)\,, \quad \partial_1 z_1 + \partial_2 z_2 = 0 \quad \text{in} \quad \Omega. \end{cases} \tag{2.9}$$

We define

$$z^n(x_1,x_2) = \int_0^L \sin\left(\frac{\pi\, n\, x_3}{L}\right) z(x_1,x_2,x_3)\, d\,x_3 \tag{2.10}$$

and

$$\sigma^n(x_1,x_2) = \int_0^L \sin\left(\frac{\pi\, n\, x_3}{L}\right) \sigma(x_1,x_2)\, d\,x_3 = \frac{L}{n\,\pi}(1-(-1)^n)\,\sigma(x_1,x_2). \tag{2.11}$$

Multiplying in (2.9) by $\sin\left(\frac{\pi\, n\, x_3}{L}\right)$ and integrating with respect to x_3 we deduce that

$$\begin{cases} -\Delta' z^n = \left[\lambda - \left(\frac{\pi\, n}{L}\right)^2\right] z^n - \nabla'\sigma^n & \text{in} \quad G \\ z^n = 0 & \text{on} \quad \partial G \\ \text{div}'\, z^n = 0 & \text{in} \quad G \end{cases} \tag{2.12}$$

where we have denoted by Δ', ∇' and div$'$ the Laplacian, gradient and divergence respectively in the variables (x_1, x_2).

Thus, (z^n, σ^n) are eigenfunctions of the two-dimensional Stokes system in G with eigenvalue $\mu_n = \lambda - \left(\frac{n\,\pi}{L}\right)^2$.

However, it is well-known that all the eigenvalues of the Stokes system are strictly positive. This implies

$$z^n = 0; \quad \nabla'\sigma^n = 0$$

except possibly if $\mu_n = \lambda - \left(\frac{n\,\pi}{L}\right)^2 > 0$ i.e.

$$n < \frac{L}{\pi}\sqrt{\lambda}$$

Then

$$z(x_1,x_2,x_3) = \frac{2}{L}\sum_{n<L/\pi\sqrt{\lambda}} \sin\left(\frac{\pi\, n\, x_3}{L}\right) z^n(x_1,x_2)$$

and

$$\sigma(x_1,x_2,x_3) = \frac{2}{L}\sum_{n<L/\pi\sqrt{\lambda}} \sin\left(\frac{\pi\, n\, x_3}{L}\right) \sigma^n(x_1,x_2).$$

Using (2.11) this last identity writes

$$\sigma = \frac{2}{\pi}\sum_{n<L/\pi\sqrt{\lambda}} (1-(-1)^n)\sin\left(\frac{n\pi x_3}{L}\right)\sigma$$

so that

$$\sigma = 0. \tag{2.13}$$

Then $\sigma^n = 0$. Therefore (2.12) reduces to

$$\begin{cases} -\Delta' z^n = \left[\lambda - \left(\frac{\pi n}{L}\right)^2\right] z^n & \text{in} \quad G \\ z^n = 0 & \text{on} \quad \partial G \\ \operatorname{div}' z^n = 0 & \text{in} \quad G. \end{cases} \tag{2.14}$$

In order to show Lemma 1, we cover the "genericity" part of that Lemma if we prove it under the hypothesis

$$\text{the spectrum of } -\Delta \text{ for Dirichlet in } G \text{ is simple} \tag{2.15}$$

(since we know, as recalled in Remarks 1.1, a), that (2.15) is generically true).

We then argue as in [LZ]. In view of (2.14), for every $n \in \mathbb{N}$, $n < L/\pi\sqrt{\lambda}$, there are two possibilities: either $\lambda - \left(\frac{\pi n}{L}\right)^2$ is not an eigenvalue of $-\Delta'$ in $H_0^1(G)$, in which case $z^n \equiv 0$, or it is an eigenvalue. If $\lambda - \left(\frac{\pi n}{L}\right)^2$ is an eigenvalue of $-\Delta'$ in $H_0^1(G)$ since we have assumed G to be such that the spectrum of $-\Delta'$ in $H_0^1(G)$ is simple, we conclude the existence of some constant $\alpha^n \in \mathbb{R}$ such that

$$z_1^n = \alpha^n z_2^n \quad \text{in} \quad G$$

But then, the condition $\operatorname{div}' z^n = 0$ implies that

$$\alpha^n \partial_1 z_2^n + \partial_2 z_2^n = 0 \quad \text{in} \quad G$$

Taking into account that $z_2^n = 0$ on ∂G this implies that $z_2^n \equiv 0$ in G.

We have proved that $z^n \equiv 0$ for all $n \in \mathbb{N}$. Thus $z \equiv 0$. This concludes the proof of Lemma 1 and Theorem 1.

Remark 2.1 We thus have proven Theorem 1 assuming (2.15). This hypothesis is by no means necessary as it is shown at the end of Section 4.

3. A COUNTEREXAMPLE

In this section we are going to show that the uniqueness property fails when G is a ball of $\mathbb{R}^2$ for any $L > 0$ and $T > 0$ and even if $\omega = \Omega$.

We claim that if G is a ball of $\mathbb{R}^2$ there exists an (actually there exist infinitely many) eigenfunction $z = (z_1, z_2)$, $z_j = z_j(x_1, x_2)$ of the Stokes system in G such that the corresponding pressure vanishes, i.e.

$$\begin{cases} -\Delta' z = \lambda z & \text{in} \quad G \\ z = 0 & \text{on} \quad \partial G \\ \operatorname{div}' z = 0 & \text{in} \quad G. \end{cases} \tag{3.1}$$

This is a known example (see, for instance, H. Lamb [La]), a simple proof of it being presented below.

Let us assume for a moment that this is true. Then, for any $n \in \mathbf{N}$,

$$\omega^n(x_1, x_2, x_3) = \left(\sin\left(\frac{n\,\pi}{L}\,x_3\right) z_1(x_1, x_2)\ ,\ \sin\left(\frac{n\,\pi}{L}\,x_3\right) z_2(x_1, x_2)\ ,\ 0\right)$$

is an eigenfunction of the Stokes system in $\Omega = G \times (0, L)$ with zero pressure and $\omega_3^n \equiv 0$. More precisely

$$\begin{cases} -\Delta\,\omega^n = \left(\lambda + \dfrac{n^2\pi^2}{L^2}\right)\omega^n & \text{in} \quad \Omega \\ \omega^n = 0 & \text{on} \quad \partial\Omega \\ \operatorname{div}\,\omega^n = 0 & \text{in} \quad \Omega \end{cases} \tag{3.2}$$

and

$$\omega_3^n \equiv 0 \quad \text{in} \quad G. \tag{3.3}$$

Then,

$$u(x,t) = e^{-(\lambda + n^2\pi^2/L^2)t}\,\omega^n(x)$$

is a solution of the Stokes system (1.1) such that $p \equiv 0$ and $u_3 \equiv 0$.

Let us now go back to the claim and show that there exists z such that (3.1) holds. Proceeding as in [LZ] we consider a radially symmetric eigenfunction $\varphi = \varphi(r)$ of the following fourth order problem:

$$\begin{cases} (\Delta')^2\,\varphi = -\lambda\,\Delta'\,\varphi & \text{in} \quad G \\ \varphi = \dfrac{\partial\,\varphi}{\partial\,\nu} = 0 & \text{on} \quad \partial\,G. \end{cases} \tag{3.4}$$

Of course, there are infinitely many radially symmetric eigenfunctions of (3.4).

We then define the vector field

$$z = \left(\frac{\partial\,\varphi}{\partial\,x_2}\ ,\ -\frac{\partial\,\varphi}{\partial\,x_1}\right). \tag{3.5}$$

Then, it is obvious that

$$z = 0 \quad \text{on} \quad \partial\,G$$

$$\operatorname{div}'\,z = 0 \quad \text{in} \quad G.$$

We have to show that

$$-\Delta'\,z = \lambda\,z \quad \text{in} \quad G. \tag{3.6}$$

In view of (3.4)-(3.5) we have

$$\frac{\partial}{\partial\,x_2}(\Delta' z_1 + \lambda z_1) - \frac{\partial}{\partial\,x_1}(\Delta' z_2 + \lambda z_2) = (\Delta')^2\varphi + \lambda\Delta'\varphi = 0 \quad \text{in} \quad G.$$

This implies the existence of $\pi = \pi(x_1, x_2)$ such hat

$$-\Delta' z = \lambda z + \nabla' \pi \quad \text{in} \quad G.$$

If is then sufficient to show that π can be chosen such that $\pi \equiv 0$.

Obviously, π satisfies

$$\begin{cases} \Delta' \pi = 0 & \text{in} \quad G \\ \dfrac{\partial \pi}{\partial \nu} = -\Delta' z \cdot \nu = \dfrac{\partial \Delta' \varphi}{\partial \tau} & \text{on} \quad \partial G \end{cases}$$

where τ is the unit tangent vector to ∂G. But since φ is radially symmetric, $\dfrac{\partial \Delta' \varphi}{\partial \tau} = 0$ on ∂G. Thus, π satisfies

$$\begin{cases} \Delta' \pi = 0 \quad \text{in} \quad G \\ \dfrac{\partial \pi}{\partial \nu} = 0 \quad \text{on} \quad \partial G \end{cases}$$

and therefore $\pi = 0$ up to an additive constant.

This concludes the proof of the counter-example.

4. SOME APPROXIMATE CONTROLLABILITY RESULTS AND FURTHER COMMENTS

4.1. Proof of Theorem 2

As we said in the Introduction, from Theorem 1 one can obtain easily Theorem 2. This can be done by means of Hahn-Banach Theorem or by the direct method introduced in [L2] and [FaPZ1,2]. We are going to briefly sketch the proof of Theorem 2 following the latter.

Let G be a smooth, bounded domain of $\mathbb{R}^2$ such that the uniqueness property (1.2) holds for all $L > 0$ and $T > 0$.

Given any $L > 0$ we consider the three-dimensional cylinder $\Omega = G \times (0, L)$ and we fix any $T > 0$.

We first observe that due to the linearity and the well posedness of the Stokes system in $L^2_{\text{div}}(\Omega)$, it is sufficient to prove the density of $R(y^0; T)$ in $L^2_{\text{div}}(\Omega)$ with $y^0 \equiv 0$.

Given any $y^1 \in L^2_{\text{div}}(\Omega)$ and $\varepsilon > 0$ we are going to construct a scalar control $v \in L^2(\Omega \times (0, T))$ such that the solution of (1.3) satisfies

$$\|y(T) - y^1\|_{L^2_{\text{div}}(\Omega)} \leq \varepsilon. \tag{4.1}$$

We consider the adjoint Stokes system

$$\begin{cases} -\varphi_t - \Delta \varphi = \nabla q & \text{in} \quad \Omega \times (0, T) \\ \varphi = 0 & \text{on} \quad \partial\Omega \times (0, T) \\ \text{div}\, \varphi = 0 & \text{in} \quad \Omega \times (0, T) \\ \varphi(T) = \varphi^0 & \text{in} \quad \Omega \end{cases} \tag{4.2}$$

with $\varphi^0 \in L^2_{\text{div}}(\Omega)$.

We then define on $L^2_{\text{div}}(\Omega)$ the functional

$$J(\varphi^0) = \frac{1}{2}\int_{\omega\times(0,T)} |\varphi_3|^2 \, dx\, dt + \varepsilon \|\varphi^0\|_{L^2_{\text{div}}(\Omega)} - \int_\Omega y^1 \cdot \varphi^0 \, dx. \tag{4.3}$$

If is easy to see that J is continous in $L^2_{\text{div}}(\Omega)$ and strictly convex.

On the other hand, proceeding as in [FaPZ1,2] we deduce easily that the uniqueness property (1.2) implies that

$$\liminf_{\|\varphi^0\|_{L^2_{\text{div}}(\Omega)}\to\infty} \frac{J(\varphi^0)}{\|\varphi^0\|_{L^2_{\text{div}}(\Omega)}} \geq \varepsilon. \tag{4.4}$$

Therefore, J is coercive in $L^2_{\text{div}}(\Omega)$ and there is a unique $\hat{\varphi}^0 \in L^2_{\text{div}}(\Omega)$ such that

$$J(\hat{\varphi}^0) = \min_{\varphi^0 \in L^2_{\text{div}}(\Omega)} J(\varphi^0). \tag{4.5}$$

If is then easy to see that, if $\hat{\varphi}$ is the solution of (4.2) and we set

$$v = \hat{\varphi}_3 \tag{4.6}$$

then, (4.1) holds.

If is also easy to see that $\hat{\varphi}_0 \equiv 0$ (and therefore $\hat{\varphi}_3 \equiv 0$) if and only if $\|y^1\|_{L^2_{\text{div}}(\Omega)} \leq \varepsilon$.

4.2. Bang-bang controls

Following the arguments of [FaPZ1,2,3] it can be proved that the approximate controllability holds with "quasi bang-bang controls".

Let us go back to the proof of Theorem 2 and let us consider instead of (4.3) the functional

$$J(\varphi^0) = \frac{1}{2}\left(\int_{\omega\times(0,T)} |\varphi_3| \, dx\, dt\right)^2 + \varepsilon \|\varphi^0\|_{L^2_{\text{div}}(\Omega)} - \int_\Omega y^1 \cdot \varphi^0 dx. \tag{4.7}$$

The functional J: $L^2_{\text{div}}(\Omega) \longrightarrow \mathbb{R}$ is continuous and strictly convex and also satisfies

$$\liminf_{\|\varphi^0\|_{L^2_{\text{div}}(\Omega)}\to\infty} \frac{J(\varphi^0)}{\|\varphi^0\|_{L^2_{\text{div}}(\Omega)}} \geq \varepsilon \tag{4.8}$$

Thus J achieves its minimum value at some $\hat{\varphi}^0 \in L^2_{\text{div}}(\Omega)$. As it was shown in [FaPZ1,2,3] there exists then some

$$\lambda(x,t) \in \left(\int_{\omega\times(0,T)} |\hat{\varphi}_3| \, dx\, dt\right) \text{sgn}\,(\hat{\varphi}_3(x,t))$$

such that the solution of (1.3) with $v = \lambda$ satisfies

$$\|y(T) - y^1\|_{L^2_{\text{div}}}(\Omega) \leq \varepsilon.$$

By sgn we have denoted the sign function:

$$\operatorname{sgn}(z) = \begin{cases} 1 & \text{if } z > 0 \\ [-1,1] & \text{if } z = 0 \\ -1 & \text{if } z < 0. \end{cases}$$

Rigorously speaking, to ensure that the control obtained in this way is of bang-bang form we would need to show that the set $\{(x,t) \in \Omega \times (0,T)\ ;\ \hat{\varphi}_3(x,t) = 0\}$ has zero Lebesgue measure. This is done in the following Theorem.

Theorem 3

Generically with respect to the smooth, bounded cross-section G, the following uniqueness property holds for all $L > 0$, $T > 0$ and for any measurable set $E \subset \Omega = G \times (0,L)$ with $\text{meas}(E) > 0$*:*

$$\textit{If } u \textit{ is a solution of (1.1) such that } u_3 = 0 \textit{ in } E \textit{ then } u \equiv 0 \textit{ in } \Omega \times (0,T). \tag{4.9}$$

When this uniqueness property holds, then system (1.3) is approximately controllable with bang-bang controls supported on E and oriented in the direction of e_3.

Proof of Theorem 3.

It is sufficent to prove the uniqueness result (4.9). The generic approximate controllability result then holds immediately, as above, by considering the functional:

$$J(\varphi^0) = \frac{1}{2}\left(\int_E |\varphi_3|\, dx\, dt\right)^2 + \varepsilon \|\varphi^0\|_{L^2_{\text{div}}(\Omega)} - \int_\Omega y^1 \cdot \varphi^0 dx. \tag{4.10}$$

Let (u,p) be a solution of (1.1) with initial data in $L^2_{\text{div}}(\Omega)$ such that

$$u_3 = 0 \quad \text{in } E.$$

(We may consider as well, equivalently, solutions (φ, q) of the adjoint system (4.2)). Then there exists a set $E_1 \subset (0,T)$ with $\text{meas}(E_1) > 0$, such that

$$\text{if } t \in E_1,\ u_3(\cdot,t) = \partial u_3/\partial t = \Delta u_3 = 0 \text{ on } F_t \subset \Omega, \quad \text{meas}(F_t) > 0. \tag{4.11}$$

On the other hand,

$$\Delta p(\cdot,t) = 0 \quad \text{in } \Omega,\ \forall t$$

and by virtue of (4.11),

$$\frac{\partial p(\cdot,t)}{\partial x_3} = 0 \text{ on } F_t,\ \forall t.$$

Then, since the zero set of non-trivial harmonic functions in Ω is of zero Lebesgue measure, we deduce that

$$\frac{\partial p(\cdot,t)}{\partial x_3} = 0 \text{ in } \Omega,\ \forall t \in E_1. \tag{4.12}$$

But $t \to u(t)$ is analytic from $t > 0 \to V = H^1_{0,\mathrm{div}}(\Omega)$, so that

$$t > 0 \to \nabla p(\cdot, t) \quad \text{is real analytic from} \, t > 0 \to V'$$

which together with (4.12) implies that

$$\frac{\partial p(\cdot, t)}{\partial x_3} = 0 \text{ in } \Omega \times (0, T). \tag{4.13}$$

(By $V = H^1_{0,\mathrm{div}}(\Omega)$ we denote the space $V = (H^1_0(\Omega))^3 \cap L^2_{\mathrm{div}}(\Omega)$ and by V' its dual.) Therefore,

$$\frac{\partial u_3}{\partial t} - \Delta u_3 = 0 \quad \text{in } \Omega \times (0, T), \quad u_3 = 0 \quad \text{in } \Gamma \times (0, T)$$

and since u_3 satisfies (4.11) it follows that (see, for instance F. H. Lin [Li])

$$u_3 = 0 \quad \text{in } \Omega \times (0, T).$$

We are in a situation in which Theorem 1 applies so that (4.9) follows.

4.3 Relaxing the assumption of simple spectrum

Our proof of Theorem 1 shows that given a cross-section G (not necessarily smooth) the uniqueness property (1.2) holds for all $L > 0$ and $T > 0$ simultaneously if and only if there is no eigenfunction of the Stokes system in G with Dirichlet boundary conditions such that the corresponding pressure vanishes, i.e. if $z = (z_1, z_2)$ satisfies

$$\begin{cases} -\Delta' z = \lambda z & \text{in} \quad G \\ z = 0 & \text{on} \quad \partial G \\ \mathrm{div}' z = 0 & \text{in} \quad G \end{cases} \tag{4.14}$$

then, necessarily, z has to be identically zero.

A sufficient condition to guarantee that this holds is to request all eigenvalues of the Laplacian in $H^1_0(G)$ to be simple. In view of the counterexample of section 3 this condition turns to be sharp in general.

The simplicity of the Dirichlet spectrum for the Laplacian is a generic property among smooth domains. We have then decided to formulate our uniqueness result as a generic property. However, it is clear that for a particular given domain G, even if it is not smooth, as soon as the spectrum of the Laplacian is simple we can conclude that the uniqueness property holds.

Furthermore, is some particular geometries, even if the spectrum of the Laplacian is not simple we can conclude, analyzing (4.14), that (1.2) holds.

Let G be the square $G = (0, \pi)^2 \subset \mathbb{R}^2$. It is well known that the spectrum of the Laplacian in $H^1_0(G)$ is not simple since there may exist two couples (n, m) and (n', m') in $\mathbb{N}^2$ such that

$$n^2 + m^2 = (n')^2 + (m')^2.$$

In particular we may take $(n', m') = (m, n)$ when $n \neq m$.

Let us consider two such couples and the corresponding eigenfunctions of the Laplacian:

$$\varphi(x_1, x_2) = \sin n\, x_1 \sin m\, x_2$$

$$\psi(x_1, x_2) = \sin n'\, x_1 \sin m'\, x_2.$$

It is clear that these functions (φ, ψ), or any linear combination of them, cannot be such that

$$\text{div}'(\varphi, \psi) = 0.$$

Indeed

$$\text{div}'(\varphi, \psi) = n \cos n\, x_1 \sin m\, x_2 + m' \sin n'\, x_1 \cos m'\, x_2$$

and it is clear that $\text{div}'(\varphi, \psi)$ can not be identically zero since since at $x_1 = 0$, $\text{div}'(\varphi, \psi) = n \sin m\, x_2$.

Thus, in this particular case, even if the spectrum of the Dirichlet Laplacian is not simple the uniqueness property (1.2) holds.

Acknowledgements The work of the second author was supported by Project PB93-1203 of the DGICYT (Spain) and Projects SC1*-CT91-0732 and CHRX-CT94-0471 of the European Community.

REFERENCES

[A] J. H. Albert, Genericity of simple eigenvalues for elliptic pde's, *Proc. AMS*, **48** (2) (1975), 413-418.

[C1] J.-M. Coron, Contrôlabilité exacte frontière de l'équation d'Euler des fluides parfaites incompressibles bidimensionnels, *C. R. Acad. Sci. Paris*, **317**, série **I**, (1993), 271-276.

[C2] J.-M. Coron, On the control of $2D$ incompressible perfect fluids, *J. Math. pures appl.* , to appear.

[DF] J. I. Díaz and A. Fursikov, Approximate controllability of the Stokes system on cylinders by external unidirectional forces, preprint.

[FaPZ1] C. Fabre, J. P. Puel and E. Zuazua, Contrôlabilité approchée de l'équation de la chaleur semilinéaire, *C. R. Acad. Sci. Paris*, **315**, série **I**, (1992), 807-812.

[FaPZ2] C. Fabre, J. P. Puel and E. Zuazua, Approximate controllability of the semilinear heat equation, Proc. Royal. Soc. Edinburgh, to appear.

[FaPZ3] C. Fabre, J. P. Puel and E. Zuazua, Contrôlabilité approchée de l'équation de la claleur linéaire avec des contrôles de norme L^∞ minimale, *C. R. Acad. Sci. Paris*, **316**, série **I** (1993), 679-684.

[La] H. Lamb, *Hydrodynamics*, 6th ed., Cambridge Univ. Press, 1932.

[L1] J. L. Lions, *Contrôlabilité exacte, stabilisation et perturbations de systèmes distribués. Tome 1. Contrôlabilité exacte*, RMA8, Masson, Paris, 1988.

[L2] J. L. Lions, Remarks on approximate controllability, *J. Analyse Math.*, **59** (1992), 103-116.

[L3] J. L. Lions, Are there connections between turbulence and controllability?, *Lect. Notes* 144 (1990), Springer Verlag, A. Bensoussan and J. L. Lions eds.

[LZ] J. L. Lions and E. Zuazua, Approximate controllability of a hydro-elastic coupled system, preprint.

[Li] F. H. Lin, Nodal Sets of Solutions of Elliptic and Parabolic Equations, Comm. pure and appl. Math., **44** (1991), 287-308.

[M] A. M. Micheletti, Perturbazione della spettro dell' operatore di Laplace in relazione ad una variazione del campo, *Ann. Scuola Norm. Sup. Pisa*, **26** (3) (1972), 151-169.

[U] K. Uhlenbeck, Generic properties of eigenfunctions, *American J. Math.*, **98** (4) (1976), 1059-1078.

22. On a Stefan Problem in a Concentrated Capacity

E. Magenes Dipartimento di Matematica, Università di Pavia, Pavia, Italy

Introduction

The Stefan problems in a "concentrated capacity" appear in the heat diffusion phenomena involving two adjoining three-dimensional bodies Ω and $\tilde{\Omega}$, such that in $\tilde{\Omega}$ a change of phase takes place and the thermal conductivity along the direction ν normal to the boundary Γ of Ω is much greater than in the other directions. Assuming to be infinite, the body $\tilde{\Omega}$, as far as heat diffusion is concerned, behaves like a manifold of dimension less tham three and can be identified with a subset of Γ. The mathematical formulation and many related results in some particular important geometrical situations have been first give in [11] (cfr. also [1], [19] and the references therein). In [14], [15] I studied the case where Ω is a regular bounded set of $\Re^n$ and $\tilde{\Omega} = \Gamma$ in a framework of Hilbert spaces. In the present paper I shall consider the case $\tilde{\Omega} = \Gamma_1$ where Γ_1 is a subset of Γ, open and regular in Γ, and on the remaining part of Γ the "thermal flux" coming from Ω is given. This case presents many different and more difficult questions that the previons one. The exact formulation and the main result are described in section 1. I express my thanks to G. Savare' for the helpful discussions on the subject of the present paper.

1. Notation and main results.

1.1 Let Ω be an open bounded regular set of $\Re^n$, $n \geq 2$, whose boundary Γ is an oriented connected C^∞ $(n-1)$-manifold. Let Γ_0 and Γ_1 be two disjoint open and regular sets in Γ such that $\Gamma = \Gamma_0 \cup \Gamma_1 \cup \Gamma_*$, where Γ_* is a $C^\infty(n\text{-}2)$-manifold. Let T be a positive number, $Q = \Omega \times \,]0,T[$,

$\Sigma = \Gamma \times]0,T[$, $\Sigma_i = \Gamma_i \times]0,T[$, $i = 0, 1$, $\Sigma_* = \Gamma_* \times]0,T[$. Let us denote by ν the "inward" (with respect to Ω) normal to Γ, by ∇ and Δ the gradient vector and the Laplace operator in $\Re^n$. Moreover we shall suppose that a (proper) C^∞-Riemannian structure g is defined on Γ_1 and we denote by Δ_g the Laplace Beltrami operator on Γ_1 with respect to g (see e g. [2]) and by $(u,v)_g$ the global scalar product with respect to g, either for u and $v \in L^2(\Gamma_1)$ or for $u \in H^{-1}(\Gamma_1)$ and $v \in H_0^1(\Gamma_1)$(1). We recall that $u \to \Delta_g u$ is a linear continuous and bijective operator from $H_0^1(\Gamma_1)$ onto $H^{-1}(\Gamma_1)$ and that

$$-(\Delta_g u, v)_g = (du, dv)_g \quad \forall u, v \in H_0^1(\Gamma_1) \tag{1.1}$$

where d is the exterior differential on Γ_1.

Let β be a function from $\Re$ to $\Re$ such that

$$\begin{aligned} &\beta(0) = 0, \ |\beta(\xi) - \beta(\eta)|^2 \le c_1(\beta(\xi) - \beta(\eta))(\xi - \eta)) \ \forall \xi, \ \eta \in \Re \\ &|\beta(\xi)| \ge c_2|\xi| - c_3 \ \forall \xi \in \Re \ (c_i, \ i = 1, 2, 3 \text{ positive numbers}) \end{aligned} \tag{1.2}$$

In order to fixe the ideas we shall take

$$\beta(\xi) = a(\xi - 1)^+ - b\xi^- \quad \forall \xi \in \Re \ \text{a, b positive numbers} \tag{1.3}$$

which has a precise physical meaning (Remark 1.1).

Our problem (P) is now formally defined by

$$(\mathrm{P}) \quad \begin{cases} \text{find } u:\Sigma_1 \to \Re \text{ and } \theta:Q \to \Re \text{ such that} \\ \dfrac{\partial \theta}{\partial t} - \Delta\theta = \varphi \text{ in } Q, \ \theta = \beta(u) \text{ on } \Sigma_1, \ \dfrac{\partial \theta}{\partial \nu} = 0 \text{ on } \Sigma_0 \\ \dfrac{\partial u}{\partial t} - \Delta_g \beta(u) = f + \dfrac{\partial \theta}{\partial \nu} \text{ on } \Sigma_1, \ \beta(u) = 0 \text{ on } \Sigma_* \\ \theta(\cdot, 0) = 0, \ u(\cdot, 0) = 0 \end{cases}$$

where φ and f are real functions defined respectively on Q and Σ_1.

REMARK 1.1: As we said in the introduction problem (P) is a mathematical modelization of a Stefan problem in the "concentrated capacity" Γ_1: u denotes the enthalpy density in Γ_1, θ the

---

(1) We shall use the L^2-Sobolev spaces H^s, $s \in \Re$, as $H^s(\Omega)$, $H^s(\Gamma_i)$, $H^s(0,T; E)$, (E Hilbert space), ...with the standard definitions and identifications (see e.g. [13], chap. 1)

temperature and $\theta = \beta(u)$ is the state equation. $\frac{\partial\theta}{\partial\nu}$ on Γ_1 is the "thermal flux" coming from Ω to Γ_1, that plays a role of an additional heat source in Γ_1. By the choice (1.3), we suppose that the temperature of the phase-change in Γ_1 is equal to 0 and that latent heat in Γ_1 is equal to 1. The Riemannian structure g on Γ_1 is related to the conductivity properties of Γ_1. The heat conduction in Ω is supposed to be modelized by the usual linear heat equation in the temperature θ and no phase-change is present in Ω.

Let us introduce the definition of "weak solution" of problem (P):

DEFINITION 1.1: Under the assumption

$$\varphi \in L^2(Q),\ f \in L^2(\Sigma_1) \tag{1.4}$$

$(u,\ \theta)$ is a "weak solution" of (P) if

$$u \in L^\infty(0,T;\ L^2(\Gamma_1)),\ \beta(u) \in L^2(0,T;\ H_0^1(\Gamma_1)),\ \theta \in L^\infty(0,T;\ L^2(\Omega)) \cap L^2(0,T;\ H^1(\Omega)) \tag{1.5}$$

$$\theta = \beta(u)\ \text{ on }\ \Sigma_1 \tag{1.6}$$

$$-\int_0^T\int_{\Gamma_1} u\frac{\partial z}{\partial t}\,d\sigma dt + \int_0^T (d\beta(u), dz)_g\,dt - \int_0^T\int_\Omega \theta\frac{\partial z}{\partial t}\,dxdt + \int_0^T\int_\Omega \nabla\theta\nabla z\,dxdt = \int_0^T\int_{\Gamma_1} fz\,d\sigma dt + \int_0^T\int_\Omega \varphi z\,dxdt \quad \forall z \in Z \tag{1.7}$$

where

$$Z = \left\{z \in H^{1,1}(Q),\ z|_\Sigma \in H^{1,1}(\Sigma),\ z(\cdot,\ T) = 0,\ z|_{\Sigma^*} = 0\right\}^{(2)} \tag{1.8}$$

REMARK 1.2: Let us remark that (1.6) is meaningful (for a.e. t in $]0,T[$ in the sense of $H^{\frac{1}{2}}(\Gamma_1)$), from the assumption (1.5). Also (1.8) is well defined by the "trace" theorems in $H^{1,1}(Q)$ and $H^{1,1}(\Sigma)$(cf. [13], chap. 4, n. .2). Moreover from (1.7), taking $z \in \mathcal{D}(Q)$, we obtain

$$\frac{\partial\theta}{\partial t} - \Delta\theta = \varphi \text{ in the sense of } \mathcal{D}'(Q),$$

---

(2) The spaces $H^{r,s}(Q)$, $H^{r,s}(\Sigma)$, $H^{r,s}(\Sigma_i)$, ..., $r,s \geq 0$, are defined as in [13]; f.i. $H^{r,s}(Q) = L^2(0,T;\ H^r(\Omega)) \cap H^s(0,T;\ L^2(\Omega))$, $H^{r,s}_{0,}(Q) = L^2(0,T;\ H^r_0(\Omega)) \cap H^s(0,T;\ L^2(\Omega))$, $H^{r,s}_{,0}(Q) = L^2(0,T;\ H^s(\Omega)) \cap H^s_0(0,T;\ L^2(\Omega))$, $H^{r,s}_{0,0}(Q) = H^{r,s}_{,0}(Q) \cap H^{r,s}_{0,}(Q)$

therefore, from (1.5) and (1.4), we have $\frac{\partial\theta}{\partial t} \in L^2(0,T;\ H^{-1}(\Omega))$ and consequently from (1.5) and a well known interpolation result we obtain

$$\theta \in C^0([0,T];\ L^2(\Omega)) \tag{1.9}$$

and we deduce from (1.7) (or more directly from the following Propostion 1.1)

$$\theta(\cdot,0) = 0 \text{ in the sense of } (1.9) \tag{1.10}$$

REMARK 1.3: In a "formal way" it is easy to deduce that if (u,θ) is a "weak solution", then (u,θ) is a solution of (P).

1.2 The main result of the present paper is now the following

THEOREM 1.1 Under the assumption (1.4) there exists a unique "weak solution" (u,θ) of (P).

We shall first prove the uniqueness property; and in order to do that let us introduce the following space

$$V = \{\eta \in H^1(\Omega),\ \eta|_\Gamma \in H^1(\Gamma),\ \eta = 0 \text{ in } \Gamma.\} \tag{1.11}$$

which, endowed with the natural structure, is a separable Hilbert space. Then, for every $\delta \in L^2(0,T)$ and every $\eta \in V$, let us define $l(t) = \int_t^T \delta(s)\, ds$ and take in (1.7) $z(x,\ t) = l(t)\eta(x)$. By an obvious integration by parts and since V is separable we obtain easily the following

PROPOSITION 1.1 If (u,θ) is a "weak solution" of (P), there exists a subset E of $]0,T[$, depending on (u,θ), with meas E=0, such that we have $\forall t \in]0,T[-E$

$$\begin{aligned} &\int_{\Gamma_1} u(t)\eta d\sigma + \int_0^t (d\beta(u(s)),\ d\eta)_g ds + \int_\Omega \theta(t)\eta dx + \int_0^t\int_\Omega \nabla\theta(s)\nabla\eta dx ds = \\ &= \int_0^t\int_{\Gamma_1} f(s)\eta d\sigma ds + \int_0^t\int_\Omega \varphi(s)\eta dx ds \quad \forall \eta \in V \end{aligned} \tag{1.12}$$

REMARK 1.4: From PROPOSITION 1.1 and (1.10), we also deduce that u verifies the initial condition $u(\cdot,\ 0) = 0$ in the following sense

$$\lim_{t\to 0,\ t\in]0,T[-E} \int_{\Gamma_1} u(t)\eta d\sigma = 0 \quad \forall \eta \in V \tag{1.13}$$

Now, by using PROPOSITION 1.1 and standard techniques as in [15] or in [16] (THEOREM 1.1), we can obtain the following

THEOREM 1.2 The "weak solution" of (P) is unique

2. A "mixed" linear problem for the heat equation

2.1 We start by studying the following "mixed" linear problem

$$\begin{gathered}\frac{\partial\theta}{\partial t}-\Delta\theta=\varphi \text{ in } Q\\ \theta=h \text{ on } \Sigma_1, \frac{\partial\theta}{\partial\nu}=0 \text{ on } \Sigma_0\\ \theta(\cdot,0)=0\end{gathered} \tag{2.1}$$

where $\varphi\in L^2(Q)$ and $h\in B^\sigma$ and B^σ is defined by

$$B^\sigma=L^2(0,T;\ H_0^{1-\sigma}(\Gamma_1))\cap H^{\frac{1}{2}-\frac{\sigma}{2}}(0,T;\ L^2(\Gamma_1)),\ 0<\sigma<\frac{1}{2} \tag{2.2}$$

endowed with the natural Hilbert structure.

Let us recall the following notations for the spaces, wich we shall use in the present paper. First of all, if E is a (real) Hilbert space, we shall denote by ${}_0H^{\frac{1}{2}}(0,T;E)$(3) the space of the restrictions to $]0,T[$ of the elements of $H^{\frac{1}{2}}(-\infty,T;\ E)$ which vanish for $t<0$ (cf. also [13], chap 3, n. 5.1 for the definition by interpolation and for properties). Similarly, interchanging the role of 0 and T we define the space ${}_TH^{\frac{1}{2}}(0,\ T;\ E)$ (cf. again [13], chap 3, n. 5.2). Moreover, denoting by P the differential operator $\frac{\partial}{\partial t}-\Delta$, we shall use the space ${}_0D_P^{2s,s}(Q)$, $0<s<1$, (cf [6], n.2 and n.4) of the restrictions v to $]0,T[$ of the elements $\tilde{v}\in L^2(\Re;H^{2s}(\Omega))\cap H^s(\Re;L^2(\Omega))$, such that $P\tilde{v}\in L^2(\Re;L^2(\Omega))$ and $\tilde{v}=0$ for $t<0$, endowed with the natural Hilbert structure.

Let us also introduce the space A^σ, $0<\sigma<\frac{1}{2}$, defined by

$$A^\sigma={}_0D_P^{1-\sigma,\frac{1}{2}-\frac{\sigma}{2}}(Q)\cap L^2(0,T;\ H^1(\Omega))\cap{}_0H^{\frac{1}{2}}(0,T;\ L^2(\Omega)) \tag{2.3}$$

(3) Sometime denoted also by ${}_{00}H^{\frac{1}{2}}(0,T;\ E)$

endowed with the natural Hilbert structure.

From [6], n. 2 and n. 4, for every $v \in A^\sigma$ we can define v and $\frac{\partial v}{\partial \nu}$ on Σ and we have

$$v|_\Sigma \in H^{\frac{1}{2}-\sigma,\frac{1}{4}-\frac{\sigma}{2}}(\Sigma),\quad \left.\frac{\partial v}{\partial \nu}\right|_\Sigma \in H^{-\frac{1}{2}-\sigma,-\frac{1}{4}-\frac{\sigma}{2}}(\Sigma)\ \ (\text{def} = (H^{\frac{1}{2}+\sigma,\frac{1}{4}+\frac{\sigma}{2}}(\Sigma))') \tag{2.4}$$

the application $v \to v|_\Sigma$ and $\left.\frac{\partial v}{\partial \nu}\right|_\Sigma$ being linear and continuous.

Now we can formulate the problem (2.1) in the following "weak sense"

find $v \in A^\sigma$ such that

$$\begin{gathered}\frac{\partial v}{\partial t} - \Delta v = \varphi \ \text{ in } Q \ (\text{in the distributional sense in } Q)\\ v = h \ \text{ on } \Sigma_1, \text{ in the sense of } H^{\frac{1}{2}-\sigma,\frac{1}{4}-\frac{\sigma}{2}}(\Sigma_1)\\ \frac{\partial v}{\partial \nu} = 0 \ \text{ on } \Sigma_0, \text{ in the sense of } H^{-\frac{1}{2}-\sigma,-\frac{1}{4}-\frac{\sigma}{2}}(\Sigma_0)\ \ (\text{def} = (H_{\Sigma_1}^{\frac{1}{2}+\sigma,\frac{1}{4}+\frac{\sigma}{2}}(\Sigma_0))')\end{gathered} \tag{2.5}$$

and we can prove the following

THEOREM 2.1: For every fixed σ, $0<\sigma<\frac{1}{2}$, and $h \in B^\sigma$, there exists one and only one $v \in A^\sigma$, "weak solution" of (2.1) in the sense of (2.5), and we have

$$\|v\|_{A^\sigma} \le c(\|h\|_{B^\sigma} + \|\varphi\|_{L^2(Q)}) \ \ (c \text{ independent of } h \text{ and } \varphi) \tag{2.6}$$

PROOF: In order to prove the existence of v we first study the following problem:

$$\frac{\partial v_1}{\partial t} - \Delta v_1 = \varphi \ \text{ in } Q,\ v_1 = \tilde{h} \ \text{ on } \Sigma,\ v_1(\cdot,0) = 0 \tag{2.7}$$

where $\tilde{h}$ denotes the function defined by

$$\tilde{h} = h \ \text{ on } \Sigma_1,\ \tilde{h} = 0 \ \text{ on } \Sigma - \Sigma_1 \tag{2.8}$$

Since $h \in B^\sigma$, we have $\tilde{h} \in H^{1-\sigma,\frac{1}{2}-\frac{\sigma}{2}}(\Sigma)$. Therefore we have (cf. [13], chap 4, n. 15.1, [6] n.2 and 4.) that there exists one and only one solution of (2.7) such that

$$v_1 \in {}_0D_P^{\frac{3}{2}-\sigma,\frac{3}{4}-\frac{\sigma}{2}}(Q) \tag{2.9}$$

and we also obtain that

$$\frac{\partial v_1}{\partial \nu} \in H^{-\sigma,-\frac{\sigma}{2}}(\Sigma)\ \ (\mathrm{def} = (H^{\sigma,\frac{\sigma}{2}}(\Sigma))') \tag{2.10}$$

Now let us consider the problem

$$\frac{\partial v_2}{\partial t} - \Delta v_2 = 0 \text{ in } Q,\ v_2 = 0 \text{ on } \Sigma_1,\ \frac{\partial v_2}{\partial \nu} = \frac{\partial v_1}{\partial \nu} \text{ on } \Sigma_0 \tag{2.11}$$

Since from (2.10) we deduce that $\frac{\partial v_1}{\partial \nu} \in H^{-\sigma,-\frac{\sigma}{2}}(\Sigma_0)$ $(\mathrm{def} = (H^{\sigma,\frac{\sigma}{2}}(\Sigma_0))')$, we can formulate (2.11) as in [13], chap 3, n. 5.1 and 5.2, choosing there $H = L^2(\Omega)$, $V = H^1_{\Gamma_1}(\Omega) = \{w \in H^1(\Omega),\ w|_{\Gamma_1} = 0\}$ and setting

$$\begin{aligned} \Phi^{\frac{1}{2}} &= L^2(0,T\ ;\ H^1_{\Gamma_1}(\Omega)) \cap {}_0H^{\frac{1}{2}}(0,T;\ L^2(\Omega)) \\ \Phi_*^{\frac{1}{2}} &= L^2(0,T;\ H^1_{\Gamma_1}(\Omega)) \cap {}_TH^{\frac{1}{2}}(0,T;\ L^2(\Omega)) \end{aligned} \tag{2.12}$$

We can indeed consider $\left.\frac{\partial v_1}{\partial \nu}\right|_{\Sigma_0}$ as an element of $(\Phi_*^{\frac{1}{2}})'$, because, using also the "trace" theorem for the space $H^{\frac{1}{2}+\sigma,\frac{1}{4}+\frac{\sigma}{2}}(Q)$ (cf. [13], chap. 4, n. 2.2) we have that $\forall z \in \Phi_*^{\frac{1}{2}}$

$$\left|<\frac{\partial v_1}{\partial \nu}, z>\right| \le c\left\|\frac{\partial v_1}{\partial \nu}\right\|_{H^{-\sigma,-\frac{\sigma}{2}}(\Sigma_0)} \|z\|_{H^{\sigma,\frac{\sigma}{2}}(\Sigma_0)} \le c\left\|\frac{\partial v_1}{\partial \nu}\right\|_{H^{-\sigma,-\frac{\sigma}{2}}(\Sigma_0)} \|z\|_{H^{\frac{1}{2}+\sigma,\frac{1}{4}+\frac{\sigma}{2}}(Q)} \le c\left\|\frac{\partial v_1}{\partial \nu}\right\|_{H^{-\sigma,-\frac{\sigma}{2}}(\Sigma_0)} \|z\|_{\Phi_*^{\frac{1}{2}}} \tag{2.13}$$

where <,> denotes the pairing between $H^{-\sigma,-\frac{\sigma}{2}}(\Sigma_0)$ and $H^{\sigma,\frac{\sigma}{2}}(\Sigma_0)$.

Therefore, from [13], chap 3, n. 5.2, we deduce that there exists one and only one $v_2 \in \Phi^{\frac{1}{2}}$ solution of (2.11).

Let us now set $v = v_1 - v_2$. Since $v_1 \in {}_0D_P^{\frac{3}{2}-\sigma,\frac{3}{4}-\frac{\sigma}{2}}(Q)$ with $Pv_1 = \varphi$ in Q and $v_2 \in \Phi^{\frac{1}{2}}$ with $Pv_2 = 0$ in Q, we obtain that $v \in A^\sigma$ and v is a "weak solution" of (2.1) and satisfies also (2.6)

In order to prove the uniqueness of such a v let us remark that if w is another "weak solution" of (2.1) then $v - w$ is a solution (in the same sense of [13], chap. 3, n. 5.2, alredy recalled for (2.11)) of the problem

$$\frac{\partial(v-w)}{\partial t} - \Delta(v-w) = 0 \text{ in } Q,\ v-w = 0 \text{ on } \Sigma_1,\ \frac{\partial(v-w)}{\partial \nu} = 0 \text{ on } \Sigma_0 \tag{2.14}$$

and then $v - w = 0$.

Since the solution v of THEOREM 2.1 belongs to A^σ, we have from (2.4) that

$$\left.\frac{\partial v}{\partial \nu}\right|_{\Sigma_1} \in H^{-\frac{1}{2}-\sigma,-\frac{1}{4}-\frac{\sigma}{2}}(\Sigma_1) \ \ (\text{def} = (H_{0,}^{\frac{1}{2}+\sigma,\frac{1}{4}+\frac{\sigma}{2}}(\Sigma_1))') \tag{2.15}$$

Moreover, if we consider φ fixed, v depends only on h; therefore we can introduce the operator $\Im$

$$h \to \Im(h) = \left.\frac{\partial v}{\partial \nu}\right|_{\Sigma_1} \ \text{from } B^\sigma \text{ into } H^{-\frac{1}{2}-\sigma,-\frac{1}{4}-\frac{\sigma}{2}}(\Sigma_1)) \tag{2.16}$$

and we have

$$\Im \text{ is continuous and maps bounded sets of } B^\sigma \text{ into bounded sets of } H^{-\frac{1}{2}-\sigma,-\frac{1}{4}-\frac{\sigma}{2}}(\Sigma_1) \tag{2.17}$$

3. The "regularized" problem (P_ε).

3.1 In order to prove the existence part of THEOREM 1.1 we introduce the "regularized" problem (P_ε), which is "formally" obtained by (P) replacing β with a suitable approximation β_ε, $\varepsilon > 0$; more precisely let β_ε be defined e.g. as follows (we fixe β given by (1.3))

$$\begin{gathered}\beta_\varepsilon(\xi) = \beta(\xi) \ \text{if} \ \xi \le 0; \ \beta_\varepsilon(\xi) = \beta(\xi) + \varepsilon \ \text{if} \ \xi \ge 1; \\ \beta_\varepsilon \in C^\infty(\Re), \ l\varepsilon \le \beta_\varepsilon'(\xi) \le L \ \ \forall\ \xi \ \ (l \text{ and } L \text{ positive numbers independent of } \varepsilon)\end{gathered} \tag{3.1}$$

We shall define the "weak solution" of (P_ε) by the same Definition 1.1, where we intend to replace β by β_ε. Let us remark that, if $(u_\varepsilon, \theta_\varepsilon)$ is a weak solution of (P_ε), then not only θ_ε satisfies (cf. REMARK 1.1)

$$\theta_\varepsilon \in C^0([0,T];\ L^2(\Omega)) \tag{3.2}$$

$$\theta_\varepsilon(0) = 0 \tag{3.3}$$

but, since β_ε verifies (3.1), also u_ε satisfies

$$u_\varepsilon \in L^\infty([0,T];\ H_0^1(\Gamma_1)). \tag{3.4}$$

Moreover we can obviously prove the analogue of PROPOSITION 1.1 and the uniqueness of the "weak solution" as in THEOREM 1.1.

3.2 Let us prove the existence of a solution of problem (P_ε). Using the operator $\Im$ introduced by (2.16), (P_ε) can be rewritten "formally" in the following sense

$$\begin{gathered}\text{given } f \text{ and } \varphi, \text{ find } u_\varepsilon \text{ such that} \\ \frac{\partial u_\varepsilon}{\partial t} - \Delta_g \beta_\varepsilon(u_\varepsilon) = f + \Im(\beta_\varepsilon(u_\varepsilon)) \ \text{on} \ \Sigma_1, \ \beta_\varepsilon(u_\varepsilon) = 0 \ \text{on} \ \Sigma_\bullet, \ u_\varepsilon(\cdot,0) = 0\end{gathered} \tag{3.5}$$

The idea for solving (3.5) is to use the Leray-Schauder method in a suitable Hilbert space, applied to the following family of problems depending on the paramenter α, $0 \le \alpha \le 1$:

given f, φ and w, find $z_{\varepsilon,\alpha}$ such that

$$\frac{\partial z_{\varepsilon,\alpha}}{\partial t} - \Delta_g \beta_\varepsilon(z_{\varepsilon,\alpha}) = f + \alpha \Im(\beta_\varepsilon(w)) \text{ on } \Sigma_1,\ \beta_\varepsilon(z_{\varepsilon,\alpha}) = 0 \text{ on } \Sigma_\bullet,\ z_{\varepsilon,\alpha}(\cdot, 0) = 0 \tag{3.6}$$

In order to do this let us take $w \in B^\sigma$ and introduce the Banach space

$$F = C^0([0,T];\ H^{-1}(\Gamma_1)) \cap L^2(0,T;\ H_0^1(\Gamma_1)) \cap {}_0H^{\frac{1}{2}}(0,T;\ L^2(\Gamma_1)) \tag{3.7}$$

endowed with the natural norm. We have the following

PROPOSITION 3.1: For every fixed $f \in L^2(\Sigma_1)$, $\varphi \in L^2(Q)$, $w \in B^\sigma$ and α, $0 \le \alpha \le 1$, there exists one and only one function $z_{\varepsilon,\alpha} \in F$ solution of (3.6) in the sense that

$$z_{\varepsilon,\alpha}(\cdot,0) = 0 \text{ in the sense of } C^0([0,T];\ H^{-1}(\Gamma_1)) \tag{3.8}$$

$$-\int_0^T\int_{\Gamma_1} z_{\varepsilon,\alpha}\frac{\partial q}{\partial t}d\sigma dt + \int_0^T (d\beta_\varepsilon(z_{\varepsilon,\alpha}), dq)_g\, dt = \int_0^T\int_{\Gamma_1} fq d\sigma dt + \alpha < \Im(\beta_\varepsilon(w)), q > \ \forall q \in U \tag{3.9}$$

where $U = \left\{q \in H^{1,1}(\Sigma_1),\ q(T) = 0, q|_{\Sigma_\bullet} = 0\right\}$ and <,> denotes, here and so on, the pairing between $H^{-\frac{1}{2}-\sigma, -\frac{1}{4}-\frac{\sigma}{2}}(\Sigma_1)$ and $H_0^{\frac{1}{2}+\sigma, \frac{1}{4}+\frac{\sigma}{2}}(\Sigma_1)$; moreover we have

$$\|z_{\varepsilon,\alpha}\|_F \le c(\varepsilon, \|\varphi\|_{L^2(Q)}, \|f\|_{L^2(\Sigma_1)}, \|w\|_{B^\sigma}) \tag{3.10}$$

where c(...), for every ε, depends continuously on the other variables.

PROOF.: By (3.1) it is easy to see that $\beta_\varepsilon(w) \in B^\sigma$ so that by (2.16) we have that $\Im(\beta_\varepsilon(w)) \in H^{-\frac{1}{2}-\sigma, -\frac{1}{4}-\frac{\sigma}{2}}(\Sigma_1)$. Let us denote for simplicity $\Im(\beta_\varepsilon(w))$ by ψ. Then we have to solve a nonlinear, non degenerate parabolic problem of the Stefan type on Σ_1 (we omit the indices ε, α, in $z_{\varepsilon,\alpha}$ when no misunderstanding is possible)

$$\frac{\partial z}{\partial t} - \Delta_g \beta_\varepsilon(z) = f + \alpha\psi \text{ on } \Sigma_1,\ \beta_\varepsilon(z) = 0 \text{ on } \Sigma_\bullet,\ z(\cdot,0) = 0,\ 0 \le \alpha \le 1 \tag{3.11}$$

with a righthand side $f + \alpha\psi$ in an unusual space, since $\psi \in H^{-\frac{1}{2}-\sigma, -\frac{1}{4}-\frac{\sigma}{2}}(\Sigma_1)$. If ψ belongs to $L^2([0,T];\ H^{-1}(\Gamma_1))$ then the problem (3.11) admits one and only one solution z in the space $H^1([0,T];\ H^{-1}(\Gamma_1)) \cap L^2([0,T];\ H_0^1(\Gamma_1))$ (cf. [8], [5], [12]) in the sense that

$$-\int_0^T\int_{\Gamma_1} z\frac{\partial q}{\partial t}d\sigma dt + \int_0^T (d\beta_\varepsilon(z), dq)_g dt = \int_0^T\int_{\Gamma_1} f d\sigma dt + \alpha << \psi, q >> \ \forall q \in U \text{ and } z(\cdot,0) = 0 \tag{3.12}$$

where <<.>> denotes the pairing between $L^2([0,T]; H^{-1}(\Gamma_1))$ and $L^2([0,T]; H_0^1(\Gamma_1))$ and moreover we have

$$\|z\|_{H^1([0,T]; H^{-1}(\Gamma_1)) \cap L^2([0,T]; H_0^1(\Gamma_1))} \leq c(\varepsilon, \|f\|_{L^2(\Sigma_1)}, \|\psi\|_{L^2([0,T]; H^{-1}(\Gamma_1))}) \tag{3.13}$$

where $c(\dots)$ is independent of α.

Let us remark that (3.11) can be written in the following way

$$\frac{\partial z}{\partial t} - \delta[\beta_\varepsilon'(z)dz] = f + \alpha\psi \text{ on } \Sigma_1,\ z = 0 \text{ on } \Sigma_\bullet,\ z(\cdot,0) = 0 \tag{3.14}$$

with

$$\beta_\varepsilon'(z) \in L^\infty(\Sigma_1) \text{ and } l\varepsilon \leq \beta_\varepsilon'(z) \leq L \text{ on } \Sigma_1 \tag{3.15}$$

Then we can apply the method and the results of [13], chap. 3, n. 5, considering z as the solution ζ of the linear problem

$$\frac{\partial \zeta}{\partial t} - \delta[\beta_\varepsilon'(z)d\zeta] = f + \alpha\psi \text{ on } \Sigma_1,\ \zeta = 0 \text{ on } \Sigma_\bullet,\ \zeta(\cdot,0) = 0 \tag{3.16}$$

with the notations of [13] (take $H = L^2(\Gamma_1)$, $V = H_0^1(\Gamma_1)$, $a(t, u, v) = (\beta_\varepsilon'(z)du, dv)_g$, $\Phi^{\frac{1}{2}} = L^2(0, T; H_0^1(\Gamma_1) \cap {}_0H^{\frac{1}{2}}(0, T; L^2(\Gamma_1))$ and $\Phi_\bullet^{\frac{1}{2}} = L^2(0, T; H_0^1(\Gamma_1)) \cap {}_TH^{\frac{1}{2}}(0, T; L^2(\Gamma_1))$

Therefore we deduce the following estimate, also taking into account that $\Phi_\bullet^{\frac{1}{2}} \subset H^{\frac{1}{2}+\sigma, \frac{1}{4}+\frac{\sigma}{2}}(\Sigma_1)$

$$\|z\|_{L^2([0,T]; H_0^1(\Gamma_1)) \cap {}_0H^{\frac{1}{2}}([0,T]; H^{-1}(\Gamma_1))} \leq c(\varepsilon)\ \|f + \alpha\psi\|_{(\Phi_\bullet^{\frac{1}{2}})'} \leq c(\varepsilon)\left\{\|f\|_{L^2(\Sigma_1)} + \|\psi\|_{H^{-\frac{1}{2}-\sigma, -\frac{1}{4}-\frac{\sigma}{2}}(\Sigma_1)}\right\} \tag{3.17}$$

with $c(\varepsilon)$ independent of α, since $0 \leq \alpha \leq 1$

Now a recent result [18] on the maximal regularity of weak solutions of abstract evolution inequalities, wich can be applied also to the Stefan type problems (cf. THEOREM 3, REMARK 1.6 of [18]), affirms that the mapping $f + \alpha\psi \to z$ is also uniformly continuous on the bounded sets of $H^{-\frac{1}{2}+\delta}(0,T; H^{-1}(\Gamma_1))$, $0 < \delta < \frac{1}{2}$ with values in $C^0([0,T]; H^{-1}(\Gamma_1))$ (independently of α)

Then if we take $\delta = \frac{1}{4} - \frac{\sigma}{2}$ we have that the mapping $f + \alpha\psi \to z$ is uniformly continuous on the bounded sets of $H^{-\frac{1}{4}-\frac{\sigma}{2}}(0,T; H^{-1}(\Gamma_1))$ with values in $C^0([0,T]; H^{-1}(\Gamma_1))$, independently of α, and we have the estimate:

$$\|z\|_{C^0([0,T]; H^{-1}(\Gamma_1))} \leq c(\varepsilon,\ \|f\|_{L^2(\Sigma_1)}, \|\psi\|_{H^{-\frac{1}{4}-\frac{\sigma}{2}}(0,T; H^{-1}(\Gamma_1))}) \leq c(\varepsilon,\ \|f\|_{L^2(\Sigma_1)}, \|\psi\|_{H^{-\frac{1}{2}-\sigma, -\frac{1}{4}-\frac{\sigma}{2}}(\Sigma_1)}) \tag{3.18}$$

independently of α. Therefore the mapping $f + \alpha\psi \to z$ can be uniquely extended to a continous mapping between these last two spaces. Hence, always if we suppose that $\psi \in L^2(0,T;\ H^{-1}(\Gamma_1))$, from (3.17) and (3.18), we obtain the final estimate:

$$\|z\|_F \leq c(\varepsilon, \|f\|_{L^2(\Sigma_1)}, \|\psi\|_{H^{-\frac{1}{2}-\sigma, -\frac{1}{4}-\frac{\sigma}{2}}(\Sigma_1)}) \tag{3.19}$$

independently of α.

Now if $\psi \in H^{-\frac{1}{2}-\sigma, -\frac{1}{4}-\frac{\sigma}{2}}(\Sigma_1)$ we can approximate ψ by a sequence $\{\psi_n\}$ of $\psi_n \in L^2(0,T;H^{-1}(\Gamma_1))$, solve (3.11) for these ψ_n finding a function z_n satisfying (3.12), and, using (3.19), the previos remarks on the mapping $f + \alpha\psi_n \to z_n$ and standard arguments of compactness and monotonicity, we can pass to the limit as $n \to \infty$ and obtain the existence of a solution z of (3.11) in the sense that $z \in F$ and satisfies

$$-\int_0^T\int_{\Gamma_1} z\frac{\partial q}{\partial t}\,d\sigma dt + \int_0^T (d\beta_\varepsilon(z), dq)_g\, dt = \int_0^T\int_{\Gamma_1} fq\,d\sigma dt + \langle \alpha\psi, q\rangle \quad \forall q \in U \text{ and } z(\cdot,0) = 0 \tag{3.20}$$

But (3.20) is exactly (3.9), if we recall that $\psi = \Im(\beta_\varepsilon(w))$. Moreover from (2.17) and (3.19) we obtain also (3.10).

Let us now consider z as a function only of w and α, with f, φ fixed:

$$z = \Phi(w,\alpha),\quad w \in B^\sigma,\quad \alpha \in [0,1] \tag{3.21}$$

Since $z \in F$ and $F \subset B^\sigma$, we can consider Φ as a family (depending on α) of operators from B^σ into itself and try to apply the Leray-Schauder method, taking in account that the injection of F into B^σ is compact. In order to do that we first prove the following

PROPOSITION 3.2: The operator Φ is continuous from B^σ into B^σ, uniformly with respect to $\alpha \in [0,1]$.

PROOF. Since the mapping $w \to \beta_\varepsilon(w)$ is continuous in B^σ, by the quoted result of [18] Φ is continuous from B^σ into $C^0([0,T];\ H^{-1}(\Gamma_1))$, uniformly with respect to α; moreover it is bounded from B^σ into F (by PROPOSITION 3.1), uniformly with respect to α. Since F is compactly embedded into B^σ, the desired continuity follows immediately.

Let us now prove the following

PROPOSITION 3.3: There exists a number $c(\varepsilon)$, indipendent of $\alpha \in [0,\ 1]$ and depending continuously on $\|f\|_{L^2(\Sigma_1)}$ and $\|\varphi\|_{L^2(Q)}$, such that for any fixed point u of Φ and $0 \leq \alpha \leq 1$ we have

$$\|u\|_{B^\sigma} \leq c(\varepsilon) \tag{3.22}$$

PROOF: Let u be a fixed point of Φ; then (PROPOSITION 3.1) we have $u \in F$ and u satisfies

$$-\int_0^T\int_{\Gamma_1} u\frac{\partial q}{\partial t}d\sigma dt + \int_0^T(d\beta_\varepsilon(u), dq)_g dt = \int_0^T\int_{\Gamma_1} fqd\sigma dt + \alpha < \Im(\beta_\varepsilon(u)), q > \ \forall q \in U \tag{3.23}$$

and

$$u(\cdot,0) = 0 \text{ in the sense of } C^0([0,T];\ H^{-1}(\Gamma_1)) \tag{3.24}$$

By defintiion of $\Im$ we have $\Im(\beta_\varepsilon(u)) = \left.\frac{\partial\theta}{\partial v}\right|_{\Sigma_1}$, where θ is the solution of the problem

$$\frac{\partial\theta}{\partial t} - \Delta\theta = \varphi \text{ in } Q,\ \theta = \beta_\varepsilon(u) \text{ on } \Sigma_1,\ \frac{\partial\theta}{\partial v} = 0 \text{ on } \Sigma_0,\ \theta(\cdot,0) = 0 \tag{3.25}$$

in the weak sense defined for problem (2.1). Therefore $\theta \in A^\sigma$ and we have $\left.\frac{\partial\theta}{\partial v}\right|_{\Sigma_1} \in H^{-\frac{1}{2}-\sigma,-\frac{1}{4}-\frac{\sigma}{2}}(\Sigma_1)$.

Moreover if $z \in Z$ (cf. (1.8)), $z|_{\Sigma_1} \in H_{0.}^{\frac{1}{2}+\sigma,\frac{1}{4}+\frac{\sigma}{2}}(\Sigma_1)$. Then, by density and "trace" arguments, following the method of [13], chap. 2 and 4 (cf. also [3] and 4]), we can obtain the "Green's formula":

$$-\int_0^T\int_\Omega \theta\frac{\partial z}{\partial t}dxdt + \int_0^T\int_\Omega \nabla\theta\nabla z dxdt = \int_0^T\int_\Omega \varphi z dxdt - < \frac{\partial\theta}{\partial v}, z > \ \forall z \in Z \tag{3.26}$$

Then, multiplying (3.26) by $\alpha \in [0,\ 1]$ and adding with (3.23) where we take $q = z|_{\Sigma_1}$, we obtain

$$\begin{aligned} -\int_0^T\int_{\Gamma_1} u\frac{\partial z}{\partial t}d\sigma dt + \int_0^T(d\beta_\varepsilon(u), dz)_g dt - \alpha\int_0^T\int_\Omega \theta\frac{\partial z}{\partial t}dxdt + \alpha\int_0^T\int_\Omega \nabla\theta\nabla z dxdt &= \\ = \int_0^T\int_{\Gamma_1} fz d\sigma dt + \alpha\int_0^T\int_\Omega \varphi z dxdt \ \ \forall z \in Z \end{aligned} \tag{3.27}$$

As in PROPOSITION 1.1, from (3.27) we can obtain that there exists a subset E of $]0,T[$, depending on $(u,\ \theta)$, with meas $E = 0$, such that $\forall t \in]0,T[\ -\ E$ we have

$$\begin{aligned} \int_{\Gamma_1} u(t)\eta d\sigma + \int_0^t (d\beta_\varepsilon(u, d\eta))_g ds + \alpha\int_\Omega \theta(t)\eta dx + \alpha\int_0^t\int_\Omega \nabla\theta(s)\nabla\eta dxds &= \\ = \int_0^t\int_{\Gamma_1} f(s)\eta d\sigma ds + \alpha\int_0^t\int_\Omega \varphi(s)\varphi dxds \ \ \forall\eta \in V \end{aligned} \tag{3.28}$$

Then we can take in (3.28) $\eta = \theta(t)$ (since $\beta_\varepsilon(u) = \theta$ on Σ_1) and finally we obtain the estimate

$$\|u\|_{L^\infty(0,T;\ L^2(\Gamma_1))} + \|\beta_\varepsilon(u)\|_{L^2(0,T;\ H_0^1(\Gamma_1))} + \alpha\|\theta\|_{L^\infty(0,T;\ H^1(\Omega))} \le c \tag{3.29}$$

(c independent of α and ε) and moreover, recalling that β_ε satisfies (3.1),

$$\|u\|_{L^2(0,T;\ H_0^1(\Gamma_1))} \le c(\varepsilon) \ \ (c(\varepsilon) \text{ independent of } \alpha) \tag{3.30}$$

Now from (3.29) we deduce that

$$\alpha\|\Delta\theta\|_{L^2(0,T;\,H^{-1}(\Omega))} \le c \text{ (c independent of } \alpha \text{ and } \varepsilon) \tag{3.31}$$

and then, since $\frac{\partial\theta}{\partial t} - \Delta\theta = \varphi$ in Q

$$\alpha\left\|\frac{\partial\theta}{\partial t}\right\|_{L^2(0,T;\,H^{-1}(\Omega))} \le c \text{ (c independent of } \alpha \text{ and } \varepsilon) \tag{3.32}$$

Therefore from (3.29) and (3.32) we obtain that $\alpha\theta \in C^0([0,T];\, L^2(\Omega)) \cap {}_0H^{\frac{1}{2}}(0,T;\, L^2(\Omega))$ with the estimate

$$\alpha\|\theta\|_{C^0([0,T];\,L^2(\Omega)) \cap {}_0H^{\frac{1}{2}}(0,T;\,L^2(\Omega))} \le c \text{ (c independent of } \alpha \text{ and } \varepsilon) \tag{3.33}$$

Moreover $\alpha\theta \in A^\sigma$ and from the "trace" theorem of [6] still used, we have $\alpha\left\|\frac{\partial\theta}{\partial\nu}\right\|_{H^{\frac{1}{2}-\sigma,\frac{1}{4}-\frac{\sigma}{2}}(\Sigma)} \le c$ (c independent of α and ε) and in conclusion

$$\alpha\|\Im(\beta_\varepsilon(u))\|_{H^{\frac{1}{2}-\sigma,\frac{1}{4}-\frac{\sigma}{2}}(\Sigma_1)} \le c \text{ (c independent of } \alpha \text{ and } \varepsilon) \tag{3.34}$$

Now we can apply the same techniques as in the proof of PROPOSITION 3.1 to the problem (3.23) and, taking in account the estimates (3.29), (3.33), (3.34) we conclude the proof.

We can also obtain the estimate

$$\|u\|_{C^0([0,T];\,H^{-1}(\Gamma_1))} \le c(\varepsilon) \text{ (}c(\varepsilon)\text{ independent of } \alpha) \tag{3.35}$$

taking into account the estimate (3.34) and the results already quoted of [18].

Now we can apply the Leray-Schauder method, since (3.23) has a unique solution for $\alpha = 0$ and reduces to (3.5) for $\alpha = 1$ and the injection of F into B^σ is compact. We conclude that there exists one weak solution of problem (P_ε). More precisely, recalling also the uniqueness already proved (cf. n. 3.1), we have the following

THEOREM 3.1: Under the assumptions $\varphi \in L^2(Q)$ and $f \in L^2(\Sigma_1)$, there exists one and only one pair $(u_\varepsilon,\, \theta_\varepsilon)$ such that

$$\begin{gathered} u_\varepsilon \in C^0([0,T];\, H^{-1}(\Gamma_1)) \cap L^\infty(0,T;\, L^2(\Gamma_1)) \cap L^2(0,T;\, H_0^1(\Gamma_1)) \cap {}_0H^{\frac{1}{2}}(0,T;\, L^2(\Gamma_1)) \\ \theta_\varepsilon \in C^0([0,T];\, L^2(\Omega)) \cap L^2(0,T;\, H^1(\Omega)) \cap {}_0H^{\frac{1}{2}}(0,T;\, L^2(\Omega)) \end{gathered} \tag{3.36}$$

$$u_\varepsilon(\cdot,0) = 0,\ \theta_\varepsilon(\cdot,0) = 0,\ \theta_\varepsilon = \beta_\varepsilon(u_\varepsilon) \text{ on } \Sigma_1 \tag{3.37}$$

$$-\int_0^T\int_\Gamma u_\varepsilon \frac{\partial z}{\partial t} d\sigma dt + \int_0^T (d\beta_\varepsilon(u_\varepsilon), dz)_g dt - \int_0^T\int_\Omega \theta_\varepsilon \frac{\partial z}{\partial t} dxdt + \int_0^T\int_\Omega \nabla\theta_\varepsilon \nabla z dxdt = \tag{3.38}$$
$$= \int_0^T\int_{\Gamma_1} fx d\sigma dt + \int_0^T\int_\Omega \varphi z dxdt \quad \forall z \in Z;$$

moreover $(u_\varepsilon, \theta_\varepsilon)$ satifies

$$\|u_\varepsilon\|_{L^\infty(0,T;\, L^2(\Gamma_1))} + \|\beta_\varepsilon(u_\varepsilon)\|_{L^2(0,T;\, H_0^1(\Gamma_1))} + \|\theta_\varepsilon\|_{L^\infty(0,T;\, L^2(\Omega))} + \|\theta_\varepsilon\|_{L^2(0,T;\, H^1(\Omega))} + \|\theta_\varepsilon\|_{{}_0H^{\frac{1}{2}}(0,T;\, L^2(\Omega))} \leq c \tag{3.39}$$

where c is indipendent of ε and depends continuously on $\|f\|_{L^2(\Sigma_1)}$ and $\|\varphi\|_{L^2(Q)}$.

4. Existence of the solution of problem (P). Concluding remarks.

4.1 By standard techniques of compactness and monotonicity well known for the usual Stefan problem, from THEOREM 3.1 we can pass to the limit as $\varepsilon \to 0$ and obtain a pair (u, θ) of functions satisfying the DEFINITION 1.1 of weak solution of problem (P). Therefore, using also THEOREM 1.2, we obtain the proof of THEOREM 1.1-

4.2 We conclude with some remarks and open questions

a) If the "concentrated capacity" coincide with Γ then the problem can be studied (cf. [14] [15]) in an easier way, since the "mixed" problem (2.1) is replaced by a problem with "Dirichlet conditions" on Σ, more precisely by the problem

$$\frac{\partial\theta}{\partial t} - \Delta\theta = \varphi \text{ in } Q, \ \theta = h \text{ on } \Sigma, \ \theta(\cdot,0) = 0 \tag{4.1}$$

which is a more regular problem.

b) In the same condition (the concentrated capacity = Γ) it is also possible (cf [16]) to replace the linear heat equation in Q by a nonlinear one, introducing a change of phase also in Ω. If we denote by v the enthalpy density in Ω and by $\theta = \gamma(v)$ the state equation in Ω (γ satisfing a condition similar to (1.2)) then formally problem (P) is replaced by the following one

$$\frac{\partial v}{\partial t} - \Delta\gamma(v) = \varphi \text{ in } Q, \ \gamma(v) = \beta(u) \text{ in } \Sigma, \ v(\cdot,0) = 0$$
$$\frac{\partial u}{\partial t} - \Delta_g\beta(u) = f + \frac{\partial\gamma(v)}{\partial v} \text{ in } \Sigma, \ u(\cdot,0) = 0 \tag{4.2}$$

c) An open question is now to extend the present paper, in the case where the "concentrated capacity" is still Γ_1, to the situation similar to b), where a phase-change is present also in Ω. "Formally" the new problem is the following one

find (u,v) such that

$$\frac{\partial v}{\partial t} - \Delta\gamma(v) = \varphi \text{ in } Q,\ \gamma(v) = \beta(u) \text{ on } \Sigma_1,\ \frac{\partial\gamma(v)}{\partial\nu} = 0 \text{ on } \Sigma_0,\ v(\cdot,0) = 0 \tag{4.3}$$

$$\frac{\partial u}{\partial t} - \Delta_g\beta(u) = f + \frac{\partial\gamma(v)}{\partial\nu} \text{ on } \Sigma_1,\ \beta(u) = 0 \text{ on } \Sigma_*,\ u(\cdot,0) = 0$$

d) The extension of the present paper to the case where the initial conditions $\theta(\cdot,0) = 0$ and $u(\cdot,0) = 0$ and the boundary condition $\beta(u) = 0\ \Sigma_*$, and $\frac{\partial\theta}{\partial\nu} = 0$ on Σ_0 are replaced by non homogeneous conditions

$$\theta(\cdot,0) = \theta_0,\ u(\cdot,0) = u_0,\ \beta(u) = \tilde{\theta}_0 \text{ on } \Sigma_*,\ \frac{\partial\theta}{\partial\nu} = F_0 \text{ on } \Sigma_0 \tag{4.4}$$

under suitable assumptions on θ_0, u_0, $\tilde{\theta}_0$, F_0, does not seem to present particular difficulties, but it is not a straightforward consequence of the present paper.

e) It is also interesting to consider the case where on Σ_0 a Dirichlet condition is given instead of a flux condition for θ, (for instance the homogeneous condition $\theta = 0$ on Σ_0). This case seems easier since it can be considered to be similar to the case a).

f) Another important question is the continuity of the temperature θ. In the usual Stefan problem this property is a classical result in the case of two phases, corresponding to β given by (1.3) [7] (cf. also [9]) and in the general case of more than two phases is a very recent result [10]. The question for the Stefan problem in a concentrated capacity has been discussed in [17] by the method of [9], when the concentrated capacity coincides with Γ, both for the cases a) and b), when β and γ are of the type (1.3). The result of [17] can also be used in our present problem for the continuity of θ at the points of Σ_1. Obviously for the points of Σ_0 or of Q the continuity of θ is a classical result related to the heat equation. On the contrary the continuity at the points of Σ_* needs to be investigated more deeply.

References

[1] D. Andreucci: Existence and uniqueness of solutions to a concentrated capacity problem with change of phase, *Europ. J. Appl. Math.*, 1 (1990), 339-351.

[2] T. Aubin: *Non linear analysis on manifolds. Monge-Ampère equations*, Springer, New York, 1982.

[3] C. Baiocchi: Problemi misti per l'equazione del calore, *Rend. Sem. Mat. Fis. di Milano*, XLI (1971), 3-38.

[4] C. Baiocchi, A. Capelo: *Variational and quasi-variational inequalities, Applications to free-boundary problems*, J. Wiley, New York, 1984.

[5] H. Brèzis: Monotonicity Methods in Hilbert Spaces and some Applications to nonlinear Partial Differential Equations, *Contributions to nonlinear functional analysis*, Pergamon Press, New York (1971), 101-155.

[6] F. Brezzi: Sul metodo di penalizzazione per operatori parabolici, *Rend. Ist. Lombardo*, Sc. A 107 (1973), 241-256.

[7] L. A. Caffarelli, L. Evans: Continuity of the temperature in the two-phase Stefan problems, *Arch. Rat. Mech. Anal.*, 81, 3 (1983), 199-220.

[8] A. Damlamian: On the Stefan problem: the variational approach and some applications, *Banach Center Publ.*, 13 (1985), 253-275.

[9] E. Di Benedetto: Continuity of weak-solutions to certain singular parabolic equations, *Ann. Mat. Pura Appl.*, IV, 130 (1982), 131-176.

[10] E. Di Benedetto, V. Vespri: Continuity for bounded solutions of multiphase Stefan problem, *Rend. Acc. Lincei Matem. e Applic.*, IX, vol. V (1994), 297-302.

[11] A. Fasano, M. Primicerio, L. Rubinstein: A model problem for heat conduction with a free boundary in a concentrated capacity, *J. Inst. Maths. Applics.*, 26 (1980), 327-347.

[12] O. Ladyzhenskaya, V. Solonnikov, N. Uraltzeva: *Linear and quasi-linear equations of parabolic type*, Trans. Amer. Math. Soc., 23, Providence, 1968.

[13] J.L. Lions, E. Magenes: *Non homogeneous boundary value problems and applications*, Vol. 1 and 2, Springer-Verlag, Berlin, 1972.

[14] E. Magenes: On a Stefan problem on the boundary of a domain. *Partial differential equations and related subjects*. (M. Miranda Ed.). Longman Scient. Techn. 1992, 209-226.

[15] E. Magenes: Some new results on a Stefan problem in a concentrated capacity. *Rend. Acc. Lincei, Matem. e Applic.*, s. 9, v. 3 (1992), 23-34.

[16] E. Magenes: The Stefan problem in a concentrated capacity. Atti Simp. Int. su *Problemi attuali dell'Analisi e della Fisica Matematica*. (P.E. Ricci editor). Dipartimento di Matematica. Università "La Sapienza", Roma. 1993, 155-182.

[17] E. Magenes: Regularity and approximation properties for the solution of a Stefan problem in a concentreted capacity, Proc. Int. Workshop *Variational Methods, Nonlinear Analysis and Differential Equations*, E.C.I.G., Genova 1994, 88-106.

[18] G. Savarè: Weak solutions and maximal regularity for abstract evolution inequalities, to appear in *Advances in Math. Sc. and Appl.*

[19] M. Shillor: Existence and continuity of a weak solution to the problem of a free boundary in a concentrated capacity, *Proc. Roy. Soc. Edinburgh*, Sect. A. 100 (1985), 271-280.

23. Total Total Internal Reflection

Keith Miller Department of Mathematics, University of California, Berkeley, California

1 INTRODUCTION

Another title for this paper might be "blocking the tunnel effect in total internal reflection for a certain wave equation", or alternatively "nonunique continuation for a reduced wave equation in self-adjoint divergence form with Hölder continuous coefficients".

In [1] the author constructed an example of backward nonuniqueness for a uniformly parabolic equation

$$u_t = \nabla \cdot (\mathcal{A}(x, y, t)\nabla u) \tag{1.1}$$

on the half space $\Lambda = R^2 \times [0, \infty)$, and an example of nonunique continuation for a uniformly elliptic equation

$$\nabla \cdot (\mathcal{A}(x, y, z)\nabla u) = 0. \tag{1.2}$$

on the half space $\Lambda = R^2 \times [0, \infty)$. The solution $u(x, y, z)$ of (1.2) is periodic in x and y, C^∞ on Λ, and dies out to be identically zero for $z \geq$ a certain finite depth Z. Moreover, this can be accomplished with a symmetric coefficient matrix $\mathcal{A}$ which stays arbitrarily close to the identity I on Λ. See [2] for more recent references.

In §2 we show that the construction for (1.2) can be extended, with only rather modest modifications, to a certain equation of the form

$$\nabla \cdot (\mathcal{A}(x,y,z)\nabla u) + 1^2 u = 0. \tag{1.3}$$

This equation (1.3) is the "reduced wave equation" associated with the wave equation

$$v_{tt} = \nabla \cdot (\mathcal{A}(x,y,z)\nabla v). \tag{1.4}$$

That is, consider solutions v of (1.4) which are sinusoidal in t with a circular frequency of $\omega = 1$, but with amplitude and phase which varies with position, i.e.,

$$\begin{aligned} v(x,y,z,t) &= a(x,y,z)\cos t + b(x,y,z)\sin t, \\ &= \mathrm{Re}(u(x,y,z)\cdot e^{-it}), \text{ where} \\ u(x,y,z) &= a(x,y,z) + i\, b(x,y,z). \end{aligned} \tag{1.5}$$

(The use of complex notation is purely a notational convenience to keep track of sine and cosine identities more easily.) Then the v in (1.5) is a solution of (1.4) if and only if the u in (1.5) (and hence the a and b separately) is a solution of (1.3).

In §3 we remind the reader of the well known phenomenae of "total internal reflection" and "the tunnel effect" for (1.4) with an "optically light" gap between two "optically dense" half spaces. That is, let

$$\begin{aligned} \mathcal{A}(x,y,z) &= I \text{ for } 0 < z < G, \\ &= c^2 I \text{ for } z < 0 \text{ or } z > G, \text{ with a constant } c < 1. \end{aligned} \tag{1.6}$$

If the gap width G is infinite then an incident plane wave coming from $z = -\infty$ with an angle of incidence θ greater than arcsin c is "totally reflected". If G is finite, however, then such an incident wave "tunnels" (with exponential decay) across the gap and yields a tiny transmitted plane wave on the other side.

In §4 we note that the "starting values" (2.2) for our $u(x,y,z)$ of §2 result from total internal reflection for the sum of two incident plane waves at angles of incidence $+\theta$ and $-\theta$. Thus, if the gap width G in (1.6) is chosen greater than the annihilation depth Z of §2, then the tunnel effect for such an incident "double wave" can be completely blocked by slightly perturbing the coefficients in the gap to those values $\mathcal{A}(x,y,z)$ constructed in §2.

It is extremely fitting that this particular paper should be dedicated to Carlo Pucci. He was my teacher and mentor in the fields of ill-posed problems and of elliptic equations. Moreover, it was he who made possible the setting (the CNR villa at Via Santa Marta 13a) where much of my best work (and that of many other young mathematicians) was done. It was there in the spring of '72, sitting in a big chair before a sunny southern window, recovering from my luncheon wine pressed in the farm yard just outside, and stimulated by frequent afternoon discussions with Carlo throughout the spring, that the nonunique continuation results of [1]—results whose opposite I had been trying to prove for eight years—first came to me.

2 NONUNIQUE CONTINUATION FOR A REDUCED WAVE EQUATION

We refer to [1] and keep our exposition here extremely close to that found there.

THEOREM 1. There exists an example of nonunique continuation on the half space $\Lambda = R^2 \times [0, \infty)$ for a uniformly elliptic equation

$$u_{zz} + ((1 + A + a)u_x)_x + (bu_y)_x + (bu_x)_y + ((1 + C + c)u_y)_y + 1^2 u = 0 \text{ in } \Lambda. \qquad (2.1)$$

(i) The solution $u(x, y, z)$ is C^∞ on Λ, identically zero for $z \geq$ a certain positive number Z, but not identically zero on any open subset of $R^2 \times [0, Z)$.
(ii) The coefficients $a(x, y, z)$, $b(x, y, z)$, $c(x, y, z)$ are C^∞ on Λ and identically zero for $z \geq Z$.
(iii) The coefficients $A(z)$, $C(z)$ are Hölder continuous of order $1/6$ on $[0, \infty)$, C^∞ on $[0, Z)$, and identically zero for $z \geq Z$.
(iv) All functions are periodic in x and y with a certain period $2L$; u is symmetric about $x = jL$ and $y = jL$ for all integers j.
(v) Moreover, u satisfies the "no flow" condition that the normal component of $\mathcal{A}\nabla u$ is zero on the boundary $\partial\Omega \times [0, \infty)$ of the cylinder $\Omega \times [0, \infty)$, $\Omega = [0, L] \times [0, L]$, since both b and the normal derivative of u are zero there.
(vi) Although the coefficient matrix $\mathcal{A}$ is only Hölder continuous at $z = Z$, ∇u dies out sufficiently rapidly as $z \to Z$ that $\mathcal{A}\nabla u$ is C^∞ on Λ. Thus u is a classical (as well as weak) solution of (2.1) on Λ.
(vii) Moreover, for any $\lambda > 1$, this example can be constructed such that u for small positive z has the form

$$u(x, y, z) = \cos \lambda x \cdot e^{-\gamma z}, \text{ where } \gamma(\lambda) \equiv (\lambda^2 - 1)^{1/2}, \qquad (2.2)$$

and such that the coefficient perturbations A, C, a, b, c stay as small as one likes (bounded less than any given positive number).

Outline of the Construction

Note that for any $\lambda > 1$, the equation (2.1) with $A = C = a = b = c = 0$ has solutions of the form (2.2). Our construction proceeds with an infinity of *steps* of successively shorter z-durations $Z_1, Z_2, Z_3, \ldots$ whose sum $Z = \sum Z_n$ is finite. The solution begins with initial values $\cos \lambda_1 x \cdot e^{-\gamma_1 z}$ and then manages to rotate itself successively into values proportional to $\cos \lambda_2 y \cdot e^{-\gamma_2 z}, \cos \lambda_3 x \cdot e^{-\gamma_3 z}, \ldots$ at the beginning of the 2nd,3rd,... steps, where $\lambda_1 = \lambda > 1$ is given and $\lambda_1, \lambda_2, \lambda_3, \ldots$ is a certain increasing sequence, to be chosen later. Each step will consist of three major *phases* sandwiched between four simpler "transition" phases.

Let us consider the nth step, in which the solution begins proportional to $\cos \lambda_n x \cdot e^{-\gamma_n z}$, ends proportional to $\cos \lambda_{n+1} y \cdot e^{-\gamma_{n+1} z}$, and at each intermediate z is always a linear combination of $\cos \lambda_n x$ and $\cos \lambda_{n+1} y$. The seven phases will have z-durations of $\lambda_n^{-1}, \lambda_n^{-1}, \lambda_n^{-1}, \lambda_n^{-1} s_n, \lambda_n^{-1}, \lambda_n^{-1}$, and λ_n^{-1} respectively, where s_n is a certain sequence tending to infinity, also to be chosen later.

We normalize by expanding the x, y, and z scales by a factor of λ_n, and let $\bar{u}$, $1+\bar{A}+\bar{a}$, $\bar{b}$, $1+\bar{C}+\bar{c}$ denote the corresponding solution and coefficients in the normalized geometry. Under normalization the final term 1^2u in (2.1) should be replaced by $\omega^2\bar{u}$, where $\omega=\omega_n=1/\lambda_n$. We define the five parameters $\mu=\mu_n\equiv\lambda_{n+1}/\lambda_n$, $\gamma=\gamma_n=\gamma_n/\lambda_n$, $\nu=\nu_n=\gamma_{n+1}/\lambda_n$, $s\equiv s_n$, and $\varepsilon^2=\varepsilon_n^2=e^{-(\gamma_{n+1}-\gamma_n)(\lambda_n^{-1}s_n)}=e^{-(\nu-\gamma)s}$. We let η and β be two fixed C^∞ functions on $[0,1]$ with the following behavior: $\eta(z)$ is $\equiv 0$ near $z=0$, monotone, and $\equiv 1$ near $z=1$; $\beta(z)$ is $\equiv 0$ near $z=0$ and $\equiv z-1$ near $z=1$.

We now proceed to list the duration, solution, and coefficients (in the order $1+\bar{A}+\bar{a},\bar{b},1+\bar{C}+\bar{c}$) for each phase. We adopt the convention of beginning the "time" z anew (with $z=0$) at the beginning of each phasae and employ the notation "$\sim$" (proportional to) to avoid having to write explicit magnitude factors to patch the solution together continuously from phase to phase. Our solution $\bar{u}$ should begin with $\bar{u}\sim\cos x\cdot e^{-\gamma z}$ and coefficients $1,0,1$ and finish after these 7 phases with $\bar{u}\sim\cos\mu y\cdot e^{-\nu z}$ and coefficients $1,0,1$.

The "transition 1" phase has duration 1, solution $\bar{u}\sim\cos x\cdot e^{-\gamma z}$, and coefficients $1,0,\psi(z)$, where $\psi(z)=1+(\mu^{-2}-1)\eta(z)$. This phase merely changes the coefficients smoothly from $(1,0,1)$ to $(1,0,\mu^{-2})$.

The "seed" phase has duration 1, solution $\bar{u}\sim\cos x\cdot e^{-\gamma z}+\varepsilon\eta(z)\cdot\cos\mu y\cdot e^{-\gamma z}$, and coefficients $1+\bar{a},\bar{b},\mu^{-2}$, where the rather complicated $\bar{a}(x,y,z)$ and $\bar{b}(x,y,z)$ will be derived later. This phase introduces a tiny $\cos\mu y$ component into the solution.

The "transition 2" phase has duration 1, solution $\bar{u}\sim\cos x\cdot e^{-\varphi(z)}+\varepsilon\cdot\cos\mu y\cdot e^{-\gamma z}$, and coefficients $1+\bar{A}(z)\equiv(\varphi'(z))^2-\varphi''(z)+\omega^2,0,\mu^{-2}$, where $\varphi(z)=\gamma z+(\nu-\gamma)\beta(z)$. This phase smoothly changes the rate of decay of the first component from an initial $e^{-\gamma z}$ rate to a final $e^{-\nu z}$ rate. It does not changes the 1 to ε ratio of the two components. Note that

$$\begin{aligned}1+\bar{A}(z)&\equiv(\varphi'(z))^2-\varphi''(z)+\omega^2\\&=(\gamma+(\nu-\gamma)\beta')^2+(\nu-\gamma)\beta''+\omega^2\end{aligned}\tag{2.3}$$

is $\approx\gamma^2+\omega^2=1$ when $(\nu-\gamma)$ is small. In fact, since $\omega=\lambda_n^{-1}\leq\lambda_1^{-1}<1$, and thus $\nu-\gamma=(\mu^2-\omega^2)^{1/2}-(1-\omega^2)^{1/2}$ is uniformly $O(\mu-1)$, we have

$$\bar{A}(z)\text{ is uniformly }O(\mu-1).\tag{2.4}$$

Note that $1+\bar{A}(z)$ begins equal $\gamma^2+\omega^2=1$ and finishes equal $\nu^2+\omega^2=\mu^2$.

The "distorted decay" phase has duration s, solution $\bar{u}\sim\cos x\cdot e^{-\nu z}+\varepsilon\cos\mu y\cdot e^{-\gamma z}$, and coefficients $\mu^2,0,\mu^{-2}$. The ratio between the two components is decaying at an $e^{-(\nu-\gamma)z}$ rate; thus, since $\varepsilon^2=e^{-(\nu-\gamma)s}$, the initial 1 to ε ratio has reversed to an ε to 1 ratio by the end of the phase.

The "transition 3" phase has duration 1, solution $\bar{u}\sim\varepsilon\cos x\cdot e^{-\nu z}+\varepsilon\cos\mu y\cdot e^{-\varphi(z)}$, and coefficients $\mu^2,0,1+\bar{C}(z)=\mu^{-2}((\varphi'(z))^2-\varphi''(z)+\omega^2)$, where $\varphi(z)=\nu z+(\nu-\gamma)\cdot\gamma(1-z)$. This phase smoothly changes the rate of decay of the second component from an initial $e^{-\gamma z}$ rate to a final $e^{-\nu z}$ rate. It does not change the ε to 1 ratio of the two components. Note that $1+\bar{C}(z)$ begins equal $\mu^{-2}(\gamma^2+\omega^2)=\mu^{-2}$ and ends equal $\mu^{-2}(\nu^2+\omega^2)=1$. Also, as in (2.4), we see that $\bar{C}(z)$ is uniformly $O(\mu-1)$.

The "removal phase" has duration 1, solution $\bar{u}\sim\varepsilon\cdot\eta(1-z)\cdot\cos x\cdot e^{-\nu z}+\cos\mu y\cdot e^{-\nu z}$, and coefficients $\mu^2,\bar{b},1+\bar{c}$ where $\bar{b}(x,y,z)$ and $\bar{c}(x,y,z)$ will be described later. This phase completely removes the first component from the solution.

The "transition 4" phase has duration 1, solution $u \sim \cos\mu y \cdot e^{-\nu z}$, and coefficients $\psi(z), 0, 1$, where $\psi(z) = \mu^2 + (1-\mu^2)\eta(t)$. This phase merely changes the coefficients smoothly from $(\mu^2, 0, 1)$ to $(1, 0, 1)$.

Derivation of $\bar{a}, \bar{b}$ for the Seed Phase

We consider $\bar{u}$ defined by

$$\bar{u} \sim \cos x \cdot e^{-\gamma z} + \varepsilon\eta(z) \cdot \cos\mu y \cdot e^{-\gamma z}, \tag{2.5}$$

which should be a solution of an equation of the form

$$((1+\bar{a})\bar{u}_x)_x + (\bar{b}\bar{u}_y)_x + (\bar{b}\bar{u}_x)_y + (\mu^{-2}\bar{u}_y)_y + \bar{u}_{zz} + \omega^2\bar{u} = 0. \tag{2.6}$$

Since $\cos x \cdot e^{-\gamma z}$ and $\cos\mu y \cdot e^{-\gamma z}$ are already solutions of the equation with coefficients $1, 0, \mu^{-2}$, the perturbations $\bar{a}(x,y,z)$ and $\bar{b}(x,y,z)$ must only take care of the $\eta'(z)$ and $\eta''(z)$ terms in the equation, and we are quickly led to the following *perturbation equation* for $\bar{a}$ and $\bar{b}$:

$$(\bar{a}\sin x)_x + (\bar{b}\varepsilon\eta(z)\mu\sin\mu y)_x + (\bar{b}\sin x)_y = \varepsilon[2\gamma\eta'(z) - \eta''(z)]\cos\mu y. \tag{2.7}$$

Let us consider z fixed for the moment and suppress it in the notation. Thus the η, η', η'' in (2.7) are just constants with respect to x and y. We want our $\bar{a}(x,y)$ and $\bar{b}(x,y)$ to be C^∞ with x period 2π and y period $2\pi/\mu$.

Erratum. There was a minor error in the exposition of this Seed Phase in [1]. There, in equation (7), we made a *further* normalization of the geometry, expanding the y scale again by a factor of μ, thus changing our $\cos\mu y$ term in $\bar{u}$ into a $\cos y$ term in a normalized function $\tilde{u}$. We then proceeded to find *symmetric* coefficients $\tilde{b}$ for $\tilde{u}$ (that is the coefficients in the $(\tilde{b}\tilde{u}_y)_x$ and $(\tilde{b}\tilde{u}_x)_y$ terms were the same). The error is that under the nonisotropic change back to unnormalized geometry the desired symmetry would be lost, i.e., the two coefficients $\bar{b}$ and $\bar{b}$ in (2.6) would *not* be the same. The correction for this minor error is simple—we should merely skip the further normalization and instead work with the original $\bar{u}, \bar{a}, \bar{b}$ with their slightly differing periods 2π in x and $2\pi/\mu$ in y, just as we have done in (2.5)–(2.7) of the present paper. The needed adjustments in equations (7)–(18), (29)–(31) and Figure 1 of [1] are then obvious. Assume henceforth that those corrections have been made.

After those minor corrections in [1] are made, we see that *our* (2.7) *is in exactly the same form as the previous perturbation equations* (9) *or* (31) *of* [1], except for the trivial substitution of the quantity $[2\gamma\eta' - \eta'']$ here for the quantity $[-\eta']$ in (9) or $[2\eta' - \eta'']$ in (31). Thus we can construct $\bar{a}$ and $\bar{b}$ as in [1], and can show that, *for* $\varepsilon \le \varepsilon_0$, *each derivative with respect to* x, y, z *of the coefficients* $\bar{a}$ *and* $\bar{b}$ *is uniformly bounded by* ε *times a constant.*

We won't repeat that construction. Nevertheless, for the sake of *insight*, we offer a few words here about a linearized version of the construction. We suspect that our $\bar{a}$ and $\bar{b}$ for (2.7), together with all their derivatives, will be $O(\varepsilon)$. Thus, dropping the one $O(\varepsilon^2)$ term in (2.7) we are led to the *linearized version* of (2.7):

$$(\bar{a}\sin x)_x + \bar{b}_y\sin x = C \cdot \cos\mu y, \tag{2.8}$$

where C is the constant $\varepsilon \cdot [2\gamma\eta' - \eta'']$. Let $R \equiv \bar{a} \sin x$. Thus we want

$$\begin{aligned}
&\text{(a) } R_x = [C \cos \mu y - \bar{b}_y \sin x],\\
&\text{(b) } R \text{ and } \bar{b} \text{ have } x \text{ period } 2\pi \text{ and } y \text{ period } 2\pi/\mu,\\
&\text{(c) } R = 0 \text{ for } x = j\pi, \text{ and}\\
&\text{(d) } \bar{b} = 0 \text{ for } x = j\pi \text{ and } y = j\pi/\mu, \text{ for all integers } j.
\end{aligned} \tag{2.9}$$

Now for x near $j\pi$ we can let R_x handle all the load in (2.9a), i.e.

$$\text{let } \bar{b}(x,y) = 0 \text{ and } R(x,y) = C \cdot \cos \mu y \cdot (x - j\pi), \text{ for } x \text{ near } j\pi. \tag{2.10}$$

Then, because of (2.9c), R_x must have mean value zero on each horizontal line between $x = j\pi$ and $x = (j+1)\pi$. Thus it suffices to enforce such a zero mean value on the right hand side of (2.9a), and this we can do by choosing $\bar{b}$ appropriately. We choose

$$\bar{b}(x,y) = f(y) \cdot h(x), \tag{2.11}$$

where $h(x)$ is a 2π periodic function with its support in the middle halves of the $(j\pi, (j+1)\pi)$ intervals and such that $h(x) \cdot \sin x$ has mean value 1 on each such interval. Thus we need $f'(y) = C \cos \mu y$, which with (2.9d) yields $f(y) = \mu^{-1} \cdot C \cdot \sin \mu y$.

In the full nonlinear construction for (2.7), (2.10) still holds, but one needs to replace $h(x)$ by a function $s(x,y)$ which is constant along certain nearly vertical characteristic curves.

Construction of the $\bar{b}$ and $\bar{c}$ for the "removal phase" is completely analogous to the "seed phase".

Putting the Pieces Together and Choice of the λ_n and s_n

One stacks the solutions and coefficients for the seven phases "end to end", so to speak, to complete the nth step. One then finally stacks the individual steps "end to end".

By proper choice of λ_n and s_n one can make $Z \equiv \sum Z_n$ finite, the solution u and a, b, c tend to 0 (in C^∞ fashion) and the functions $A(z)$ and $C(z)$ tend to 0 (in Hölder continuous fashion) as $z \to Z$. The choices of λ_n and s_n for *large* n can be taken exactly as in [1]. This is because the values $\gamma_n = (\lambda_n^2 - 1)^{1/2}$ are $\approx \lambda_n$ for large values of λ_n, and thus the various factors of $\varepsilon_n \equiv e^{-(\gamma_{n+1}-\gamma_n)\lambda_n^{-1} \cdot s_n/2}$, with which one would bound the coefficients a, b, c and their derivatives here, are nearly the same for large n as the factors $\varepsilon_n \equiv e^{-(\lambda_{n+1}-\lambda_n)\lambda_n^{-1} s_n/2}$ used in (34)–(41) of [1]. One can for example choose λ_n and s_n proportional to $(n+N)^6$ and $(n+N)^4$ for large n, where N is some large integer.

Recall, see (2.4) for example, that the perturbations A, C, are uniformly $O(\mu_n - 1)$ in the nth step, and the perturbations a, b, c are uniformly $O(\varepsilon_n)$. Thus in order to fulfill our promise of Theorem 1(vii) to keep these perturbations uniformly small, one may need to keep $\mu_n - 1$ and ε_n quite small in the early steps (perhaps considerably smaller than was necessary in [1]).

Our u, a, b, c are all periodic in the nth step with x period $2\pi/\lambda_n$ and y period $2\pi/\lambda_{n+1}$. Clearly we can pick our sequence λ_n such that these periods are all integral divisors of some one large period $2L$, as promised in Theorem 1(vi).

3 TOTAL INTERNAL REFLECTION AND THE TUNNEL EFFECT

Consider the wave equation (1.4) and its reduced wave equation (1.3) with the "gap" coefficients of (1.6). We begin with the case of an *infinite* gap width $G = \infty$. Consider the solution caused by an incident plane wave of complex amplitude $E_1 = 1$ in the lower half space $z < 0$, coming in from $z = -\infty$ at an angle θ from the normal. This incident plane wave on $z < 0$ has the form

$$u_1(x, y, z) = 1 \cdot e^{i(\lambda x + \alpha z)}, \text{ with } \lambda = \sin\theta/c, \ \alpha = \cos\theta/c. \tag{3.1}$$

One also gets a reflected plane wave on $z < 0$ of the form

$$u_2(x, y, z) = E_2 \cdot e^{i(\lambda x - \alpha z)}. \tag{3.2}$$

Now if $\lambda < 1$, i.e. $\theta < \arcsin c$, then in $z > 0$ one gets a transmitted plane wave at the Snell's Law angle. But, if $\lambda > 1$, i.e. if θ exceeds the critical angle $\arcsin c$, then in $z > 0$ one gets an exponentially decaying wave of the form

$$u_3(x, y, z) = E_3 \cdot e^{i\lambda x} \cdot e^{-\gamma z}, \text{ where } \gamma(\lambda) = (\lambda^2 - 1)^{1/2}. \tag{3.3}$$

The total solution then has the form

$$\begin{aligned} u &= u_1 + u_2 && \text{in } z < 0, \\ &= u_3 && \text{in } z > 0, \end{aligned} \tag{3.4}$$

where the complex constants E_2, E_3 are uniquely determined by requiring that both u and the normal component of $\mathcal{A}\nabla u$ patch together continuously at the interface. This yields the two linear equations

$$\begin{aligned} 1 + E_2 &= E_3, \\ ic^2(\alpha \cdot 1 - \alpha \cdot E_2) &= -\gamma \cdot E_3. \end{aligned} \tag{3.5}$$

Because the solution of (3.5) yields the identity $|E_2|^2 = |1|^2$, one sees that the total energy, $|1|^2$, of the incident wave is carried back into the lower half space by the reflected wave. Hence the terminology "*total internal reflection*". Nevertheless, the solution does "*tunnel*" into the upper half space with an $e^{-\gamma z}$ exponential decay.

Consider next the case of a *finite* gap width G in (1.6). The energy of the incident plane wave u_1 of (3.1) is then partially transmitted across the gap even when θ is greater than $\arcsin c$. To see how this "*tunnel effect*" occurs, one needs to add two further waves u_4 and u_5 to the previous u_1, u_2, u_3. In the gap we also have a tiny exponentially growing wave

$$u_4(x, y, z) = E_4 \cdot e^{i\lambda x} \cdot e^{+\gamma z}, \tag{3.6}$$

and beyond the gap we have a tiny transmitted plane wave at the original angle θ from the normal,

$$u_5(x, y, z) = E_5 \cdot e^{i(\lambda x + \alpha z)}. \tag{3.7}$$

The total solution is then

$$\begin{aligned} u &= u_1 + u_2 && \text{in } z < 0, \\ &= u_3 + u_4 && \text{in } 0 < z < G, \\ &= u_5 && \text{in } z > G, \end{aligned} \tag{3.8}$$

where the 4 complex constants E_2, E_3, E_4, E_5 are uniquely determined by requiring that both u and the normal component of $\mathcal{A}\nabla u$ patch together continuously on the two interfaces at $z = 0$ and $z = G$. This yields 4 linear equations, similar to those of (3.5), which we won't bother writing down.

Because the solution of these 4 equations yields the identity $|E_2|^2 + |E_5|^2 = |1|^2$, we see that the fraction $|E_5|^2$ (very tiny if $\gamma G \gg 1$) of the incident energy is carried forward in the transmitted wave while the remainder is reflected back in the reflected wave.

4 BLOCKING THE TUNNEL EFFECT FOR CERTAIN DOUBLE WAVES

Consider a "double wave" solution v caused by two incident plane waves of the same magnitude and phase (i.e. $E_1 = 1$ for both), but with opposite angles of incidence, $+\theta$ and $-\theta$.

In the case of an *infinite* gap width $G = \infty$, with the coefficients (1.6) for (1.4), the desired "double wave" solution v is then the sum of two of the solutions discovered in §3; i.e.,

$$v(x,y,z,t) = \text{Re}([u_+(x,y,z) + u_-(x,y,z)] \cdot e^{-it}) \text{ for } -\infty < z < \infty, \tag{4.1}$$

where u_+ and u_- are the solutions of form (3.4) with $+\theta$ and $-\theta$, i.e. with $+\lambda$ and $-\lambda$. Thus on $z > 0$ this has the formula

$$v(x,y,z,t) = (\cos \lambda x \cdot e^{-\gamma z}) \cdot \text{Re}(2E_3 \cdot e^{-it}) \text{ on } z > 0. \tag{4.2}$$

The function $\cos \lambda x \cdot e^{-\gamma z}$ appearing in (4.2) is the solution to the reduced wave wave equation (1.3) with the coefficients $\mathcal{A}(x,y,z) = I$ with these particular "starting values" for small positive z. However, we discovered in Theorem 1 how to construct a solution $u(x,y,z)$ of (1.3) on $z > 0$ with these same starting values for small positive z (see (2.2)), but with slightly perturbed coefficients $\mathcal{A}(x,y,z)$, in such a way that u vanishes identically for z greater than some infinite value Z.

We patch together the resulting coefficients and solutions on the lower and upper half spaces; that is, we let

$$\begin{aligned} \mathcal{A}(x,y,z) &= c^2 I \text{ on } z < 0, \\ &= \text{ the coefficients of Theorem 1 on } z > 0, \end{aligned} \tag{4.3}$$

and

$$\begin{aligned} v(x,y,z,t) = &\text{ the formula of (4.1) on } z < 0, \\ = &\ \text{Re}([2E_2 \cdot u(x,y,z)] \cdot e^{-it}) \text{ on } z > 0, \\ &\text{where } u(x,y,z) \text{ is the solution of Theorem 1.} \end{aligned} \tag{4.4}$$

This v is then a solution of the wave equation (1.4) on all $-\infty < z < \infty$, and moreover it "tunnels" only a finite depth into the upper half space Λ; for $z > Z$ it is identically zero.

In the case of a *finite* gap width G, the desired "double wave" solution v for the coefficients (1.6) is also given by the sum (4.1), but where u_+ and u_- are now the solutions of form (3.8) with $+\theta$ and $-\theta$, i.e. with $+\lambda$ and $-\lambda$. Here the solution has tunneled across the gap, and on $z > G$ is the sum of two tiny transmitted plane waves, at angles $+\theta$ and $-\theta$, carrying energy beyond the gap.

This tunnel effect is completely *blocked*, however, if the gap width G is greater than the Z of Theorem 1 and if we switch in the gap to the slightly perturbed coefficients $\mathcal{A}(x, y, z)$ of that theorem. The desired "double wave" solution is then still the *same* $v(x, y, z, t)$ given by (4.4) for the infinite gap case. This is because v is $\equiv 0$ for $z > Z$, and hence one can alter the coefficients $\mathcal{A}$ arbitrarily on $z > Z$ and still have that v is a solution.

Note that the two incident waves have to be precisely aligned in frequency, amplitude, phase and position to accomplish this complete blocking. Also, this example is for a *scalar* wave equation. The author does not know whether a similar construction could be accomplished, by slight variations of the optical properties in the gap, for the *system* of Maxwell's equations for electromagnetic waves.

Paolo Manselli has recently pointed out to the author that, because the A, C are functions of z only and because the a, b, c inside the x and y derivatives are C^∞ on Λ, the parabolic and elliptic equations (5) and (6) in *divergence* form in [1] are simultaneously also in *nondivergence* form with smooth first order coefficients (thus obviating the need for the italicized comment on page 116 there). The same clearly also applies to equation (2.1) here.

REFERENCES

1. K. Miller, Nonunique continuation for uniformly parabolic and elliptic equations in self-adjoint divergence form with Hölder continuous coefficients, *Archive for Rational Mechanics and Analysis, 54*: 105–117 (1974).
2. T. Wolff, Recent work on sharp estimates in second order elliptic unique continuation problems, *Journal of Geometric Analysis, 3*: 621–650 (1993).

24. The Reflector Problem for Closed Surfaces

Vladimir Oliker Department of Mathematics and Computer Science, Emory University, Atlanta, Georgia

1 STATEMENT OF THE PROBLEM

In [7], problem 21, S.T. Yau described the following problem. Let R be a surface in Euclidean space $\mathbb{R}^3$. For each point on a unit sphere S^2 consider a ray from the origin through that point that strikes the surface R and reflects off according to the laws of geometric optics. The direction of the reflected ray defines a point on S^2 and, thus, we have a map from S^2 into S^2. How much information about the surface R is contained in this map?

The purpose of this note is to show that a partial answer to this problem can be obtained by applying some results of [4]. Let us formulate the problem in more

The research was partially supported by AFOSR under contract F49620-95-C0009 and NSF Grant DMS-9405808 specific form and in more detail.

In $\mathbb{R}^3$ we fix a Cartesian coordinate system with the origin at a point $\mathcal{O}$. Denote by $R \hookrightarrow \mathbb{R}^3$ a smooth, closed, convex surface, star-shaped relative to $\mathcal{O}$. Suppose that a point-source of light is positioned at $\mathcal{O}$ and is radiating with intensity $g(\mathbf{m}) > 0, \mathbf{m} \in S^2$, where S^2 is a unit sphere centered at $\mathcal{O}$. Assume, further, that the inner side of the surface R is a perfect reflector. In such circumstances, according to Snell's law, every ray of direction $\mathbf{m}$ emanating from $\mathcal{O}$ is reflected off R in direction $\mathbf{y}$ given by

$$\mathbf{y} = \mathbf{m} - 2\langle \mathbf{m}, \mathbf{n} \rangle \mathbf{n}, \tag{1}$$

where $\langle , \rangle$ denotes the scalar product in $\mathbb{R}^3$ and $\mathbf{n}$ is the unit (exterior) normal vector to R at the point of incidence of the ray in direction $\mathbf{m}$. Thus, we have a map $\gamma_R : S^2 \to S^2$, $\mathbf{y} = \gamma_R(\mathbf{m})$, associated with the reflector R.

Denote by $d\sigma$ the area element on S^2 and let $f(\mathbf{y})$ be the intensity of light in a reflected direction $\mathbf{y}$. By definition,

$$f(\mathbf{y}) = f(\gamma_R(\mathbf{m})) = \frac{g(\mathbf{m})d\sigma(\mathbf{m})}{d\sigma(\gamma_R(\mathbf{m}))}. \tag{2}$$

We rewrite this expression as

$$f(\gamma_R(\mathbf{m}))|J(\gamma_R(\mathbf{m}))| = g(\mathbf{m}), \tag{3}$$

where $J(\gamma_R)$ is the Jacobian determinant.

The "reflector" problem considered in this note is to determine a closed convex reflector R, star-shaped relative to $\mathcal{O}$, such that γ_R is a diffeomorphism and for given functions $g > 0$ and $f > 0$ on S^2 the equation (3) is satisfied on S^2.

Since it is assumed that no energy is lost in the process, and it is required that γ_R be a diffeomorphism it follows immediately from (2) that the functions f and g must satisfy the "total energy conservation" condition

$$\int_{S^2} f(\mathbf{y})d\sigma(\mathbf{y}) = \int_{S^2} g(\mathbf{m})d\sigma(\mathbf{m}). \tag{4}$$

Thus, this is a necessary condition on the data.

The expression (3) can be transformed into a differential expression for one scalar function describing the reflector R. Let $\rho(\mathbf{m}), \mathbf{m} \in S^2$, denote the radial function of the surface R, that is, $\rho(\mathbf{m})$ is the distance from $\mathcal{O}$ to the point of intersection of the ray of direction $\mathbf{m}$ from $\mathcal{O}$ with R. It was shown in [5] that

$$J(\gamma_R) = \frac{det(2b_{ij}\langle \mathbf{m}, \mathbf{n} \rangle + \rho e_{ij})}{\rho^2 det(e_{ij})}, \tag{5}$$

where (b_{ij}) is the matrix of coefficients of the second fundamental form of R and (e_{ij}) is the matrix of coefficients of the metric of S^2 in some local coordinates (u^1, u^2). In these coordinates we have the following expressions for b_{ij} and for $\langle \mathbf{m}, \mathbf{n} \rangle$ [3]:

$$b_{ij} = \frac{\rho \tilde{\nabla}_{ij}\rho - \rho^2 e_{ij} - 2\rho_i\rho_j}{\sqrt{\rho^2 + |\tilde{\nabla}\rho|^2}},$$

$$\langle \mathbf{m}, \mathbf{n} \rangle = \frac{\rho}{\sqrt{\rho^2 + |\tilde{\nabla}\rho|^2}},$$

where $\tilde{\nabla}_{ij}$ denote the second covariant derivatives in the metric $e_{ij}du^i du^j, \rho_i = \partial\rho/\partial u^i$, $\tilde{\nabla}\rho = e^{ij}\rho_i \mathbf{m}_j, \mathbf{m}_j = \partial \mathbf{m}/\partial u^j, (e^{ij}) = (e_{ij})^{-1}$. Here and everywhere else in the paper the range of latin indices is $1 \leq i, j, k, ... \leq 2$ and summation over repeated lower and upper indices is assumed. It follows now from (3 and (5) that the radial function of a reflector must satisfy the equation (see [5], formula (22))

$$M(\rho) \equiv f(\gamma_R(\mathbf{m})) \frac{\det[2\rho\tilde{\nabla}_{ij}\rho - (\rho^2 - |\tilde{\nabla}\rho|^2)e_{ij} - 4\rho_i\rho_j]}{(\rho^2 + |\tilde{\nabla}\rho|^2)^2 \det(e_{ij})} = g \text{ on } S^2. \tag{6}$$

Thus, analytically, the reflector problem requires solving equation (6) with given functions f and g satisfying (4), and the reflector R constructed from solution ρ must be such that γ_R is a diffeomorphism onto S^2.

Instead of formulating the problem as a "forward" problem (as it was done above) one can formulate it as a "dual" problem by working with the inverse map γ_R^{-1}. A curious observation is that the equation for the dual formulation can be obtained from (6) by the change of the unknown function $p = 1/\rho$ and it has the form

$$f(\mathbf{y}) \frac{\det[\tilde{\nabla}_{ij}p + (p - \rho)e_{ij}]}{\rho^2 \det[e_{ij}]} = g(\gamma_R^{-1}(\mathbf{y})) \quad \text{on } S^2, \tag{7}$$

where $\rho = (p^2 + |\tilde{\nabla}p|^2)/2p$ (see [2] for more details). Existence of weak solutions to this equation was proved in [1].

2 MAIN RESULT

For the rest of the paper we consider the special case of equation (6) when the function $f \equiv 1$.

We say that the operator M is elliptic on a function $\rho \in C^2(S^2)$ if

(8) the matrix $[2\rho\tilde{\nabla}_{ij}\rho - (\rho^2 - |\tilde{\nabla}\rho|^2)e_{ij} - 4\rho_i\rho_j]$ is negative definite.

If a smooth function ρ on S^2 is positive and satisfies (6) then at the point of maximum of ρ the condition (8) is satisfied. This fact and the simple observation that the operator M is elliptic on $\rho \equiv const$ show that the class of functions on which M is elliptic is not empty.

We are interested in positive elliptic solutions of (6). It will be convenient to introduce a new unknown function P by setting $\rho = e^{-P}$. A computation shows that

$$M(\rho) = \Phi(P) \equiv \frac{\det[\tilde{\nabla}_{ij}P + P_iP_j + (1-Q)e_{ij}]}{Q^2 \det(e_{ij})}, \tag{9}$$

where

$$P_i = \partial P/\partial u^i, \; Q = \frac{1}{2}(1 + |\tilde{\nabla}P|^2).$$

Put

$$E = det(e_{ij}), \; (G_{ij}) = (\tilde{\nabla}_{ij}P + P_iP_j + (1-Q)e_{ij}), \; (G^{ij}) = cof(G_{ij}).$$

Lemma 1. *Let $\bar{\rho} \in C^3(S^2)$ be a positive elliptic solution of (6) and $\bar{P} = -ln\bar{\rho}$. For any $h \in C^2(S^2)$ the Frechet derivative of Φ at $\bar{P}$ is given by*

$$\Phi'(\bar{P})h = \frac{1}{\sqrt{E}}\frac{\partial}{\partial u^i}\left(\frac{G^{ij}}{Q^2\sqrt{E}}\frac{\partial h}{\partial u^j}\right). \tag{10}$$

Proof. This lemma has been established in [4].

For a function $\bar{P} = -ln\bar{\rho}$, as in Lemma 1, put

$$\mathcal{P} = \{P \in C^{2,\alpha}(S^2), 0 < \alpha < 1 \mid \|P - \bar{P}\|_{C^{2,\alpha}} < \epsilon\},$$

where $\epsilon > 0$ is such that for any $P \in \mathcal{P}$ the operator M is elliptic on $\rho = e^{-P}$.

Next, we observe that for any $P \in \mathcal{P}$ the map γ_R is defined and in terms of P is given by

$$\mathbf{y} = \gamma_R(m) = \mathbf{m} - \frac{1}{Q}(\mathbf{m} + \tilde{\nabla}P)$$

In order to indicate the dependence on P we will write the map γ_R as γ_P.

Note that since $J(\gamma_{\bar{P}}) > 0$, the same is true for any $P \in \mathcal{P}$ and the map γ_P is a local diffeomorphism for every $P \in \mathcal{P}$.

Let us show that the map γ_P maps S^2 onto itself. Indeed, let $\hat{\mathbf{y}} \in S^2$. Consider a paraboloid of revolution with focus at $\mathcal{O}$ and axis of revolution $\hat{\mathbf{y}}$ directed away from the apex of the paraboloid. By rescaling the paraboloid homothetically with respect to $\mathcal{O}$ we can arrange that the corresponding to P reflector R is contained inside the convex set bounded by the paraboloid and the latter has at least one point of contact with R. At such contact point the paraboloid and reflector have a common normal. Consequently, the ray of direction $\mathbf{m}$ going into the point of contact will be mapped to $\hat{\mathbf{y}}$ by both, the paraboloid and the reflector R. Thus, γ_P is a covering projection of S^2 onto S^2.

Since S^2 is compact, γ_P is a diffeomorphism onto S^2. This implies that for all $P \in \mathcal{P}$

$$\int_{S^2} J(\gamma_P)d\sigma = 4\pi. \tag{11}$$

Finally, we note that the surface with the radial function $\bar{\rho} = e^{-\bar{P}}$ is convex. Indeed, since $\bar{\rho}$ is positive and elliptic we have from (8)

$$b_{ij}\xi^i\xi^j = \frac{\rho\tilde{\nabla}_{ij}\rho - \rho^2 e_{ij} - 2\rho_i\rho_j}{\sqrt{\rho^2 + |\tilde{\nabla}\rho|^2}}\xi^i\xi^j < -\frac{\sqrt{\rho^2 + |\tilde{\nabla}\rho|^2}}{2}e_{ij}\xi^i\xi^j \ for\ any\ \xi \in \mathcal{R}^2 - \{0\}.$$

This implies that the second fundamental form of the surface described by $\bar{\rho}$ is negative definite everywhere. Consequently, the surface is strictly convex. Furthermore, the corresponding surfaces generated by functions from $\mathcal{P}$ are also strictly convex.

Put

$$H = \{g \in C^{\alpha}(S^2) \mid g > 0 \ and \int_{S^2} g d\sigma = 4\pi\}.$$

Theorem 2. *Let $\bar{P} \in C^3(S^2)$ and let*

$$\Phi(\bar{P}) = \bar{g} \ on \ S^2. \tag{12}$$

Then there exists a $\delta > 0$ such that for any $g \in H$ such that $\|g - \bar{g}\|_{C^\alpha} < \delta$ the equation

$$\Phi(P) = g \ on \ S^2 \tag{13}$$

admits a solution of class $C^{2,\alpha}$. This solution is unique up to an additive constant. Geometrically, this family of solutions corresponds to a family of homothetic convex reflector surfaces each of which is a graph of a radial function over S^2 and redistributes the reflected rays so that the intensity in the reflected directions is uniform over S^2.

Proof. Put

$$\mathcal{P}_0 = \{P \in \mathcal{P} | \int_{S^2} P d\sigma = 0\}.$$

It follows immediately from the expression (9) that $\Phi(P + c) = \Phi(P)$ for any $c = const$. Consequently, we can normalize $\bar{\rho}$ so that

$$\int_{S^2} \bar{P} d\sigma = 0. \tag{14}$$

It follows from the discussion above and (11) that $\Phi : \mathcal{P}_0 \to H$. Note also that $\bar{g} \in H$. We wish to apply the inverse function theorem to the map Φ. For that we need to consider the map $\Phi'(\bar{P}) : \mathcal{P}_0 \to H_0$, where

$$H_0 = \{\eta \in C^{3,\alpha}(S^2) | \int_{S^2} \eta d\sigma = 0\}.$$

Since $\bar{P} = -ln\bar{\rho}$ and $\bar{\rho}$ is a positive elliptic solution to (6), the matrix (G^{ij}) is positive on $\bar{P}$. This implies that the operator $\Phi'(\bar{P})$ is positively elliptic. It follows from Lemma 1 and the maximum principle that $ker\Phi'(\bar{P}) = \{const\}$ in space $\mathcal{P}$. Then, on the space $\mathcal{P}_0$ $ker\Phi'(\bar{P}) = \{0\}$.

On the other hand, since Φ' is selfadjoint (Lemma 1) and H_0 consist of functions orthogonal to 1 , it follows from the general theory of elliptic equations that the equation

$$\Phi' h = \eta$$

is uniquely solvable in $\mathcal{P}_0$ for any $\eta \in H_0$. By the inverse function theorem there exists some $\delta > 0$ such that the equation (13) is uniquely solvable in $\mathcal{P}_0$. The rest of the statements in the Theorem follow from the discussion preceding the statemnt of the theorem. This completes the proof of Theorem 2.

Remark Let $0 \leq \phi \leq 2\pi, 0 \leq \theta \leq \pi$ be the spherical coordinates on S^2. Suppose that we are given a function $\tilde{g} \in H \bigcap C^1(S^2)$ such that $\tilde{g}(\phi, \theta) \equiv \tilde{g}(\theta)$. With minor modifications of the arguments in [6] and of Lemma 4.1 in [4] it can be shown that for such $\tilde{g}$ one can always construct a radially symmetric, $C^3(S^2)$, positive elliptic solution of the equation (13). Consequently, one can use $\tilde{g}$ and the corresponding solution of (13) as the pair $\bar{g}$, $\bar{P}$ in Theorem 2 and construct solutions of (13) that are close to the radially symmetric solution $\bar{P}$ in $C^{2,\alpha}(S^2)$. Thus, there is a sufficiently large class of functions on which the hypotheses of Theorem 2 are satisfied.

References

[1] L. Caffarelli and V. Oliker, *Weak solutions of one problem in geometric optics*, preprint

[2] E. Newman and V. Oliker, *Differential-geometric methods in design of reflector antennas*, Symposia Mathematica, Ed. by A. Alvino, E. Fabes, G. Talenti, v. 35, 205-223, 1994

[3] V. Oliker, *Hypersurfaces in R^{n+1} with prescribed Gaussian curvature and related equations of Monge-Ampère type* Comm. in PDE's, 9(8):807-838, 1984

[4] V. I. Oliker, *Near radially symmetric solutions of an inverse problem in geometric optics*, Inverse Problems, 3, 743-756, 1987.

[5] V. Oliker and E. Newman, *On the energy conservation equation in the reflector mapping problem*, Applied Mathematics Letters 6(1):91-95, 1993

[6] V. Oliker and P. Waltman, *Radially symmetric solutions of a Monge-Ampère equation arising in a reflector mapping problem*, Lecture Notes in Math., Springer-Verlag, v. 1285, 361-374, 1986.

[7] S.T. Yau, *Open problems in geometry*, Proceedings of Symposia in Pure Mathematics, Ed. by R.Greene and S.T. Yau, v.54, part 1, 1-28, AMS, 1993

25. Upper Bounds for Eigenvalues of Elliptic Operators

Murray H. Protter Department of Mathematics, University of California, Berkeley, California

1. BOUNDS FOR THE LAPLACE OPERATOR

Let Ω be a bounded domain in R^m. We consider the eigenvalue problem

$$\begin{aligned} \Delta u + \lambda u = 0 \quad &\text{in } \Omega \\ u = 0 \quad &\text{on } \partial\Omega \, . \end{aligned} \tag{1}$$

It is well known that the eigenvalues $\lambda_1 < \lambda_2 \leq \lambda_3 \leq \ldots$ form an increasing sequence of positive numbers tending to infinity.

We are interested in inequalities between the eigenvalues, ones which are independent of the size and shape of the domain Ω. The starting point for this problem is the inequality for regions in R^2

$$\frac{1}{\lambda_{n+1} - \lambda_n} \sum_{i=1}^{n} \lambda_i \geq \tfrac{1}{2} n \, , \tag{2}$$

valid for $n = 1, 2, \ldots$. This is an extension by Payne, Polya and Weinberger [8] of the inequality

$$\lambda_{n+1} \leq 3\lambda_n \tag{3}$$

which was obtained earlier. The above results in R^2 have simple extensions to R^m:

$$\frac{1}{\lambda_{n+1} - \lambda_n} \sum_{i=1}^{n} \lambda_i \geq \frac{mn}{4} \tag{4}$$

and

$$\lambda_{n+1} \leq \frac{m+4}{m} \lambda_n \, . \tag{5}$$

Inequality (4) is a special case of the following result obtained by Hile and Protter:

$$\sum_{i=1}^{n} \frac{\lambda_i}{\lambda_{n+1} - \lambda_i} \geq \frac{mn}{4} \, , \qquad n = 1, 2, \ldots \, . \tag{6}$$

We observe that if we weaken (6) by replacing each λ_i in the denominator by λ_n then we get (4). Then setting all these λ_i in (6) equal to λ_n yields (5).

Included in (6) is a series of explicit bounds for eigenvalues of the Laplacian which have not as yet been explored. A start in this direction was made in [3] where quadratic inequalities were obtained.

We begin by considering the function

$$F(x) \equiv \sum_{i=1}^{n} \frac{\lambda_i}{x - \lambda_i} - \frac{mn}{4}$$

where we fix m and n temporarily. We are interested in values of x in the range $\lambda_1 < x < \infty$. We note that for $\lambda_{i-1} < x < \lambda_i$, $i = 2, 3, \ldots, n$, F is a decreasing function from $+\infty$ to $-\infty$ with vertical asymptotics at $x = \lambda_i$, $i = 1, 2, \ldots, n$. Hence F has exactly one zero between λ_{i-1} and λ_i for $i = 2, \ldots, n$. We label these zeros $r_1, r_2, \ldots r_{n-1}$. For $x > \lambda_n$ the function is also decreasing and has one zero between λ_n and $+\infty$. We label this zero r_n and it follows from [10] that $\lambda_{n+1} < r_n$.

We form the function

$$G(x) \equiv \sum_{i=1}^{n} \lambda_i \prod_{\substack{j=1 \\ j \neq i}}^{n} (x - \lambda_j) - \frac{mn}{4} \prod_{j=1}^{n} (x - \lambda_j) \, .$$

The zeros of G are the same as those of F. We introduce the symmetric functions of $\lambda_1, \lambda_2, \ldots, \lambda_n$, denoted $\sigma_1, \sigma_2, \ldots, \sigma_n$. That is, $\sigma_1 = \sum_{i=1}^{n} \lambda_i$, $\sigma_n = \prod_{i=1}^{n} \lambda_i$ and σ_k consists of the sum of the products of the λ_i taken k at a time. We write

$$G(x) = \sum_{i=1}^{n} \lambda_i \prod_{\substack{j=1 \\ j \neq i}}^{n} (x - \lambda_j) - \frac{mn}{4} [x^n - \sigma_1 x^{n-1} + \sigma_2 x^{n-2} + \cdots + (-1)^n \sigma_n]$$

Also,

$$\begin{aligned} G(x) &= \sigma_1 x^{n-1} - 2\sigma_2 x^{n-2} + 3\sigma_3 x^{n-3} + \cdots + (-1)^n n \sigma_n \\ &\quad - \frac{mn}{4} [x^n - \sigma_1 x^{n-1} + \sigma_2 x^{n-2} + \cdots + (-1)^n \sigma_n] \\ &= -\frac{mn}{4} x^n + \left(\frac{mn}{4} + 1\right) \sigma_1 x^{n-1} - \left(\frac{mn}{4} + 2\right) \sigma_2 x^{n-2} + \cdots + (-1)^{n-1} \left(\frac{mn}{4} + n\right) \sigma_n \, . \end{aligned}$$

Thus if the symmetric functions of G are written $S_1, S_2, \ldots, S_n$ we find

$$S_1 = \frac{mn+4}{mn} \sigma_1 \, , \qquad S_2 = \frac{mn+8}{mn} \sigma_2, \ldots \, , \qquad S_n = \frac{mn+4n}{mn} \sigma_n = \frac{m+4}{m} \sigma_n \, . \tag{7}$$

The inequalities (7) yield new and useful relations among the eigenvalues (1). For example, we have

$$r_1 + r_2 + \cdots + r_n = \frac{mn+4}{mn} (\lambda_1 + \lambda_2 + \cdots + \lambda_n) \, .$$

Since $\lambda_{n+1} < r_n$ and $\lambda_i < r_i < \lambda_{i+1}$ for $i = 1, 2, \ldots, n-1$, we find

$$\lambda_{n+1} \leq \frac{mn+4}{mn} (\lambda_1 + \lambda_2 + \cdots + \lambda_n) - (r_1 + r_2 + \cdots + r_{n-1}) \, . \tag{8}$$

The simplest lower bound for each r_i is λ_i and hence

$$\begin{aligned}\lambda_{n+1} &\leq \frac{mn+4}{mn}(\lambda_1 + \cdots + \lambda_n) - (\lambda_1 + \lambda_2 + \cdots + \lambda_{n-1}) \\ &\leq \frac{mn+4}{mn}\,\lambda_n + \frac{4}{mn}\sum_{i=1}^{n-1}\lambda_i \,.\end{aligned}$$

This bound is the one obtained by Payne, Polya and Weinberger. However, if a lower bound for each r_i can be obtained which is larger than λ_i then (8) yields immediately an improved bound. Because of the specific nature of the function F it is possible to get an improvement for r_i. We set $\chi_\varepsilon = (1-\varepsilon)\lambda_1 + \varepsilon\lambda_2$ with $0 < \varepsilon < 1$. If we can show that F is positive for some small value of ε (and it will be since $F \to +\infty$ as $\varepsilon \to 0$) then this value χ_ε which is larger than λ_1 is a lower bound for r_1. Similar lower bounds can be obtained for each r_i, $i = 2, \ldots, n-1$. For example, we set $m = 2$, $n = 3$ and consider

$$F(\chi_\varepsilon) = \frac{\lambda_1}{(1-\varepsilon)\lambda_1 + \varepsilon\lambda_2 - \lambda_1} + \frac{\lambda_2}{(1-\varepsilon)\lambda_1 + \varepsilon\lambda_2 - \lambda_2} + \frac{\lambda_3}{(1-\varepsilon)\lambda_1 + \varepsilon\lambda_2 - \lambda_3} - \tfrac{3}{2}\,.$$

If for some value of ε, the function F is positive then χ_ε is a lower bound for r_1 which is larger than λ_1. We desire

$$\frac{(1-\varepsilon)\lambda_1(\lambda_1 - \lambda_2) + \varepsilon\lambda_2(\lambda_2 - \lambda_1)}{\varepsilon(\varepsilon - 1)(\lambda_2 - \lambda_1)(\lambda_2 - \lambda_1)} > \tfrac{3}{2} + \frac{\lambda_3}{\lambda_3 - \varepsilon\lambda_2 + (\varepsilon - 1)\lambda_1}$$

or

$$\frac{\lambda_1 - (\lambda_2 - \lambda_1)\varepsilon}{\varepsilon(1-\varepsilon)(\lambda_2 - \lambda_1)} > \tfrac{3}{2} + \frac{\lambda_3}{\lambda_3 - \varepsilon\lambda_2 + (\varepsilon - 1)\lambda_1}\,.$$

We observe that as $\varepsilon \to 0$ the left side of the above inequality tends to $+\infty$ while the right side tends to $\frac{3}{2} + \frac{\lambda_3}{\lambda_3 - \lambda_1}$. Thus choosing ε sufficiently small, say $\varepsilon < \frac{1}{2}$ and, in addition, if necessary

$$\varepsilon < \frac{\lambda_1}{\lambda_3 + \frac{7}{2}\,\lambda_2}\,,$$

we get

$$\chi_0 = \lambda_1 + \frac{\lambda_1}{\lambda_3 + \frac{7}{2}\,\lambda_2}\,(\lambda_2 - \lambda_1)\,.$$

Thus the zero r_1 has the lower bound

$$r_1 > \lambda_1 + \frac{2\lambda_1(\lambda_2 - \lambda_1)}{3\lambda_3 + 7\lambda_2} \equiv \lambda_1 + t_1\,. \tag{9}$$

Similar estimates may be obtained for the other zeros of F. We get the estimate

$$\lambda_4 \leq \tfrac{5}{3}\lambda_3 + \tfrac{2}{3}(\lambda_1 + \lambda_2) - (t_1 + t_2 + t_3)\,. \tag{10}$$

Since each t_i is positive and expressible explicitly in terms of $\lambda_1, \lambda_2, \lambda_3$, inequality (10) yields a new upper bound for λ_4 which is independent of the domain and is only marginally related to the dimension m.

In the general case we consider the function

$$F(x_\varepsilon) = \sum_{i=1}^{n} \frac{\lambda_i}{(1-\varepsilon)\lambda_1 + \varepsilon\lambda_2 - \lambda_i} - \frac{mn}{4} .$$

Choose $\eta = \varepsilon(\lambda_2 - \lambda_1)/\lambda_1$ and as before let $\varepsilon < \frac{1}{2}$. We wish to choose η so small that

$$\frac{1}{\eta} > \frac{\lambda_2}{(1-\varepsilon)(\lambda_2 - \lambda_1)} + \sum_{i=3}^{n} \frac{\lambda_i}{\lambda_i - \varepsilon\lambda_2 - (1-\varepsilon)\lambda_1} + \frac{mn}{4} .$$

The above inequality certainly holds when

$$\frac{1}{\eta} = \frac{2\lambda_2}{\lambda_2 - \lambda_1} + \sum_{i=3}^{n} \frac{\lambda_i}{\lambda_i - \lambda_2} + \frac{mn}{4} .$$

Thus $t_1 = \lambda_1 + \eta\lambda_2$ is larger than λ_1 and smaller than r_1 since $F(t_1) > 0$. In a similar way we obtain $t_2, t_3 \ldots t_{n-1}$ with the result

$$\lambda_{n+1} \leq \frac{mn+4}{mn} \sum_{i=1}^{n} \lambda_i - \sum_{i=1}^{n-1} t_i . \tag{8}$$

By the same technique every symmetric function of the eigenvalues combined with (6) and (7) yields new upper bounds. For example the fact that $s_n = \prod_{i=1}^{n} r_i$ and $\sigma_n = \prod_{i=1}^{n} r_i$ shows at once that

$$\lambda_{n+1} \leq \frac{m+4}{m} \lambda_n \prod_{i=1}^{n-1} \lambda_i / \prod_{i=1}^{n-1} r_i .$$

By using the quantities t_i which are lower bounds for r_i and greater than λ_i we get an improvement over the standard inequality $\lambda_{n+1} \leq \frac{m+4}{m} \lambda_n$.

Of course, more precise results could be obtained by solving the polynomial equation of the n-th degree explicitly. But since only numerical results can be obtained for $n \geq 5$, the technique outlined here yields explicit inequalities with little additional effort.

2. HOOK'S INEQUALITIES AND EXTENSIONS

The results yielding universal inequalities between eigenvalues of the Laplacian described in the previous section have been extended to inequalities for entire classes of elliptic operators. The work of Hile and Yeh [4] and Chen [2] develop the theory for the biharmonic and polyharmonic operators. Ashbaugh and Benguria [1] obtain not only upper bounds for eigenvalues but show how to obtain inequalities relating eigenvalues of two different elliptic operators, a problem considered previously by many. The literature by now has become vast and the article by Ashbaugh and Benguria [1] contains a comprehensive bibliography.

The most far-reaching results obtained thus far for universal inequalities between eigenvalues of general operators are those of S. M. Hook which we now describe. He considers linear operators in a general inner product space, obtains a variety of inequalities for such operators, and shows how these may be applied to yield many of the previously known results for partial differential operators as well as an abundant supply of new ones.

Some of the results of Hook are amenable to the methods described in section 1 for obtaining explicit bounds which are improvements over the current ones. We begin with the following general result on operators.

Theorem 1 (Hook [5]). *Let $A : \Omega \subset H \to H$ be a self-adjoint operator in a real or complex Hilbert space H with an inner product which is semibounded below. Suppose the spectrum of A is discrete with eigenvalues $\lambda_1 \leq \lambda_2 \leq \dots$. Let $\{T_j : \Omega \to H\}_{j=1}^{M}$ be any collection of skew-symmetric operators and $\{B_j : T_j(\Omega) \to H\}_{j=1}^{M}$ any collection of symmetric operators which leave Ω invariant. Denote the normalized eigenvectors of A by $\{u_j\}_{j=1}^{\infty}$ where u_j corresponds to λ_j. Suppose that $\lambda_{n+1} > \lambda_n$. Let $\langle \cdot \rangle$ denote the inner product. If for some i, j we have $T_j u_i \neq 0$ and if there is a $\gamma > 0$ and $\beta > \lambda_n$ so that*

$$\gamma \sum_{i=1}^{n} \sum_{j=1}^{M} \langle [T_j, B_j] u_i, u_j \rangle + \gamma^2 \sum_{i=1}^{n} \sum_{j=1}^{M} \frac{1}{\beta - \lambda_i} \langle T_j u_i, T_j u_i \rangle = \tfrac{1}{2} \textstyle\sum_{i=1}^{n} \sum_{j=1}^{M} \langle [[A, B_j], B_j] u_i, u_i \rangle$$

then $\beta \geq \lambda_{n+1}$. Moreover, if $T_j u_i \neq \Omega$ for all $i = 1, 2, \dots, n$ and $j = 1, 2, \dots, M$ we have $\beta > \lambda_{n+1}$.

The above result is an abstract version of a generalization of the theorems of Payne, Polya and Weinberger [8] and Hile and Protter [3]. Effective applications of special cases were given by Hile and Yeh [4] and Chen [2].

As one application of the above theorem we consider the general second order linear elliptic equation in $\mathbb{R}^m$ with Dirichlet boundary conditions:

$$\begin{aligned} Au &= -e^{\mathbf{w}\cdot\mathbf{x}} \mathrm{div}(M^2 e^{\mathbf{w}\cdot\mathbf{x}} \mathrm{grad}\, u) - \tfrac{1}{4} \|M\mathbf{w}\|^2 u = \lambda u \quad \text{in } \Omega \\ u &= 0 \quad \text{on } \partial\Omega\,. \end{aligned} \tag{1}$$

Here Ω is an arbitrary bounded domain in $\mathbb{R}^m$, M is a constant $m \times m$ real symmetric postive definite matrix and $\mathbf{w} \in \mathbb{R}^m$ is a given constant vector. Actually (1) is the same as a general second order linear elliptic equation with constant coefficients. We have

Theorem 2. *If equation (1) above with Dirichlet boundary conditions has a discrete spectrum $\lambda_1 \leq \lambda_2 \leq \dots \leq \lambda_n < \lambda_{n+1} < \dots$, and is self-adjoint with the inner product*

$$\langle u, v \rangle = \int_\Omega uv\ e^{-\mathbf{w}\cdot\mathbf{x}}$$

then the eigenvalues satisfy

$$\sum_{i=1}^{n} \frac{\lambda_i}{\lambda_{n+1} - \lambda_i} > \frac{mn}{4}\,. \tag{2}$$

The proof is given in [6]. The inequality (2) is identical with the one considered in section 1 and by forming the function

$$F(x) = \sum_{i=1}^{n} \frac{\lambda_i}{x - \lambda_i} - \frac{mn}{4}$$

we can obtain explicit upper bounds such as (8) for each eigenvalue in terms of lower eigenvalues which are improvements of those obtained earlier.

3. HIGHER ORDER OPERATORS

Hook is able to adapt Theorem 1 so that it applies to higher order elliptic operators. The general theorem on abstract elliptic operators is given in [5]. As an application he shows how eigenvalue inequalities can be obtained for a class of higher order operators subject to Dirichlet boundary conditions. We let $p(x) = \sum_{i=0}^{\ell} a_i x^i$ be a monic polynomial and $q(x) = \sum_{i=0}^{r} b_i x^i$ a polynomial where generally we suppose $r < \ell$ and $a_i \geq 0$, $b_i \geq 0$ for all i. We introduce the quantities $\alpha_{i,j}$ as follows: $\alpha_{i,0} = 1$ and for $j \geq 1$ $\alpha_{i,j}$ is the largest positive root of

$$a_0 + \sum_{k=j}^{\ell} a_k x^k - q(x) - \lambda_i = 0 \quad \text{if } r \leq j \leq \ell \tag{1}$$

and of

$$a_0 + \sum_{k=j+1}^{r} b_k \alpha_{i,k}^k - \sum_{k=j}^{\ell} a_k x^k - \sum_{k=0}^{j} b_k x^k - \lambda_i = 0 \quad \text{if } 1 \leq j \leq r-1\,, \tag{2}$$

where the quantities λ_i are defined below.

Theorem 3. *Let $p(x), q(x)$ be as above and let $\Omega \subset \mathbb{R}^m$ be a bounded domain. Suppose the problem*

$$[p(-\Delta) - q(-\Delta)]u = \lambda u \quad \text{in } \Omega \tag{3}$$

$$u = \frac{\partial u}{\partial n} = \frac{\partial^2 u}{\partial n^2} = \cdots = \frac{\partial^{\ell-1} u}{\partial n^{\ell-1}} = 0 \quad \text{on } \partial\Omega \tag{4}$$

is self-adjoint and has a discrete spectrum $\lambda_1 \leq \lambda_2 \leq \cdots \leq \lambda_n \leq \ldots$. If $\lambda_n < \lambda_{n+1}$ then the eigenvalues of (3),(4) satisfy the inequality

$$\left(\sum_{i=1}^{n} \frac{\alpha_{i,1}}{\lambda_{n+1} - \lambda_i}\right)\left(\sum_{i=1}^{n}\sum_{j=1}^{\ell}(m+2j-2)j a_j \alpha_{i,j-1}^{j-1}\right) > \frac{m^2 n^2}{4} \tag{5}$$

where the $\alpha_{i,j}$ are defined by (1),(2).

The proof of Theorem 3 is given in [6]. We see that if we define the function

$$F_1(x) = \left(\sum_{i=1}^{n} \frac{\alpha_{i,1}}{x - \lambda_i}\right)\left(\sum_{i=1}^{n}\sum_{j=1}^{\ell}(m+2j-2)j a_j \alpha_{i,j-1}^{j-1}\right) - \frac{m^2 n^2}{4}$$

then the zeros of F_1 lie between the vertical asymptotes given by the lines $x = \lambda_i$, $i = 1, 2, \ldots, n$ while the $(n+1)$-st zero is greater than λ_{n+1}, as the function $F_1(x) \to -m^2 n^2/4$ as $x \to \infty$. Thus the argument in section 1 is applicable for obtaining an improved explicit upper bound for λ_{n+1}. Of course the quantities $\alpha_{i,j}$ are not easily computable except in special cases. Nevertheless the polynomial equation obtained from $F_1(x)$ by multiplying through by $\prod_{i=1}^{n}(x - \lambda_i)$ and examining the roots of $F_1(x)\prod_{i=1}^{n}(x - \lambda_i) = 0$ allows us to use the symmetric functions of the roots to obtain estimates as before.

If, in Theorem 3 we choose $p(x) = x^\ell$ and $q = 0$, then (5) becomes

$$\left(\sum_{i=1}^{n} \frac{\lambda_i^{1/\ell}}{\lambda_{n+1}-\lambda_i}\right)\left(\sum_{i=1}^{n} \lambda_i^{(\ell-1)/\ell}\right) > \frac{m^2n^2}{4\ell(m+2\ell-2)}\,. \tag{6}$$

This is an inequality obtained by Hile and Yeh and by Chen. An explicit inequality for λ_{n+1} can be obtained by weakening (6) so that each λ_i in the first term is replaced by λ_n. Then we find

$$\lambda_{n+1} < \lambda_n + \frac{4\ell(m+2\ell-2)}{m^2n^2}\left(\sum_{i=1}^{n} \lambda_i^{1/\ell}\right)\left(\sum_{i=1}^{n} \lambda_i^{(\ell-1)/\ell}\right)\,. \tag{7}$$

By employing the function

$$F_2(x) = \sum_{i=1}^{n} \frac{\lambda_i^{1/\ell}}{x-\lambda_i} \sum_{i=1}^{n} \lambda_i^{(\ell-1)/\ell} - \frac{m^2n^2}{4\ell(m+2\ell-2)}$$

and getting lower bounds for each of the zeros of F_2 which are larger than the corresponding eigenvalue, we obtain not only an improvement of (7) but also a set of n universal inequalities relating the first $n+1$ eigenvalues through their symmetric functions.

REFERENCES

1. M. S. Ashbaugh and R. D. Benguria, Isoperimetric inequalities for eigenvalue ratios, in *Partial Differential Equations of Elliptic type*, A.Alvino, E.Fabes and G.Talenti, editors, pp. 1–36 (1994).
2. Z.-C. Chen, Inequalities for a class of polyharmonic operators, *Appl. Anal.* **27** (1988), 289–314.
3. G. N. Hile and M. H. Protter, Inequalities for the eigenvalues of the Laplacian, *Indiana Univ. Math. J.* **29** (1980), 523–538.
4. G. N. Hile and R. Z. Yeh, Inequalities for eigenvalues of the biharmonic operator, *Pacific J. Math.* **112** (1984), 115–133.
5. S. M. Hook, Inequalities for eigenvalues of self-adjoint operators, *Trans. Amer. Math. Soc.* **318** (1990), 237–259.
6. S. M. Hook, Domain-independent upper bounds for eigenvalues of elliptic operators, *Trans. Amer. Math. Soc.* **318** (1990), 615–642.
7. S. M. Hook, More inequalities for eigenvalues of self-adjoint operators, to appear in *Trans. Amer. Math. Soc.*
8. L. E. Payne, G. Polya and H. F. Weinberger, On the ratio of consecutive eigenvalues, *J. Math. and Physics* **35** (1956), 289–298.
9. M. H. Protter, Can one hear the shape of a drum, revisited, *SIAM Review* **29** (1987), 185–197.
10. M. H. Protter, Universal inequalities for eigenvalues, in *Maximum Principles and Eigenvalue Problems in Partial Differential Equations*, P.W.Schaefer, editor, Pitman Research Notes in Math. **175**, (1988), 111–120.

26. Stability for Abstract Evolution Equations

P. Pucci Dipartimento di Matematica, Università di Perugia, Perugia, Italy

James Serrin Department of Mathematics, University of Minnesota, Minneapolis, Minnesota

1. Introduction

In a recent paper [6] we studied the question of asymptotic stability for non-autonomous dissipative wave systems. Earlier work in the same direction is due to Marcati [2,3] and Nakao [4], who treated particularly the case of abstract evolution equations. In this note we give a new asymptotic stability theorem which extends the analysis in [2,3,4] by taking into account the techniques introduced in [6].

We focus on abstract equations of the form

$$[P(u'(t))]' + A(u(t)) + Q(t, u'(t)) + F(u(t)) = 0, \tag{1.1}$$

where A, F, P and Q are nonlinear operators on appropriate Banach spaces. We understand P to be the evolution operator, A as a differential operator of divergence form, Q as a damping term, and F as a restoring force. Concrete examples of (1.1) include the principal case of wave systems, where $P = I$, $A = -\Delta$, and also, more generally, the p-Laplacian, where $A = -\Delta_p$, $p > 1$, as well as the polyharmonic operator $A = (-\Delta)^L$, $L \geq 1$.

As an important feature of the present work, we allow the damping $Q = Q(t, v)$ to be strongly non-autonomous in t and nonlinear in v. Our main theorem is given in Section 3, while Section 2 is devoted to preliminary results. In particular, in Section 2 we formulate a careful definition of solution of (1.1), which clarifies and generalizes the corresponding definitions in [2,3,4], and moreover resolves the principal difficulty in treating the abstract case, namely that an appropriate energy balance for (1.1) cannot be derived directly, but must instead be inferred from analogy with concrete equations and systems.

2. Notation and Preliminary Results

Let $V = (V, \|\cdot\|_V)$, $W = (W, \|\cdot\|_W)$ and $X = (X\|\cdot\|_X)$ be real Banach spaces, V', W', X' their dual spaces, and $\langle\cdot,\cdot\rangle_V$, $\langle\cdot,\cdot\rangle_W$, $\langle\cdot,\cdot\rangle_X$ the natural dual pairings. Moreover, we suppose that the spaces V, W, X have a common subspace $G \neq \{0\}$.*

$$A : W \to W', \quad F : X \to X' \quad \text{and} \quad P : V \to V'$$

be functions satisfying the following structural conditions:

(S1) A, F and P are the Fréchet derivatives of real valued C^1 potentials

$$\mathcal{A} : W \to \mathbf{R}, \quad \mathcal{F} : X \to \mathbf{R}, \quad \mathcal{P} : V \to \mathbf{R},$$

respectively, where without loss of generality we assume that $\mathcal{A}(0) = 0$, $\mathcal{F}(0) = 0$ and $\mathcal{P}(0) = 0$;

(S2) $\langle A(u), u\rangle_W + \langle F(u), u\rangle_X \geq 0$ for all $u \in G$;

(S3) $\mathcal{P}^*(v) = \langle P(v), v\rangle_V - \mathcal{P}(v) \geq 0$ in V;

(S4) for every $d > 0$ the sets

$$D = \{u \in G : \mathcal{A}(u) + \mathcal{F}(u) \leq d\}, \quad E = \{v \in V : \mathcal{P}^*(v) \leq d\}$$

are bounded in X and V, respectively, and the sets $F(D)$, $P(E)$ are bounded in X' and V'.

It is easily seen that

$$\mathcal{A}(u) = \int_0^1 \langle A(su), u\rangle_W ds, \quad \mathcal{F}(x) = \int_0^1 \langle F(sx), x\rangle_X ds;$$

consequently condition (S2) implies that

(S2)′ $$\mathcal{A}(u) + \mathcal{F}(u) \geq 0 \quad \text{for } u \in G.$$

Denote the time set $[0, \infty)$ by J. Let S be a given subset of $J \times V$ and

$$Q : S \to X'$$

a continuous function.

---

* If V, W, X are themselves subspaces of a common vector space Z, then we can take $G = V \cap W \cap X$, this being the maximal subspace contained in all three spaces. As an example, references [3],[4] deal with the special case $W = X \subset V$ (in our notation), for which the appropriate subspace G is simply X itself.

We consider the abstract evolution equation

$$(2.1)\qquad [P(u'(t))]' + A(u(t)) + Q(t, u'(t)) + F(u(t)) = 0, \quad t \in J.$$

Let K denote the subset of all functions $u : J \to G$ such that

$$u \in C^1(J \to V) \cap C(J \to W) \cap L^\infty_{\text{loc}}(J \to X).$$

We say that u is a *strong solution* of (2.1) if

(a) $u \in K$ and $(t, u'(t)) \in S$ for a.a. $t \in J$;

(b) u verifies (2.1) in the following distribution sense:

$$\langle P(u'(s)), \phi(s)\rangle_V \,|_0^t = \int_0^t \{\langle P(u'(s)), \phi'(s)\rangle_V - \langle A(u(s)), \phi(s)\rangle_W - \langle Q(s, u'(s)), \phi(s)\rangle_X - \langle F(u(s)), \phi(s)\rangle_X\} ds$$

for all $\phi \in K$ and $t \in J$.

Note that the first two integrands on the right hand side of (b) are well-defined and integrable on $[0, t]$, $t \in J$. We show at the end of this section, with the help of further structural conditions, that the last two terms also are meaningful.

Let $\mathcal{E} : J \to \mathbf{R}$ be the *total energy* of the field $u \in K$, that is

$$(2.2)\qquad \mathcal{E}(t) = \mathcal{P}^*(u'(t)) + \mathcal{A}(u(t)) + \mathcal{F}(u(t)).$$

(The energy function $\mathcal{E}$, of course, arises naturally for classical conservation laws.) For simplicity in printing, we shall write

$$\mathcal{H}(t) = -\mathcal{E}(t).$$

We now postulate the following crucial *connection conditions* between Q, u' and $\mathcal{E}'$ (see [1,6]):

(S5) For every strong solution u of (2.1) the corresponding function $\mathcal{H}$ is *non-decreasing and absolutely continuous on* J. Moreover there are exponents $q > m > 1$ and a non-negative function $\delta \in L^1_{\text{loc}}(J)$ - independent of the solution u - such that

$$(2.3)\qquad \|Q(t, u'(t))\|_{X'} \le [\delta(t)]^{1/m}[\mathcal{H}'(t)]^{1/m'} + [\delta(t)]^{1/q}[\mathcal{H}'(t)]^{1/q'} \quad \text{a.e. in } J,$$

where m' and q' respectively denote the Hölder conjugates of m and q.

(S6) There is a non-negative function σ on J, with $1/\sigma \in L^{m-1}_{\text{loc}}(J)$, and a function $\omega = \omega(\tau)$, $\tau \ge 0$, with $\omega(0) = 0$, ω increasing in $[0, 1]$ and $\omega(\tau) = 1$ for $\tau \ge 1$, such that

$$\sigma(t)\omega(\|u'(t)\|_V) \le \mathcal{H}'(t) \quad \text{a.e. in } J$$

for every solution u of (2.1).

The regularity and monotonicity conditions expressed in (S5) are motivated by the classical conservation balance

$$\mathcal{H}(t) - \mathcal{H}(0) = \int_0^t \langle Q(s, u'(s)), u'(s)\rangle ds, \tag{2.4}$$

where $\langle \cdot, \cdot \rangle$ denotes an appropriate pairing between Q and v. That Q represents a damping is expresed by the requirement that the integrand in (2.4) be non-negative; we are thus led directly to the *axiomatic assumption* that $\mathcal{H}$ be absolutely continuous and non-decreasing in J along any solution u of (2.1).

In concrete subcases of (2.1), when V, W and X are standard Lebesgue or Sobolev spaces, the pairing $\langle \cdot, \cdot \rangle$ reduces to an appropriate Lebesgue integration (see [6]).

An abstract case which further clarifies (2.4) occurs when $S = J \times Y$ and Y is a Banach space with continuous inclusions $X \subset Y \subset V$, and $Q : S \to Y'$. Here the pairing $\langle \cdot, \cdot \rangle$ can be specified explicitly as $\langle \cdot, \cdot \rangle_Y$. Even more, in this case (2.3) can be derived, up to a multiplicative factor, from the direct conditions

$$\|Q(t, v)\|_{Y'} \leq \delta(t)(\|v\|_Y^{m-1} + \|v\|_Y^{q-1})$$

and

$$\|Q(t, v)\|_{Y'} \cdot \|v\|_Y \leq \gamma \langle Q(t, v), v\rangle_Y, \quad \gamma = \text{Const.} \geq 1$$

(reverse pairing inequality) as one easily checks. It is essentially this case which was discussed in [3,4]; see also the remarks at the end of the paper.

We can now show that the third and fourth integrands in (b) are meaningful for all $t \in J$. First, by (S5), Hölder's inequality, and the fact that $\phi \in L^\infty_{\text{loc}}(J \to X)$ we have

$$\begin{aligned} \int_0^t |\langle Q(s, u'(s)), \phi(s)\rangle_X| ds \leq \sup_{[0,t]} \|\phi(s)\|_X & \left[\left(\int_0^t \delta(s) ds \right)^{1/m} \left(\int_0^t \mathcal{H}'(s) ds \right)^{1/m'} \right. \\ & \left. + \left(\int_0^t \delta(s) ds \right)^{1/q} \left(\int_0^t \mathcal{H}'(s) ds \right)^{1/q'} \right] < \infty, \end{aligned} \tag{2.5}$$

as required. That the final integral in (b) is well-defined is a consequence of the first part of (2.9) in the following lemma, which will also be useful for the main result of the next section.

LEMMA. *Let* (S1)-(S5) *hold. Then for any strong solution* u *of* (2.1) *we have*

$$0 \leq \mathcal{E}(t) \leq \mathcal{E}(0) \quad \textit{in } J, \quad \mathcal{E}'(t) \leq 0 \quad \text{a.e. } \textit{in } J, \tag{2.6}$$

$$\mathcal{H}' \in L^1(J), \tag{2.7}$$

$$(2.8) \qquad 0 \le \mathcal{A}(u(t)) + \mathcal{F}(u(t)) \le \mathcal{E}(0), \qquad 0 \le \mathcal{P}^*(u'(t)) \le \mathcal{E}(0) \qquad \text{in } J.$$

Moreover, there exists a constant $C > 0$ *such that*

$$(2.9) \qquad \|u(t)\|_X,\ \|F(u(t))\|_{X'} \le C, \quad \|u'(t)\|_V,\ \|P(u'(t))\|_{V'} \le C \quad \text{in } J.$$

PROOF. From (S2)′, (S3) and (2.2) we get $\mathcal{E}(t) \ge 0$ in J. The condition $(2.6)_2$ follows directly from (S5), and in turn also $\mathcal{E}(t) \le \mathcal{E}(0)$. By (2.6) and the fact that $\mathcal{H} \in AC(J)$, we get (2.7).

Condition (2.8) is a consequence of $(2.6)_1$, (S2)′ and (S3). Relation (2.9) follows immediately from (S4), with $d = \mathcal{E}(0)$.

A final structure condition will be required for our main result (see [6], Lemma 3.4).

(S7) For all $\ell > 0$ there exists $\bar{\alpha} = \bar{\alpha}(\ell) > 0$ such that $u \in G$ and $\mathcal{A}(u) + \mathcal{F}(u) \ge \ell$ implies $\langle A(u), u\rangle_W + \langle F(u), u\rangle_X \ge \bar{\alpha}(\ell)$.

3. Asymptotic Stability

We now turn to the main result of the paper.

THEOREM. *Let* (S1)-(S7) *hold. Suppose that there is a non-negative function* $k \not\equiv 0$ *of class* $AC(J)$ *such that*

$$(3.1) \qquad \lim_{t\to\infty} \int_0^t |k'(s)|ds \Big/ \int_0^t k(s)ds = 0,$$

$$(3.2) \qquad \liminf_{t\to\infty} \int_0^t (\delta + \sigma^{1-m})k^m ds \Big/ \left(\int_0^t k ds\right)^m < \infty.$$

Then along any strong solution u *of* (2.1) *we have*

$$(3.3) \qquad \lim_{t\to\infty} \mathcal{E}(t) = 0.$$

REMARK. It is worth noting that (3.1) implies $k \notin L^1(J)$. [Otherwise, we would have $k'(t) = 0$ on J and $k(t) =$ Const. But then k must be identically zero on J, in contradiction with the assumption $k \not\equiv 0$.] On the other hand, (3.1) obviously holds if $k \notin L^1(J)$ and $k' \in L^1(J)$. The last two conditions were in fact principal requirements in the main stability Theorem 3.1 of [6], so that (3.1) gives a better condition for stability than that result.

Various applications of condition (3.2) are given in Section 5 of [6], to which the reader is referred.

PROOF OF THEOREM. Suppose for contradiction that (3.3) fails along some strong solution u of (2.1). Then by (2.6) there exists $\ell > 0$ such that $\mathcal{E}(t) \searrow 2\ell$ as $t \to \infty$.

In what follows, by a standard approximation procedure we can suppose without loss of generality that $k \in C^1(J)$. Then since $u \in K$, we also have $\phi = ku \in K$. Put

$$U = U(t) = \langle P(u'), \phi\rangle_V = k\langle P(u'), u\rangle_V. \tag{3.4}$$

By (b) with $\phi = ku \in K$, it follows that

$$\begin{aligned} U(s)\Big|_T^t = \int_T^t \{&k'\langle P(u'), u\rangle_V + k[\langle P(u'), u'\rangle_V + \mathcal{P}^*(u')] \\ &- k[\mathcal{P}^*(u') + \langle A(u), u\rangle_W + \langle F(u), u\rangle_X] - k\langle Q(s,u'), u\rangle_X\}ds \\ = \int_T^t \{&I_1 + I_2 + I_3 + I_4\}ds, \end{aligned} \tag{3.5}$$

this being valid for all $t \geq T \geq 0$ in J.

By (2.9) we have

$$\int_T^t I_1 ds \leq \int_T^t |k'(s)| \cdot \|P(u'(s))\|_{V'} \cdot \|u(s)\|_V ds \leq C^2 \int_T^t |k'(s)|ds. \tag{3.6}$$

Next, note that

$$I_2 \leq [C^2 + \mathcal{E}(0)]k \quad \text{in } J,$$

by (2.8) and (2.9). Now fix $\vartheta > 0$. By (S1) and (S3) the functions P and $\mathcal{P}^*$ are continuous on V, and $\mathcal{P}^*(0) = 0$. Hence there exists $\Lambda(\vartheta) > 0$ such that

$$\langle P(v), v\rangle_V + \mathcal{P}^*(v) \leq \vartheta$$

for all $v \in V$ with $\|v\|_V \leq \Lambda(\vartheta)$. Consequently

$$I_2 \leq k \begin{cases} \vartheta & \text{in } J_1 \\ C + \mathcal{E}(0) & \text{in } J_2, \end{cases}$$

where $J_1 = \{t \in J : \|u'(t)\|_V \leq \Lambda(\vartheta)\}$ and $J_2 = J \backslash J_1$. Now by (S6)

$$\frac{\mathcal{H}'(t)}{\sigma(t)} \geq \omega(\Lambda(\vartheta)) \quad \text{for a.a. } t \in J_2.$$

Thus

$$I_2 \leq k \begin{cases} \vartheta \\ \gamma(\vartheta)[\mathcal{H}'/\sigma]^{1/m'} \end{cases} = \begin{cases} \vartheta k & \text{in } J_1 \\ \gamma(\vartheta)[\sigma^{1-m}k^m]^{1/m}[\mathcal{H}']^{1/m'} & \text{in } J_2 \text{ (a.e.)}, \end{cases}$$

where

$$\gamma(\vartheta) = [C + \mathcal{E}(0)]\, \omega(\Lambda(\vartheta))^{-1/m'}.$$

Therefore, for all $t \geq T \geq 0$,

$$\int_T^t I_2 ds \leq \vartheta \int_T^t k(s)ds + \gamma(\vartheta)\epsilon(T)\left(\int_0^t \sigma^{1-m}k^m ds\right)^{1/m} \tag{3.7}$$

in view of Hölder's inequality and (2.7), where

$$\epsilon(T) = \left(\int_T^\infty \mathcal{H}'(s)ds\right)^{1/m'} \to 0 \quad \text{as } T \to \infty. \tag{3.8}$$

Next, if $\mathcal{P}^*(u'(t)) \geq \ell$ at some $t \in J$, then by (S2)

$$I_3(t) \leq -\ell k(t).$$

On the other hand, if $\mathcal{P}^*(u'(t)) \leq \ell$, then

$$\mathcal{A}(u(t)) + \mathcal{F}(u(t)) \geq \ell,$$

since $\mathcal{E} = \mathcal{P}^*(u') + \mathcal{A}(u) + \mathcal{F}(u) \geq 2\ell$ on J. Hence by (S7) and the fact that $u(t) \in G$, we get

$$\langle A(u(t)), u(t)\rangle_W + \langle F(u(t)), u(t)\rangle_X \geq \bar{\alpha}(\ell).$$

But $\mathcal{P}^*(u'(t)) \geq 0$ from (S3), which therefore gives $I_3(t) \leq -\bar{\alpha}(\ell)k(t)$. Consequently

$$I_3 \leq -\alpha k, \quad \alpha = \alpha(\ell) = \min\{\ell, \bar{\alpha}(\ell)\}. \tag{3.9}$$

Let $M(t) = \sup\{k(s) : s \in [T,t]\}$. By Young's inequality and the fact that $q > m > 1$ we obtain

$$\left(\int_T^t \delta k^q ds\right)^{1/q} \leq M(t)^{(q-m)/q}\left(\int_T^t \delta k^m ds\right)^{1/q} \leq M(t) + \left(\int_T^t \delta k^m ds\right)^{1/m}.$$

Thus, as in (2.5) with $\phi = u \in K$, we have

$$\begin{aligned}\int_T^t I_4 ds \leq \sup_{[0,t]}\|u(s)\|_X &\left[\left(\int_T^t \delta k^m ds\right)^{1/m}\cdot\left(\int_T^t \mathcal{H}'ds\right)^{1/m'}\right.\\ &\left.+\left(\int_T^t \delta k^q ds\right)^{1/q}\cdot\left(\int_T^t \mathcal{H}'ds\right)^{1/q'}\right]\\ \leq \epsilon_1(T)&\left[\left(\int_0^t \delta k^m ds\right)^{1/m} + M(t)\right],\end{aligned} \tag{3.10}$$

where

$$\epsilon_1(T) = C\left[\left(\int_T^\infty \mathcal{H}'ds\right)^{1/m'} + \left(\int_T^\infty \mathcal{H}'ds\right)^{1/q'}\right] \to 0 \quad \text{as } T \to \infty. \tag{3.11}$$

Combining (3.5)-(3.7), (3.9), (3.10) yields, for all $t \ge T \ge 0$,

$$\begin{aligned} U(s)|_T^t \le C^2 \int_T^t |k'(s)|ds + \theta \int_T^t k(s)ds + \gamma(\vartheta)\epsilon(T)\left(\int_0^t \sigma^{1-m}k^m ds\right)^{1/m} \\ - \alpha \int_T^t k(s)ds + \epsilon_1(T)\left[\left(\int_0^t \delta k^m ds\right)^{1/m} + M(t)\right]. \end{aligned} \tag{3.12}$$

In view of (3.2) there is a number $B > 0$ and a sequence $t_i \nearrow \infty$ as $i \to \infty$ such that, for all i,

$$\int_0^{t_i} (\delta + \sigma^{1-m})k^m ds \le \left(B \int_0^{t_i} k(s)ds\right)^m. \tag{3.13}$$

Moreover by (3.1) we can suppose as well that

$$\int_0^{t_i} |k'(s)|ds \le \frac{\alpha}{8C^2} \int_0^{t_i} k(s)ds. \tag{3.14}$$

Now choose $\vartheta = \alpha/8$ and T so large that

$$\epsilon(T) \le \alpha/8B\gamma(\vartheta), \quad \epsilon_1(T) \le \min\{\alpha/8M,\ C^2\}, \tag{3.15}$$

which can be done by (3.8) and (3.11). Then from (3.12)-(3.15) there results, for all i such that $t_i > T$,

$$\begin{aligned} U(s)\,|_T^{t_i} &\le \frac{\alpha}{8}\int_0^{t_i} k(s)ds + \frac{\alpha}{8}\int_T^{t_i} k(s)ds + \frac{\alpha}{8}\int_0^{t_i} k(s)ds \\ &\quad - \alpha \int_T^{t_i} k(s)ds + \frac{\alpha}{8}\int_0^{t_i} k(s)ds + C^2 M(t_i) \\ &= \frac{7}{8}\alpha \int_0^T k(s)ds - \frac{\alpha}{2}\int_0^{t_i} k(s)ds + C^2 M(t_i). \end{aligned} \tag{3.16}$$

On the other hand, for $t_i > T$,

$$|U(t_i)| \le k(t_i)\ \|P(u'(t_i))\|_{V'}\ \|u(t_i)\|_V \le C^2 M(t_i) \tag{3.17}$$

and

$$M(t_i) \le k(T) + \int_T^{t_i} |k'(s)|ds \le k(T) + \frac{\alpha}{8C^2}\int_0^{t_i} k(s)ds \tag{3.18}$$

by (3.14). Hence by (3.16)-(3.18) one easily obtains, for all i sufficiently large,

$$0 \le U(T) + 2C^2 k(T) + \frac{7}{8}\alpha \int_0^T k(s)ds - \frac{\alpha}{4}\int_0^{t_i} k(s)ds.$$

Consequently, since k is non-negative, we find that $k \in L^1(J)$.

This provides an immediate contradiction since (3.1) implies that $k \notin L^1(J)$, and so completes the proof.

4. Concluding Remarks

1. The special case $S = J \times Y$ discussed in Section 2 was treated earlier by Nakao [4]. For his main result it was assumed that V is a Hilbert space, that $W = X$, and that $P = I$, $\mathcal{A} + \mathcal{F} = F_A : X \to \mathbf{R}_0^+$. In addition, (S5) and (S6) were formulated with the restrictions

$$0 < \text{Const.} \leq \delta(t) = \text{Const.}\ \sigma(t) \quad \text{in } J,$$

$$\omega(\tau) = \tau^m, \quad m \geq 2;$$

and (S7) for the explicit choice $\bar{a}(\ell) = \text{Const.}\,\ell$. Finally, Nakao considered exactly the special function $k \equiv 1$, for which (3.1) and (3.2) are automatically satisfied.

The case when $S = J \times Y$ was also discussed by Marcati [3], but with autonomous damping. The hypotheses in [3] are similar to those in [4], with the exceptions that P need not be the identity, nor V a Hilbert space.

2. If W is continuously imbedded in X, slightly generalizing the assumption $W = X$ in [3,4], it is easy to see that the solution space K reduces to $C(J \to W) \cap C^1(J \to V)$. In this case, from the relation $F \in C(X \to X')$ we get

$$F \circ u \in C(J \to X') \quad \text{for all } u \in K,$$

which is enough to make the final integral in (b) well-defined. Hence the assumption in (S4) that $F(D)$ be bounded in X' can be omitted, since its only purpose was to make the relation (b) meaningful. (Of course, when (S4) is thus modified, the consequent implication in (2.9), that $\|F(u(t))\|_{X'} \leq C$ along a strong solution of (2.1), must also be dropped.)

3. Condition (S6) need not hold in the entire interval J, but in fact can be restricted to a measurable *control subset* $I \subset J$. The main theorem then needs to be revised only in two places. First, (3.1) should be modified to include the condition

$$k = 0 \quad \text{in } J \setminus I, \tag{3.1$'$}$$

and second, (3.2) should be replaced by

$$\liminf_{t \to \infty} \int_{[0,t] \cap I} (\delta + \sigma^{1-m}) k^m ds \bigg/ \left(\int_0^t k ds \right)^m < \infty. \tag{3.2$'$}$$

The proofs remain almost the same.

The importance of control subsets in studying asymptotic stability is illustrated in [5], in the context of ordinary differential systems.

Acknowledgment. P. Pucci is a member of *Gruppo Nazionale di Analisi Funzionale e sue Applicazioni* of the *Consiglio Nazionale delle Ricerche*. This research has been partly supported by the Italian *Ministero dell'Università e della Ricerca Scientifica e Tecnologica.*

REFERENCES

[1] H.A. Levine & J. Serrin, *A Global Nonexistence Theorem for Quasilinear Evolution Equations with Dissipation*, to appear.

[2] P. Marcati, *Decay and stability for nonlinear hyperbolic equations*, J. Diff. Equations **55** (1984), 30-58.

[3] P. Marcati, *Stability for second order abstract evolution equations*, Nonlinear Anal. **8** (1984), 237-252.

[4] M. Nakao, *Asymptotic stability for some nonlinear evolution equations of second order with unbounded dissipative terms*, J. Diff. Equations **30** (1978), 54-63.

[5] P. Pucci & J. Serrin, *Asymptotic stability for intermittently controlled nonlinear oscillators*, SIAM J. Math. Anal. **25** (1994), 815-834.

[6] P. Pucci & J. Serrin, *Asymptotic stability for non-autonomous wave systems*, to appear.

27. New Techniques in Critical Point Theory

Martin Schechter Department of Mathematics, University of California, Irvine, California

1 INTRODUCTION.

Critical point theory is one of the important methods of solving nonlinear problems. The method can be applied if one wishes to solve the Euler-Lagrange system corresponding to a C^1 functional G on a Banach space E. If the functional is semibounded one can attempt to find an extremum. Otherwise one is required to search for local critical points. Several techniques have been developed for finding such points, or at least Palais-Smale sequences. These are sequences satisfying

$$G(u_k) \to c, \quad G'(u_k) \to 0. \tag{1}$$

Not every such sequence leads to a critical point, but it does so if it has a convergent subsequence (the Palais-Smale condition). In this case one obtains a critical point satisfying

$$G(u) = c, \quad G'(u) = 0. \tag{2}$$

To date, these techniques can be summarized by the following result due to K. Tintarev and the author [9].

THEOREM 1. *Let G be a C^1 functional on a Banach space E which is bounded on bounded sets. Let A, B be subsets of E such that A is bounded and*

$$A \text{ links } B \tag{3}$$

and

$$a_0 = \sup_A G \le b_0 = \inf_B G. \tag{4}$$

Then G has a Palais-Smale sequence satisfying (1) with $b_0 \le c < \infty$.

The definition of linking used in this theorem is new and much more inclusive then those used in the past. Roughly speaking, we say that A links B if $A \cap B = \phi$ and one cannot shrink A to a point without intersecting B. Not only is the definition easier to verify, but it applies to a much wider class of sets. The older concept of linking required A to be a compact set which is the "boundary" of a manifold S. For A to link B with respect to this definition, one needs $A \cap B = \phi$ and every continuous map φ from S to E which is the identity on A must satisfy $\varphi(S) \cap B \neq \phi$. The new definition has the advantage that it is symmetric in A and B in a wide variety of cases (i.e., A links B implies B links A). In such cases Theorem 1 leads to two critical points, whereas older methods produced only one.

In this paper we turn our attention to condition (4). Without this hypothesis, Theorem 1 is not true. However, we shall show that there are weaker conditions which can produce critical points.

2 CRITICAL SEQUENCES.

We suppose that instead of a set A satisfying (3) and (4) there is a sequence $\{A_k\}$ of subsets of E such that A_k links B for each k and $d_k = d(A_k, B) \to \infty$. Suppose also that

$$\limsup_{k\to\infty} d_k^{-\beta} \sup_{A_k} G \le 0, \quad b_0 = \inf_B G > -\infty \tag{5}$$

for some $\beta \ge 1$. Under these circumstances one cannot, in general, expect (5) to produce a Palais-Smale sequence. However, it does produce a sequence satisfying

$$G(u_k) \to c, \quad b_0 \le c \le \infty, G(u_k) = o(\rho_k^\beta + 1), G'(u_k) = o(\rho_k^{\beta-1} + 1) \tag{6}$$

where $\rho_k = \|u_k\|$. At first glance, such a sequence does not appear to accomplish much since $\beta \ge 1$. However, if one can show that the sequence is bounded, it becomes a Palais-Smale sequence. Moreover, in applications, the same techniques that are used to show that a Palais-Smale sequence is bounded can be used here.

However, there are situations in which one cannot find a sequence $\{A_k\}$ of sets linking B such that $d_k \to \infty$ and (5) holds. For such cases one can look for sets $\{A_k\}$ which link B and such that each A_k can be decomposed into the union of two sets A'_k, A''_k where

$$\sup_{A''_k} G \le b_0$$

while $d'_k = d(A'_k, B) \to \infty$ and

$$\limsup_{k\to\infty} d_k^{-\beta} \sup_{A'_k} G \le 0. \tag{7}$$

This leads to the same conclusion as before, namely, that there is a sequence satisfying (6). Consequences of this include the following.

THEOREM 2. *Assume that*

$$G(0) \le b_0 \le G(u), \quad u \in \partial B_\delta, \delta > 0 \tag{8}$$

and that there is a $\varphi_0 \in \partial B_1$ *such that*

$$\limsup_{R\to\infty} G(R\varphi_0)/R^\beta \le 0 \tag{9}$$

where $B_r = \{u \in E : \|u\| < r\}$*. Then there is a sequence satisfying (6).*

THEOREM 3. *Let* M, N *be closed subspaces of* E *such that* $E = M \oplus N$, $M \ne E$, $N \ne E$ *and* N *is finite dimensional. Assume that*

$$G(v) \le m_R, v \in \partial B_R \cap N \tag{10}$$

$$G(w) \ge b_0, \quad w \in M \tag{11}$$

and

$$m_R/R^\beta \to 0 \text{ as } R \to \infty. \tag{12}$$

Then the same conclusion holds.

THEOREM 4. *Let* M, N *be as above with either being finite dimensional. Assume that*

$$G(v) \le b_0, \quad v \in N \tag{13}$$

$$G(w) \ge b_0, \quad w \in \partial B_\delta \cap M \tag{14}$$

$$G(sw_0 + v) \le m_R, \quad s \ge 0, \quad v \in N, \quad \|sw_0 + v\| = R \tag{15}$$

for some $w_0 \in \partial B_1 \cap M$*, where* $0 < \delta < R$*. If (12) holds, then there is a sequence satisfying (6).*

THEOREM 5. *Let* M, N *be as above with one of them finite dimensional. Assume*

$$m_0 = \sup_{v\in N} \inf_{w\in M} G(v + w) > -\infty \tag{16}$$

and

$$m_1 = \inf_{w\in M} \sup_{v\in N} G(v + w) < \infty. \tag{17}$$

Then there is a sequence satisfying

$$G(u_k) \to c, \quad m_0 \le c \le m_1, \quad G'(u_k) \to 0. \tag{18}$$

3 SOME APPLICATIONS.

We present applications of the theorems of Section 2. Let Ω be a smooth, bounded domain in $\mathbf{R}^n$, and let $A \ge \lambda_0 > 0$ be a selfadjoint operator on $L^2(\Omega)$ such that

$$C_0^\infty(\Omega) \subset D := D(A^{1/2}) \subset H^m(\Omega), \quad m > 0. \tag{19}$$

We assume that A has eigenvalues

$$\lambda_0 < \lambda_1 < \cdots < \lambda_k < \cdots$$

and eigenfunctions that are in $L^\infty(\Omega)$. Let $f(x,t)$ be a Caratheodory function on $\Omega \times \mathbf{R}$ such that

$$|f(x,t)| \le C(|t|+1). \tag{20}$$

Let

$$F(x,t) = \int_0^t f(x,s)ds.$$

We assume that there is an eigenvalue $\lambda_\ell, \ell \ge 0$, such that

$$\lambda_\ell t^2 \le 2F(x,t), \quad |t| \le \delta \tag{21}$$

$$2F(x,t) \le \lambda_{\ell+1} t^2, \quad t \in \mathbf{R} \tag{22}$$

$$2F(x,t) \le \lambda_\ell t^2 + V(x)^2 h(t) + W(x), \quad t \in \mathbf{R} \tag{23}$$

where $\delta > 0, W \in L^1(\Omega), V(x) \in L^2(\Omega)$ maps D into $L^2(\Omega)$ and $h(t)$ is a locally bounded function satisfying

$$h(t)/t^2 \to 0 \text{ as } |t| \to \infty. \tag{24}$$

Finally, we assume that

$$f(x,t)/t \to \alpha_\pm(x) \text{ a.e. as } t \to \pm\infty \tag{25}$$

and the only solution of

$$Au = \alpha_+ u^+ - \alpha_- u^- \tag{26}$$

is $u \equiv 0$, where $u^\pm = \max\{\pm u, 0\}$. We have

THEOREM 6. *Under the above hypotheses the equation*

$$Au = f(x,u), \quad u \in D \tag{27}$$

has at least one nontrivial solution.

In proving Theorem 6 one introduces the functional

$$G(u) = \|u\|_D^2 - 2\int_\Omega F(x,u)dx \tag{28}$$

where

$$\|u\|_D = \|A^{1/2}u\|.$$

It is readily verified under hypothesis (20) that G is a C^1 functional on D and that

$$(G'(u),v) = 2(u,v)_D - 2(f(u),v), u,v \in D \tag{29}$$

where we write $f(u)$ for $f(x,u)$. It therefore follows that u is a solutions of (27) if it satisfies (2). We let N_ℓ denote the subspace of $L^2(\Omega)$ spanned by the eigenfunctions of A corresponding to the eigenvalues $\lambda_0, \cdots, \lambda_\ell$, and we take $M_\ell = N_\ell^\perp \cap D$. By (22)

$$G(w) \ge 0, \quad w \in M_\ell. \tag{30}$$

Also (21) implies

$$G(v) \le -\epsilon\|v'\|_D^2, \quad v \in N_\ell, \quad \|v\|_D \le \rho \tag{31}$$

for some $\epsilon > 0, \rho > 0$, where $v = v' + y, v' \in N_{\ell-1}, y \in E(\lambda_\ell)$, the eigenspace of λ_ℓ. Moreover, it follows from (23) that there is an m_R such that

$$G(w) \geq -m_R, w \in M_{\ell-1}, \ \|w\|_D = R \tag{32}$$

and

$$m_R = o(R^2) \text{ as } R \to \infty. \tag{33}$$

If we examine (31), we see that it implies that either (a) (27) has a nontrivial solution in $E(\lambda_\ell)$ or (b) $G(v) \leq -\epsilon_1 < 0, v \in N_\ell, \|v\|_D = \rho$. In the former case we have accomplished our goal. In the latter case we can apply Theorem 4 to conclude that there is a sequence $\{u_k\} \subset D$ satisfying (6) with $\beta = 2$. I claim that the $\rho_k = \|u_k\|_D$ are bounded. For if $\rho_k \to \infty$, let $\tilde{u}_k = u_k/\rho_k$. Then there is a $\tilde{u} \in D$ and a renamed subsequence such that $\tilde{u}_k \to \tilde{u}$ weakly in D, strongly in $L^2(\Omega)$ and a.e. in Ω. Hence

$$(G'(u_k), u_k)/2\rho_k^2 = 1 - (f(u_k), u_k)/\rho_k^2 \to 1 - \alpha(\tilde{u}) = 0$$

where

$$\alpha(u) = \int_\Omega \{\alpha_+(x)(u^+)^2 + \alpha_-(x)(u^-)^2\}dx.$$

This shows that $\tilde{u} \not\equiv 0$. Moreover,

$$(G'(u_k), v)/2\rho_k = (\tilde{u}_k, v)_D - (f(u_k)/\rho_k, v) \to (\tilde{u}, v) - \alpha(\tilde{u}, v) = 0, v \in D$$

where

$$\alpha(u, v) = \int_\Omega \{\alpha_+ u^+ - \alpha_- u^-\} v dx.$$

Thus $\tilde{u}$ is a solution of (26). Consequently it vanishes by hypothesis. This contradiction shows that the ρ_k are bounded. This implies the existence of a renamed subsequence such that $u_k \to u$ weakly in D, strongly in $L^2(\Omega)$ and a.e. in Ω. Hence

$$(G'(u_k), v) = 2(u_k, v)_D - 2(f(u_k), v) \to 2(u, v)_D - 2(f(u), v) = 0, v \in D.$$

This shows that u is a solution of (27). We have also

$$\|u_k\|_D^2 = (G'(u_k), u_k)/2 + (f(u_k), u_k) \to (f(u), u) = \|u\|_D^2.$$

Thus $u_k \to u$ strongly in D and

$$0 < \epsilon_1 \leq G(u_k) \to G(u)$$

showing that $u \not\equiv 0$.

By rearranging the proof slightly, we can obtain the same conclusion if the inequalities in (21) - (23) are reversed, i.e., if $\ell \geq 1$ and

$$2F(x, t) \leq \lambda_\ell t^2, \quad |t| \leq \delta \tag{34}$$

$$\lambda_{\ell-1} t^2 \leq 2F(x, t), \quad t \in \mathbf{R} \tag{35}$$

$$\lambda_\ell t^2 \leq 2F(x, t) + V(x)^2 h(t) + W(x), t \in \mathbf{R}. \tag{36}$$

References

1. H. Brezis and L. Nirenberg, Remarks on finding critical points, Comm. Pure Appl. Math. 44 (1991) 939-964.

2. P.H. Rabinowitz, Minimax methods in critical point theory with applications to differential equations, Conf. Board of Math. Sci. Reg. Conf. Ser. in Math., No. 65, Amer. Math. Soc., 1986.

3. M. Schechter, New saddle point theorems. Preceedings of an International Symposium on Generalized Functions and their applications, Varanasi, India, December 23-26, 1991.

4. M. Schechter, A generalization of the saddle pont method with applications, Annales Polnici Mathematici 57 (1992) 269-281.

5. M. Schechter, Critical points over splitting subspaces, Nonlinearity, 6(1993) 417-727.

6. M. Schechter, Splitting subspaces and critical points, Applicable Analysis, 49 (1993) 33-48.

7. M. Schechter, The intrinsic mountain pass, Pacific Journal of Math., to appear.

8. E.A. de B.e. Silva, Linking theorems and applications to semilinear elliptic problems at resonance, Nonlinear Analysis TMA, 16 (1991) 455-477.

9. M. Schechter and K. Tintarev, Pairs of critical points produced by linking subsets with applications to semilinear elliptic problems, Bull. Soc. Math. Belg. 44 (1992) 249-261.

28. Detecting Underground Gas Sources

Giorgio G. Talenti Dipartimento di Matematicà, Università di Firenze, Florence, Italy

F. Tonani Istituto di Mineralogia, Perrografia e Geochimica dell'Università, Palermo, Italy

1 INTRODUCTION

This paper is about the following problem. Suppose gas moves out of some extended plane subsurface source and travels toward the Earth's surface with vertical velocity through some homogeneous porous-permeable medium; suppose bulk gas velocity is sampled at the surface over time, and subsurface gas concentration at and beneath the surface is measured at a particular time. Can the gas source be located?

Such a problem originates from volcanology. As a matter of fact, our work was under way on, and further spurred by, the preliminary alert condition declared at springtime 1988 over the Island of Vulcano and the ensuing Crash Programme by the Italian Gruppo Nazionale di Vulcanologia. (Vulcano I. is the southernmost among Aeolian Islands, located about $20km$ northward from the coast of Sicily in the Mediterranean Sea. A volcanic activity takes place there, involving the crater with cone — named La Fossa — and the surrounding caldera. Eruptions are largely explosive. Steady emission of carbon dioxide is apparent in hot fumaroles at the central crater, and also occurs as unapparent diffuse gas emission over the whole La Fossa caldera. Such an emission was deemed worrisome in 1984 and kept under scrutiny ever since.)

In this paper we propose a model for general gas emitting systems, and summarize related mathematical results — thus demonstrating that the problem above has a positive answer, under appropriate hypotheses. Although data required by the present theory are not at hand up till now, surrogates and estimates of these data do result from observations carried on in Vulcano Island: fed such a set of data, the present theory locate the source of carbon dioxide at about $12m$ depth — consistent with the expected depth of the ground water table in La Fossa caldera.

2 MODEL DESCRIPTION

2.1 The system of concern is modeled here as the connected gas filled pore space in a layer of some porous-permeable matrix, bounded by two parallel planes. The upper plane represents the Earth's surface and the bottom plane represents a gas source, i.e., the roof of a gas reservoir. The thickness of such a layer — ***depth of the subsurface gas source***, indicated throughout by

$$L \tag{2.1a}$$

— completely defines the geometry of the model. Estimating L amounts to locating the gas source — the primary goal of our investigations. Notice that L may or may not depend upon time.

The layer includes no sources or sinks of gas, but is sandwiched by two natural bodies of gases: the atmosphere over the Earth's surface, consisting of *air*, and the gas reservoir beneath, containing mainly *carbon dioxide*. Hence the layer is filled up by a mixture of both air and carbon dioxide. The latter — denser — underlies the former and both migrate from and to either gas reservoir.

We suppose the porous-permeable matrix is homogeneous — i.e., porosity and permeability are constant throughout — the equation of perfect gases holds, and the system is isothermal.

Our model is intended to point out the two basic and distinctive process at play: *diffusion and movement.* Other processes, which concur in determining the configuration of our system and should be reckoned with when modeling out as precisely as possible, are ignored here.

Let us name P the *total gas pressure* and call u the *concentration of carbon dioxide.* (For technical convenience, we assume P is the ratio between the actual total pressure and the reference pressure of $1 bar$ — dimensionless. Notice that u is a *molar fraction.*) As shown in Talenti & Tonani (1993), equations governing P and u read

$$\frac{\partial P}{\partial t} = \Delta\left(D \ln P + \frac{\Pi}{2} P^2\right), \tag{2.2a}$$

$$P\frac{\partial u}{\partial t} = D\Delta u + grad\left(D \ln P + \frac{\Pi}{2} P^2\right) \times grad\ u \tag{2.2b}$$

in the present setting, where $\Delta = div\ grad$, D and Π are *positive constants.* D is a diffusion coefficient; Π includes permeability, porosity, tortuosity and viscosity. Equation (2.2a) describes *change in bulk gas distribution*, equation (2.2b) accounts for *chemical composition.*

Boundary conditions can be appended. On the one hand, our method applies when pressure at ground surface is in the range of atmospheric pressure and bulk gas flow rate is low enough that atmospheric circulation is able to dispose of it. On the other hand, we define a subsurface gas source as the place where gas composition is externally fixed. Hence pure air is assumed to occur at the upper boundary, pure subsurface gas is assumed to occur at the lower boundary. In other words, pressure P satisfies

$$P = 1 \quad \text{if } x = 0; \tag{2.3}$$

concentration u satisfies

$$u = 0 \quad \text{if } x = 0, \quad u = 1 \quad \text{if } x = L \tag{2.4}$$

— x, a space variable confined to the range

$$0 \le x \le L, \tag{2.1b}$$

will denote *depth* throughout.

2.2 The model at hand becomes more tractable if *plane symmetry* prevails — i.e., pressure P and concentration u depend upon time t and depth x only. Under this hypothesis, P and u are governed by the following system of partial differential equations

$$\frac{\partial P}{\partial t} = \frac{\partial^2}{\partial x^2}\left(D \ln P + \frac{\Pi}{2} P^2\right), \tag{2.5a}$$

$$P\frac{\partial u}{\partial t} = D\frac{\partial^2 u}{\partial x^2} + \left\{\frac{\partial}{\partial x}\left(D \ln P + \frac{\Pi}{2} P^2\right)\right\}\frac{\partial u}{\partial x}. \tag{2.5b}$$

Moreover, the same hypothesis implies that the bulk gas velocity is purely vertical and

$$-(\text{vertical component of bulk gas velocity}) = \Pi\frac{\partial P}{\partial x}. \tag{2.6}$$

2.3 We concentrate on situations where the considered layer is *fairly permeable* (e.g., made up by medium to fine grained terrains), the subsurface gas source is *shallow*, bulk gas velocity is *low*. These situations entail that *Π is much larger than D*, and *total gas pressure P is nearly constant*. They are conveniently covered by a borderline form of equations (2.5) and (2.6), which will be presented now.

Let us think of D and Π as parameters, take a limit as Π tends to $+\infty$ and D is kept constant, and assume P approaches 1 in the limit. As shown in Talenti & Tonani (1993), the following situation holds in the limit. Firstly, total gas pressure P becomes a constant. Secondly, the equation governing subsurface gas concentration u takes the following form

$$u_t = Du_{xx} + F(t)u_x \,, \tag{2.7}$$

where D is a *positive constant* and F is a *function of time t only*. Thirdly,

$$-\,(\text{vertical component of bulk gas velocity}) = F(t). \tag{2.8}$$

Thus, the propagation of pressure waves is disregarded in the limit, affairs are governed by the chemical composition only and flow is described as if the gas were incompressible.

2.4 Equation (2.8) has a crucial consequence: bulk gas velocity is the same at any depth, hence must be the same as the bulk gas velocity at the Earth's surface — *observable*. In other words, function F can be viewed as a *datum*.

Equations (2.7) and (2.8), together with boundary conditions (2.4), will be our working model. Albeit changes in the depth of the subsurface gas source cannot be ruled out a priori, in our model L will be regarded constant in time.

3 PROBLEM STATEMENT AND SUMMARY OF RESULTS

3.1 Based on the previous discussion, we deal with the following mathematical model

$$u_t = Du_{xx} + F(t)u_x \qquad \text{for } 0 < x < L \text{ and } 0 < t < T, \tag{3.1}$$

$$u(0,t) = 0 \text{ and } u(L,t) = 1 \qquad \text{for } 0 < t < T, \tag{3.2}$$

a boundary value problem for a second order partial differential equation of evolution type. Observe that the boundary conditions are incomplete — the initial condition is missing.

Here t and x denote time and depth, respectively. D is a diffusion coefficient, F stands for the bulk velocity of gas — D is a *positive constant*, F is a *function of time t only*. u denotes the subsurface gas concentration, thus is bound to satisfy the constraint

$$0 \le u \le 1. \tag{3.3}$$

T is the time span of the observation — a positive parameter.

L represents the depth of the subsurface gas source: the *key unknown*. Our goal is precisely *to estimate L having available the following a priori information and the following data*. Firstly: L is a positive constant parameter. Secondly:

(i) D is given.

(ii) F is observed over the interval $[0,T]$, i.e., $F(t)$ is given for every t such that $0 \le t \le T$.

(iii) Some solution u to equation (3.1), satisfying boundary conditions (3.2) and constraint (3.3), exists such that the value of u_x at depth 0 and time T,

$$u_x(0,T), \tag{3.4}$$

is given. (Clearly, the concentration gradient at the soil surface can be monitored by analyzing the gas sitting at different depth in the ground.)

3.2 Two methods for estimating L are presented in sections 4 and 5. Under a reasonable hypothesis on F, the first method allows to sandwich L between rudimentary lower and upper bounds. The second method produces an estimate of L: a peculiarity of this estimate is that, under an appropriate hypothesis on F, the relevant error can be made smaller than any prescribed tolerance provided L is bounded from above and T can be made large enough. The former method is instrumental in using the latter, but the methods are virtually independent of each other.

Our methods are designed so as to fit cases where T takes a physically reasonable, comparatively small value — the time span an observation may spread over. Estimating L under the assumption that T equals $+\infty$ is unrealistic from our point of view and will not be considere here.

Both the bounds produced by the first method and the estimate supplied by the second method are given not in a closed form, but as roots of trascendental equations that must be handled by appropriate numerical techniques. Details, proofs, algorithms, codes and examples are presented in Talenti & Tonani (1993), Sgheri et alii (1993).

4 SANDWICHING L

Lower and upper bounds of L can be obtained by the following method.

Assumptions. Assume hypotheses (i), (ii) and (iii) from section 3, plus the following one. Function F is bounded, i.e., two constants m and M exist satisfying

$$m \leq F(t) \leq M \tag{4.1}$$

for every t such that $0 \leq t \leq T$.

Procedure. Let h and H be given by

$$h(t,L) = \frac{m/D}{1-exp(-Lm/D)} - \frac{2}{L} exp\Big(\frac{Lm}{2D} - \frac{m^2 t}{4D}\Big) \sum_{n=1}^{\infty} \frac{(-1)^{n-1}}{1+\Big(\frac{Lm}{2\pi Dn}\Big)^2} exp\Big[-n^2 \frac{\pi^2 Dt}{L^2}\Big] \tag{4.2}$$

$$H(t,L) = \frac{M/D}{1-exp(-LM/D)} + \frac{2}{L} exp\Big(-\frac{M^2 t}{4D}\Big) \sum_{n=1}^{\infty} \frac{1}{1+\Big(\frac{LM}{2\pi Dn}\Big)^2} exp\Big[-n^2 \frac{\pi^2 Dt}{L^2}\Big] \tag{4.3}$$

— h is a function of time t and depth L, also depending upon diffusion coefficient D and lower bound m for gas velocity F; H is a function of t and L, also depending upon D and upper bound M for F.

Define L_{min} to be the positive root of the following equation

$$h(T, L_{min}) = u_x(0,T) \, . \tag{4.4}$$

Define L_{max} in the following way. If

$$u_x(0,T) > \frac{M}{2D} + \frac{1}{\sqrt{\pi DT}} exp\Big(-\frac{M^2 T}{4D}\Big) + \frac{|M|}{2D} erf\Big(\sqrt{\frac{M^2 T}{4D}}\Big), \tag{4.5}$$

let L_{max} be the positive root to the equation

$$H(T, L_{max}) = u_x(0,T) \, ; \tag{4.6}$$

otherwise, let $L_{max} = +\infty$. It can be shown that $h(T,L)$ decreases strictly from $+\infty$ to 0 as L increases from 0 to $+\infty$, hence equation (3.4) has exactly one positive root. $H(T,L)$ decreases strictly as L increases from 0 to $+\infty$, moreover $H(T,0+) = +\infty$ and $H(T,+\infty)$ equals the right-hand side of (3.5). Hence equation (3.6) has exactly one positive root if inequality (3.5) holds. Graphically, L_{min} is the abscissa of the point where a horizontal straight line meets a plot of h and L_{max} is either $+\infty$ or the abscissa of the point where the same horizontal straight line meets a plot of H, if such a point exists.

The roots of equations (4.4) and (4.6) can be computed either by numerical procedures or graphically — alternative representation formulas for h and H, and algoritms for sampling and plotting h and H are relevant here. *Such roots are the output of the present procedure.*

THEOREM 1. The following inequalities hold

$$L_{min} \leq L \leq L_{max} \,. \tag{4.7}$$

In other words, L is bounded from below by a positive computable quantity L_{min} which depends on diffusion coefficient D, on lower bound m for gas velocity F, on time T and on datum (3.4). An upper bound for L may or may not be found: if datum (3.4) is large enough, then depth L can be bounded from above by a finite quantity L_{max} which depends on D, on upper bound M for F, on T and on (3.4); if datum (3.4) is too small, then the present method fails to produce an upper bound for L.

5 ESTIMATING L

5.1 An accurate estimate of L can be obtained by the following method.

Assumptions: (i), (ii), (iii) from section 3, plus the following one. A constant M exists such that

$$sup\{\,|F(t)|: 0 \leq t \leq T\} \leq M \tag{5.1}$$

for any value of parameter T which affairs may call into play — as it will become clear presently, this assumption ensures that any desirable accuracy can be obtained by aptly adjusting the value of T.

Procedure: Let U be a solution to the following problem

$$U_t = DU_{xx} + F(t)U_x \qquad \text{for } 0 < x < L \text{ and } 0 < t < T, \tag{5.2a}$$

$$U(0,t;L) = 0,\ U(L,t;L) = 1 \qquad \text{for } 0 \leq t \leq T, \tag{5.2b}$$

$$U(x,0;L) = x/L \qquad \text{for } 0 \leq x \leq L, \tag{5.2c}$$

and let l be the positive root of the following equation

$$U_x(0,T;l) = u_x(0,T). \tag{5.3}$$

Here we temporarily abuse notations and think of L as a dummy variable. Observe that (5.2) is nothing but the original model made up by (3.1) and (3.2), with an initial condition added consistent with condition (3.3). System (5.2) is a standard boundary value problem for a standard partial differential equation. Thus U is well defined as a function of three variables — x, t and L — and can be computed by routine procedures, albeit it has no closed form. An algorithm and a FORTRAN code for computing and plotting solution U to problem (5.2) are presented in Sgheri et alii (1993). Lemma A below shows that $U_x(0,t;L)$ decreases strictly from $+\infty$ to 0 as L increases from 0 to $+\infty$ and t is fixed. Thus equation (5.3) has exactly one positive root, which is the abscissa of the point where a plot of $U_x(0,T;L)$ versus L meets a horizontal straight line, and which can be routinely computed. *Such a root is the output of the present procedure.*

THEOREM 2. l is an estimate of gas source depth L which gets more and more accurate as the time span of physical observation, T, gets larger and larger. Let L_{max} be an upper bound for L — L_{max} may be obtained from available data via methods of section 4, must be determined via extra information in case these methods fail. If l is smaller than, or equals, L_{max} and T is large enough, e.g.,

$$T \geq 0.051\, D^{-1} L^2_{max}\,, \tag{5.4a}$$

then $L^{-1}_{max}\,|\,l - L\,|$, a relative error, obeys

$$L^{-1}_{max}\,|\,l - L\,| \leq min\Big\{1\,,\ 146.54\, exp\Big(- \pi^2 D L^{-2}_{max} T + 2.3\, D^{-1} L_{max} M\Big)\Big\}. \tag{5.4b}$$

Proof of theorem 2. Statement (iii) from lemma A below ensures that

$$|\,U_x(0,T;l) - U_x(0,T;L)\,| \geq \left[\frac{D^{-1}M}{exp\Big(D^{-1} max\{l,L\} M\Big) - 1}\right]^2 |\,l - L\,|\,.$$

Equation (5.3) and lemma B below tell us that

$$|U_x(0,T;l)-U_x(0,T;L)| \le 146.54\frac{1}{L}exp\left(-\frac{\pi^2DT}{L^2}+0.3\frac{LM}{D}\right),$$

provided T exceeds $0.286\,L^2/(\pi^2D)$. Thus we have that

$$\frac{|l-L|}{max\{l,L\}} \le 146.54\left[\frac{exp\left(D^{-1}max\{l,L\}M\right)-1}{D^{-1}max\{l,L\}M}\right]^2 max\left\{1,\frac{l}{L}\right\} exp\left(-\frac{\pi^2DT}{L^2}+0.3\frac{LM}{D}\right),$$

provided T exceeds $0.286\,L^2/(\pi^2D)$. Inequalities (5.4) follow via easy manipulations, since

$$max\{l,L\} \le L_{max}$$

by hypotheses. □

5.2 The following lemmas were used in the proof of theorem 2. (Lemma A rests on comparison techniques and representation formulas of suitable barriers. Lemma B — crucial — shows that the system made up by partial differential equation (3.1) and boundary conditions (3.2) has a *fading memory*, i.e., the status of any solution to such a system at any large time is negligibly influenced by remote past history of the same solution.)

LEMMA A. Suppose a positive constant M exists satisfying

$$|F(t)| \le M$$

for every t such that $0 \le t \le T$. Then solution U to problem (5.2) enjoys the following properties:

(i) $U_x(0,t;L) \longrightarrow +\infty$ as $0 < L \longrightarrow 0$ and t is fixed in $[0,T]$.

(ii) $U_x(0,t;L) \longrightarrow 0$ as $L \longrightarrow +\infty$ and t is fixed in $[0,T]$.

(iii) The following inequality

$$-\frac{\partial}{\partial L}U_x(0,t;L) \ge \left[\frac{M/D}{exp(LM/D)-1}\right]^2$$

holds if $0 \le t \le T$ and $L > 0$.

LEMMA B. Suppose a positive constant M exists satisfying

$$|F(t)| \le M$$

for every t such that $0 \le t \le T$. Let u be the difference of any two solutions to problem (3.1) and (3.2), i.e., let u satisfy

$$u_t = Du_{xx} + F(t)u_x \quad \text{for } 0<x<L \text{ and } 0<t<T\,,$$
$$u(0,t) = u(L,t) = 0 \quad \text{for } 0 \le t \le T$$

— note that the boundary conditions are both zero here. Then the following inequalities hold:

$$\left[\int_0^L u^2(x,t)\,dx\right]^{\frac{1}{2}} \le \left[\int_0^L u^2(x,0)\,dx\right]^{\frac{1}{2}} exp\left[-\frac{\pi^2Dt}{L^2}\right]$$

for $0 \le t \le T$;

$$max\{|u(x,t)|: 0 \le x \le L\} \le 4.35\left[L^{-1}\int_0^L u^2(x,0)\,dx\right]^{\frac{1}{2}} exp\left[\frac{LM}{4D}-\frac{\pi^2Dt}{L^2}\right]$$

for $0.143 \times L^2/(\pi^2 D) \leq t \leq T$;

$$|u_x(0,t)| \leq \frac{65.26}{L}\left[L^{-1}\int_0^L u^2(x,0)\,dx\right]^{\frac{1}{2}} max\left\{1,\frac{LM}{4D}\right\} exp\left[\frac{LM}{4D}-\frac{\pi^2 Dt}{L^2}\right]$$

for $0.286 \times L^2/(\pi^2 D) \leq t \leq T$.

REFERENCES

Sgheri L., Talenti G., Tonani F. (1993). Complementary techniques and results on soil gas distribution. Istituto di Analisi Globale e Applicazioni, Consiglio Nazionale delle Ricerche, Firenze.

Talenti G., Tonani F. (1993). Sounding for underground gas sources by surface physico-chemical methods: a mathematical basis. Istituto di Analisi Globale e Applicazioni, Consiglio Nazionale delle Ricerche, Firenze, 1993.

29. Conservative Operators

Edoardo Vesentini Scuola Normale Superiore, Pisa, Italy

Abstract. Densely defined, m-dissipative linear operators in a Banach space, whose opposite operators are dissipative, are characterized as the infinitesimal generators of all strongly continuous semigroups of linear isometries. This characterization extends the classical result by M.H.Stone on self-adjoint operators in Hilbert spaces.

According to M.G.Kreĭn [4], a linear operator X acting on a complex Hilbert space $\mathcal{H}$ endowed with an inner product $(|)$, is called a conservative operator if

$$\mathrm{Re}(Xx|x) = 0 \tag{1}$$

for every $x \in \mathcal{D}(X)$, the domain of X. Thus, X is conservative if, and only if, both operators X and $-X$ are dissipative.

Let now X be densely defined. Then iX is symmetric if, and only if, its numerical range is contained in the imaginary axis [7], i.e., if, and only if, (1) holds for all $x \in \mathcal{D}(X)$; that is, if, and only if, X is conservative. Hence, a necessary and sufficient condition for a densely defined, linear operator X on $\mathcal{H}$ to be conservative and maximal dissipative is that iX is maximal symmetric, i.e., either the open left half-plane $\Pi_l = \{\zeta \in \mathbf{C} : \mathrm{Re}\,\zeta < 0\}$ or the open right half-plane $\Pi_r = \{\zeta \in \mathbf{C} : \mathrm{Re}\,\zeta > 0\}$ (or their union) is contained in the resolvent set $r(X)$ of X.

Let $\mathcal{L}(\mathcal{H})$ be the Banach algebra of all bounded linear operators in $\mathcal{H}$. According to the (generalized) Stone theorem [4], a closed densely defined, linear operator X in $\mathcal{H}$ is the infinitesimal generator of a strongly continuous semigroup $T : \mathbf{R}_+ \to \mathcal{L}(\mathcal{H})$ of linear isometries of $\mathcal{H}$ if, and only if, X is conservative and maximal dissipative. The semigroup T is the restriction to $\mathbf{R}_+$ of a strongly continuous group $\mathbf{R} \to \mathcal{L}(\mathcal{H})$ if, and only if, iX is self-adjoint, i.e., both X and $-X$ are maximal dissipative.

These results will now be extended to strongly continuous semigroups of linear isometries of a complex Banach space $\mathcal{E}$ into itself, showing that they are characterized by the property of being closed, densely defined, conservative and m-dissipative. If, and only if, also the opposite of such an operator is m-dissipative, then the corresponding semigroup is the restriction to $\mathbf{R}_+$ of a strongly continuous group of surjective isometries of $\mathcal{E}$[1]. The spectral structure of the infinitesimal generator will also be investigated.

1. If λ is a linear form on $\mathcal{E}$, $< x, \lambda >$ will denote the value of λ at a point $x \in \mathcal{E}$. Let $\mathcal{E}'$ be the strong dual of $\mathcal{E}$, and, for $x \in \mathcal{E}$, let

$$\Lambda(x) = \{\lambda \in \mathcal{E}' :< x, \lambda >= \|x\|, \|\lambda\| = 1\}.$$

By the Hahn-Banach theorem, $\Lambda(x) \neq \emptyset$ for all $x \in \mathcal{E}$. As is well known [6], the linear operator $X : \mathcal{D}(X) \subset \mathcal{E} \to \mathcal{E}$ is dissipative (i.e., such that

$$\text{Re} < Xx, \lambda > \leq 0 \tag{2}$$

for every $x \in \mathcal{D}(X)$ and some $\lambda \in \Lambda(x)$) if, and only if,

$$\|(\zeta I - X)x\| \geq \zeta\|x\| \text{ for all } \zeta > 0.$$

This implies that $\Pi_r \subset r(X) \cup r\sigma(X)$, where $r\sigma(X)$ is the residual spectrum of X.

If the closed, dissipative operator X is such that $\Pi_r \cap r(X) \neq \emptyset$, then $\Pi_r \subset r(X)$, and (2) holds for all $x \in \mathcal{D}(X)$ and all $\lambda \in \Lambda(x)$. The operator X is then said to be m-dissipative, and, by the Lumer-Phillips theorem, if moreover $\mathcal{D}(X)$ is dense in $\mathcal{E}$, then X is the infinitesimal generator of a strongly continuous semigroup $T : \mathbf{R}_+ \to \mathcal{L}(\mathcal{E})$ of contractions of $\mathcal{E}$. Viceversa, the infinitesimal generator of a strongly continuous semigroup of linear contractions of $\mathcal{E}$ is m-dissipative [5], [6], and (2) holds for every $x \in \mathcal{D}(X)$ and all $\lambda \in \Lambda(x)$.

If both X and $-X$ are dissipative, i.e., if

$$\text{Re} < Xx, \lambda >= 0 \text{ for all } x \in \mathcal{D}(X) \text{ and some } \lambda \in \Lambda(x),$$

then X is called a *conservative operator*. Note that, if X is a bounded, linear, conservative operator, then both X and $-X$ are m-dissipative, and $\sigma(X) \subset i\mathbf{R}$.

Now, let $X : \mathcal{D}(X) \subset \mathcal{E} \to \mathcal{E}$ be a linear conservative operator.

[1]The case of surjective isometries has been investigated by several authors; cf. [1] and bibliographical references therein.

Lemma 1 *If* $u \in \mathcal{D}(X)$ *and* $\|u\| = 1$, *then*

$$\text{liminf}_{t\to 0}\frac{\|u + t\,Xu\| - 1}{|t|} \geq 0.$$

If $v \in \mathcal{D}(X^2) = \{x \in \mathcal{D}(X) : Xx \in \mathcal{D}(X)\}$ *and* $\|v\| = 1$, *then*

$$\text{limsup}_{t\to 0}\frac{\|v + t\,Xv\| - 1}{|t|} \leq 0.$$

Proof Let $\lambda \in \Lambda(u)$ be such that $\text{Re} < Xu, \lambda >= 0$. Since

$$\|u + t\,Xu\|^2 \geq | < u + t\,Xu, \lambda > |^2 = 1 + (t\,\text{Im} < Xu, \lambda >)^2,$$

the first part of the lemma is proved.

Let now $y = v + t\,Xv \in \mathcal{D}(X)$, and let $\mu \in \Lambda(y)$ be such that $\text{Re} < Xy, \mu >= 0$. Then

$$\begin{aligned}\|y - t\,Xy\| &\geq | < y - t\,Xy, \mu > | = | \|y\| - t < Xy, \mu > | = \\ &= ((\|y\| - t\,\text{Re} < Xy, \mu >)^2 + (t\,\text{Im} < Xy, \mu >)^2)^{\frac{1}{2}} \geq \|y\|.\end{aligned}$$

Since, on the other hand,

$$\|y - t\,Xy\| = \|v - t^2X^2v\| = 1 + o(t)$$

as $t \to 0$, then, for $t \neq 0$,

$$\frac{\|v + t\,Xv\| - 1}{|t|} = \frac{\|y\| - 1}{|t|} \leq \frac{\|y - t\,Xy\| - 1}{|t|} = o(1)$$

as $t \to 0$, and the second part of the lemma follows.

Corollary 1 *If* X *is a linear conservative operator, then*

$$\lim_{t\to 0}\frac{\|v + t\,Xv\| - \|v\|}{t} = 0$$

for all $v \in \mathcal{D}(X^2)$.

Assume now that X generates a strongly continuous semigroup $T : \mathbf{R}_+ \to \mathcal{L}(\mathcal{E})$. For all $t \geq 0$, $x \in \mathcal{D}(X)$ and all $h \in \mathbf{R}$ such that $t + h \in \mathbf{R}_+$, then $T(t)x \in \mathcal{D}(X)$, and

$$T(t+h)x = T(t)x + \int_t^{t+h} XT(s)x\,ds.$$

Since the map $s \mapsto XT(s)x = T(s)Xx$ is continuous, then

$$T(t+h)x = T(t)x + hXT(t)x + o(h) \tag{3}$$

as $h \to 0$.

The set $\mathcal{D}^\infty$ of all differentiable vectors of the semigroup T is a dense linear subvariety of $\mathcal{E}$, which is defined [1] by $\mathcal{D}^\infty = \cap_{n\geq 1}\mathcal{D}(X^n)$.

Hence, by Corollary 1 and by (3),

$$\lim_{h\to 0}\frac{\|T(t+h)x\| - \|T(t)x\|}{h} = 0$$

for all $t \geq 0$ and all $x \in \mathcal{D}^\infty$. (If $t = 0$, the limit as $h \to 0$ shall be replaced by the limit as $t \downarrow 0$). Since $T(t) = I$, this proves that $T(t)$ is a linear isometry for all $t \geq 0$.

2. It will be shown now that the converse holds.

For $x \in \mathcal{E}\setminus\{0\}$, $y \in \mathcal{E}$, $t > 0$, let $\lambda \in \Lambda(x)$, $\mu_t \in \Lambda(x + ty)$. Then, since $\mathrm{Re} < x, \mu_t > \leq | < x, \mu_t > | \leq \|x\|$,

$$\begin{aligned}\mathrm{Re} < y, \lambda > &= \frac{\mathrm{Re} < ty, \lambda >}{t} = \frac{\mathrm{Re} < x + ty, \lambda > - \|x\|}{t} \leq \\ &\leq \frac{\|x + ty\| - \|x\|}{t} \leq \\ &\leq \frac{< x + ty, \mu_t > - \mathrm{Re} < x.\mu_t >}{t} = \\ &= \frac{\mathrm{Re} < x + ty, \mu_t > - \mathrm{Re} < x, \mu_t >}{t} = \mathrm{Re} < y, \mu_t > \end{aligned} \tag{4}$$

Let $\{t_\nu\}$ be a sequence of positive numbers converging to 0. Since the closed unit ball of $\mathcal{E}'$ is compact for the weak* topology, the set $\{\mu_{t_\nu}\}$ has a weak* cluster point $\mu \in \mathcal{E}'$ with $\|\mu\| \leq 1$. If $< x, \mu > \neq \|x\|$, there exist some $\epsilon > 0$ and a subsequence $\{t'_\nu\}$ of the sequence $\{t_\nu\}$ such that $| < x, \mu_{t'_\nu} > - \|x\| \, | > \epsilon$ for all ν. Since $| < y, \mu_{t_\nu} > | \leq \|y\|$ for all ν, then

$$< x, \mu_{t'_\nu} > = < x + t'_\nu y, \mu_{t'_\nu} > - t'_\nu < y, \mu_{t'_\nu} > = \|x + t'_\nu y\| - t'_\nu < y, \mu_{t'_\nu} >,$$

which converges to $\|x\|$ as $\nu \to \infty$. This contradiction shows that $< x, \mu >$ $= \|x\|$. Hence $\|\mu\| = 1$, and, in conclusion, $\mu \in \Lambda(x)$.

Suppose now that X generates a strongly continuous semigroup $T : \mathbf{R}_+ \to \mathcal{L}(\mathcal{E})$ of linear isometries of $\mathcal{E}$. For $x \in \mathcal{D}(X)$ and $t > 0$,

$$T(t)x = x + t\,Xx + o(t)$$

as $t \downarrow 0$. Then

$$\|x + t\,Xx\| - \|x\| = \|T(t)x\| - \|x\| + o(t) = o(t) \tag{5}$$

as $t \downarrow 0$, and, setting $y = Xx$, (4) shows that - as is well known - Re $< Xx, \lambda > \leq 0$ for all $\lambda \in \Lambda(x)$.

If Re $< Xx, \mu > < 0$, arguing as before, one shows that there exist some $\epsilon > 0$ and a subsequence $\{t'_\nu\}$ of the sequence $\{t_\nu\}$ such that Re $< Xx, \mu_{t'_\nu} >$ $< -\epsilon$ for all ν. On the other hand, by (4),

$$\text{Re} < Xx, \mu_{t'_\nu} > \geq \frac{\|x + t'_\nu Xx\| - \|x\|}{t'_\nu},$$

which, by (5), tends to 0 as $\nu \to \infty$. Thus Re $< Xx, \mu > = 0$, and the following theorem holds, which extends to all Banach spaces the classical Stone theorem.

Theorem 1 *Let X be a closed, densely defined, linear operator on $\mathcal{E}$. Then X is the infinitesimal generator of a strongly continuous semigroup of linear isometries of $\mathcal{E}$ into $\mathcal{E}$, if, and only if, X is conservative and m-dissipative. If, and only if, also $-X$ is m-dissipative, then T is the restriction to $\mathbf{R}_+$ of a strongly continuous group $\mathbf{R} \to \mathcal{L}(\mathcal{E})$ of surjective isometries of $\mathcal{E}$.*[2]

[2]According to a different terminology [1], in this latter case iX is called a (generalized) hermitian operator.

Remark In the case of groups of isometries, the proof of the theorem can be simplified. Indeed, if the closed, densely defined, linear operator X is the infinitesimal generator of a strongly continuous group $T : \mathbf{R} \to \mathcal{L}(\mathcal{E})$, then - by the Lumer-Phillips theorem - $T(t)$ is a contraction for every $t \in \mathbf{R}$ if, and only if, X and $-X$ are m-dissipative. But, if $T(t)$ is a contraction for every $t \in \mathbf{R}$, then, for all $x \in \mathcal{E}$,

$$\begin{aligned} \|x\| &= \|T(-t)T(t)x\| \leq \|T(-t)\|.\|T(t)x\| \\ &\leq \|T(t)x\| \leq \|T(t)\|.\|x\| \leq \|x\|, \end{aligned}$$

showing that $T(t)$ is an isometry.

Also in the general case of semigroups, the proof of the theorem can be simplified and brought closer to that holding for Hilbert spaces, when the norm on $\mathcal{E}$ is assumed to be Gateaux differentiable, i.e. [2], when $\mathcal{E}$ is smooth.

According to a result of G.Lumer and R.S.Phillips [5], [6], if the closed operator X is densely defined and both X and the dual operator X' are dissipative, then X and X' are m-dissipative. Hence the above theorem yields

Corollary 2 *If X is closed, densely defined and conservative, and if the dual operator X' is dissipative, then X generates a strongly continuous semigroup of linear isometries of $\mathcal{E}$.*

If X is densely defined and m-dissipative, since $\Pi_r \subset r(X) \subset r(X')$ and, for all $\zeta \in r(X)$, $(\zeta I - X)^{-1'} = (\zeta I - X')^{-1}$, then also X' is m-dissipative. Thus, denoting by $\mathcal{F}'$ the closure of $\mathcal{D}(X')$ in $\mathcal{E}'$, the part $\tilde{X}'$ of X' in $\mathcal{F}'$, i.e., the restriction of X' to the linear space $\mathcal{D}(\tilde{X}') = \{\lambda \in \mathcal{D}(X') : X'\lambda \in \mathcal{F}'\}$, generates a strongly continuous semigroup of contractions $\tilde{T}' : \mathbf{R}_+ \to \mathcal{L}(\mathcal{F}')$.

Note that the additional hypothesis that X be conservative, whereby $\Pi_l \subset r(X) \cup r\sigma(X)$, does not imply that $-X'$ is dissipative. However the stronger additional condition that also $-X$ be m-dissipative *does* imply that also $-X'$ is m-dissipative. In conclusion, the following proposition holds.

Proposition 1 *If X is densely defined and X and $-X$ are m-dissipative, then $\tilde{X}'$ is the infinitesimal generator of a strongly continuous group $\tilde{T}'$: $\mathbf{R} \to \mathcal{L}(\mathcal{F}')$ of isometries of $\mathcal{F}'$ onto $\mathcal{F}'$.*

If $\mathcal{E}$ is reflexive, then $\mathcal{D}(X')$ is dense in $\mathcal{E}'$.

Corollary 3 *Under the hypotheses of Proposition 1, if moreover the Banach space $\mathcal{E}$ is reflexive, then X' generates a strongly continuous group T' of isometries of $\mathcal{E}'$ onto $\mathcal{E}'$.*

3. If the closed linear operator X is conservative, then Π_r belongs entirely either to $r\sigma(X)$ or to $r(X)$, and a similar conclusion holds for Π_l. If X is also m-dissipative, then

$$r(X) \subset \Pi_r,$$

and, if moreover $\Pi_l \subset r(X)$, also $-X$ is m-dissipative. If furthermore X is densely defined, then it generates a strongly continuous group $T : \mathbf{R} \to \mathcal{L}(\mathcal{E})$ of surjective isometries of $\mathcal{E}$. Therefore $\sigma(X) \neq \emptyset$ ([1], pp.212-214), and the following theorem holds.

Theorem 2 *If the closed operator X is densely defined, conservative and m-dissipative, the spectrum of X is non-empty.*

If, under the hypotheses of this latter theorem, $\sigma(X)$ is compact, then $\sigma(X) \subset i\mathbf{R}$, and X generates a strongly continuous group of surjective isometries. Hence ([1], p.214), X is a continuous operator and $T(t) = \exp tX$ for all $t \in \mathbf{R}$.

As is well known, every self-adjoint linear operator in a complex Hilbert space has an empty residual spectrum. This result will now be extended to infinitesimal generators of strongly continuous groups of isometries in reflexive, strictly convex Banach spaces.

Lemma 2 *If the closed, densely defined operator X generates a strongly continuous semigroup $T : \mathbf{R}_+ \to \mathcal{L}(\mathcal{E})$ of contractions of a reflexive, strictly convex, complex Banach space $\mathcal{E}$* [3]*, then $i\mathbf{R} \cap r\sigma(X) = \emptyset$.*

Proof Let $0 \in r\sigma(X)$. There is some $\lambda \in \mathcal{E}'$, with $\|\lambda\| = 1$, for which $< Xx, \lambda > = 0$ for all $x \in \mathcal{D}(X)$, and therefore also $< X(\zeta I - X)^{-1}y, \lambda > = 0$

[3] A reflexive Banach space $\mathcal{E}$ is strictly convex if, and only if, $\mathcal{E}'$ is smooth [2].

for all $y \in \mathcal{E}$ and all $\zeta \in r(X)$. For $\zeta > 0$, let $X_\zeta = \zeta X(\zeta I - X)^{-1}$ be a Yosida approximation of X. Since $< X_\zeta y, \lambda > = 0$ for all $y \in \mathcal{E}$, then $< \exp(t X_\zeta)\, y, \lambda > = < y, \lambda >$, and therefore also

$$< T(t)\, y, \lambda > = \lim_{\zeta \to +\infty} < \exp(t X_\zeta)\, y, \lambda > = < y, \lambda >$$

for all $t \geq 0$ and all $y \in \mathcal{E}$. If $x \in \mathcal{E}$, with $\|x\| = 1$, is such that $< x, \lambda > = \|\lambda\| = 1$, then

$$1 \geq \|T(t)\, x\| \geq | < T(t)\, x, \lambda > | = < x, \lambda > = 1,$$

and therefore $< T(t)\, x, \lambda > = < x, \lambda > = \|T(t)\, x\| = 1$.

Since $\mathcal{E}$ is strictly convex, then $T(t)\, x = x$ for all $t \geq 0$. Hence $x \in \mathcal{D}(X)$ and $Xx = 0$, showing that 0 is an eigenvalue of X, and thus contradicting the fact that $0 \in r\sigma(X)$.

Since, for any $a \in \mathbf{R}$, $X - ia\, I$ satisfies the hypotheses of the lemma, and since $r\sigma(X - ia\, I) = \{\zeta - ai : \zeta \in r\sigma(X)\}$, the proof is complete.

If X generates a strongly continuous group of isometries, then $\sigma(X) \subset i\mathbf{R}$. Thus Lemma 2 yields

Theorem 3 *If X is the infinitesimal generator of a strongly continuous group $T : \mathbf{R} \to \mathcal{L}(\mathcal{E})$ of isometries of a reflexive, strictly convex, complex Banach space $\mathcal{E}$, then the residual spectrum of X is empty.*

By the spectral mapping theorem, this implies that $r\sigma(T(t)) = \emptyset$ for every $t \geq 0$. This conclusion follows also from the fact (which can be established by a similar argument to the proof of Lemma 2) that, if A is a surjective isometry of a reflexive, strictly convex, complex Banach space $\mathcal{E}$ onto itself, then $r\sigma(A) = \emptyset$.

Going back once more to the general case in which $\mathcal{E}$ is any complex Banach space, let the linear operator X be densely defined, conservative and m-dissipative on $\mathcal{E}$. If the strongly continuous semigroup $T : \mathbf{R}_+ \to \mathcal{L}(\mathcal{E})$ of linear isometries generated by X is the restriction to $\mathbf{R}_+$ of a bounded holomorphic semigroup, then there is some $\theta \in [0, \frac{\pi}{2})$ such that $\sigma(X)$ is contained in the angle $\{z \in \mathbf{C} : \mathrm{Re}\, z \leq 0, \frac{|\mathrm{Im}\, z|}{|\mathrm{Re}\, z|} \leq \tan\theta\}$. On the other hand, the fact that X is conservative implies that either $\Pi_l \subset r\sigma(X)$ or

$\Pi_l \subset r(X)$. Thus $\sigma(X) = \{0\}$, and therefore $X \in \mathcal{L}(\mathcal{E})$. Since the norm of any bounded conservative linear operator equals the spectral radius ([3]; [1], p. 217), then $X = 0$, and, in conclusion, the following proposition holds.

Proposition 2 *There is no bounded holomorphic semigroup defined in an angular neighbourhood of* $\mathbf{R}_+$, *whose restriction to* $\mathbf{R}_+$ *is a non-trivial, strongly continuous semigroup of linear isometries of* $\mathcal{E}$.

In particular, there are no non-trivial holomorphic semigroups of linear isometries.

References

[1] E.B.Davies, One-parameter semigroups, Academic Press, London-New York, 1980.

[2] J.Diestel, Geometry of Banach spaces - Selected topics, Lecture Notes in Marhematics, n. 485, Springer-Verlag, Berlin-Heidelberg-New York, 1975.

[3] V.E.Kacnelson, *A conservative operator has norm equal to its spectral radius*, Mat.Issled. 5, 3 (17) (1970), 186-189.

[4] M.G.Kreĭn, Linear differential equations in Banach spaces, Amer.Math.Soc.Translations, Vol. 29, Providence, R.I., 1971.

[5] G.Lumer and R.S.Phillips, *Dissipative operators in a Banach space*, Pacific J. Math., 11 (1959), 679-698.

[6] A.Pazy, Semigroups of linear operators and applications to partial differential equations, Springer-Verlag, NewYork-Berlin-Heidelberg-Tokyo, 1983.

[7] M.H.Stone, Linear transformations in Hilbert space, Amer.Math. Soc. Colloquium Publications, Vol. 15, Providence, R.I., 1932.

30. On the Regularization of the Antenna Synthesis Problem

Giovanni Alberto Viano Departimento di Fisica, Università di Genova, Genoa, Italy

Introduction

The synthesis problem consists of finding the "sources" which produce a desired effect: for instance prescribed boundary values. We could keep, as a typical example, the antenna synthesis: to determine the current density which produces a prescribed (within a certain degree of approximation) radiation pattern. This problem, as we shall see below, can be solved by the use of a Fredholm integral equation of first kind; therefore, it presents the typical pathology of ill-posed problems: widely different sources can produce almost the same radiation pattern. Furthermore, if we take into account that the data must be discretized, handled by a computer, and can be specified only within a certain degree of approximation (related to the sensitivity of the apparatus) then we clearly see that the antenna synthesis problem must be regularized [1].

The method of regularization which is now widely used is precisely based on the ideas of Carlo Pucci [2,3] of restricting the class of admitted solutions by the use of a-priori bounds. Now let us remark that in the synthesis problem the a-priori bounds are intrinsic to the problem itself and do not represent an additional "ad-hoc" assumption (in this sense synthesis problems differ in a significant way from the inverse problems). In the specific case of the antenna synthesis a bound on the ohmic losses, associated with the current density, is necessary and it is therefore a natural constraint intrinsic to the problem. Therefore, we can formulate the regularization method as a variational problem: to find the minimum of a functional, which can be derived by two bounds. One is a bound on the solution, which is a constraint on the ohmic losses, the other is a bound on the data, related to the fact that the radiation pattern can be specified only within a certain degree of approximation.

1 Solution of the Antenna Synthesis Problem by the Use of a Fredholm Integral Equation of First Kind

The starting point is the so-called radiation problem: to look for a solution of the inhomogeneous Helmholtz equation, which satisfies also the Sommerfeld radiation condition. The source term in the Helmholtz equation is here related to the density of current. This problem has been deeply explored in the framework of partial differential equations and we do not intend to return on these questions. We simply recall that the asymptotic behaviour at large distance of the solution of the radiation problem is given by the product of the spherical wave times the Fourier transform of the source. Then the synthesis problem can be posed as follows: given the radiation pattern (i.e. the Fourier transform of the source at fixed value of $k = \frac{\omega}{c}$, ω being the frequency and c the light velocity), derive the density of current. Let us notice that in this problem both uniqueness and continuity are usually missing: we may have oscillating charge-current distribution which do not radiate. This non-uniqueness is precisely due to those non-zero sources whose Fourier transform, at $k = \frac{\omega}{c}$ fixed, is zero. This problem is very delicate and we intend to return on it elsewhere [4]. Here we limit ourselves to consider a linear aperture extending from -1 to +1 on the y-axis. In this particular case we are not involved with the non-uniqueness problem. Let us represent the current density by a function f(y) whose support is contained in the closed interval [-1,+1]. Then we introduce the integral operator L as follows:

$$Lf = \int_{-1}^{1} e^{-iky \cos\theta} f(y)\, dy = \hat{f}\,(\cos\theta) \tag{1}$$

k being fixed. The operator L transforms a function f, which is supposed to belong to $L^2(-1,+1)$, into a function, defined on the unit sphere $\mathbf{S}^2 \subset \mathbf{R}^3$, and which belongs to $L^2\,(\mathbf{S}^2)$. We can then determine the adjoint L^* of L $\left(L^* :\ L^2(\mathbf{S}^2) \to L^2(-1,+1)\right)$, integrating on the sphere $\mathbf{S}^2$, whose measure is given by $ds = \sin\theta\, d\theta\, d\varphi$. We have:

$$L^* Lf = 2\pi \int_0^{\pi} e^{ik\,x\,\cos\theta}\, \hat{f}(\cos\theta)\, \sin\theta\, d\theta =$$

$$= 4\pi \int_{-1}^{1} \frac{\sin\left[k\,(x-y)\right]}{k\,(x-y)}\, f\,(y)\, dy \tag{2}$$

Indeed one can easily verify that:

$$(\mathcal{L}f, \hat{f})_{L^2(S^2)} = (f,\ \mathcal{L}^*\hat{f})_{L^2(-1,+1)} \tag{3}$$

We are thus led to the following integral equation:

$$(Af)(x) \equiv \int_{-1}^{1} \frac{\sin\,[k(x-y)]}{k\,(x-y)}\, f\,(y)\, dy = g\,(x) \quad (-1 \leq x \leq 1) \tag{4}$$

where $g \equiv \frac{\mathcal{L}^*\mathcal{L}f}{4\pi}$ represents the data.

The equation (4) is a Fredholm integral equation of first kind. The operator $A : L^2(-1,+1) \rightarrow L^2(-1,+1)$ is self-adjoint and compact, and we can associate with it an eigenfunction expansion. Indeed the following equations hold true [5,6]:

$$\int_{-1}^{1} \frac{\sin\,[k(x-y)]}{\pi\,(x-y)}\, \psi_n\,(k,y)\, dy = \lambda_n\, \psi_n\,(k,x)\ , (n = 0,1,2..) \tag{5}$$

where the eigenvalues λ_n form a decreasing sequence: $1 > \lambda_0 > \lambda_1 > \lambda .... > 0$. Their asymptotic behaviour for $n \rightarrow \infty$ is given by [5,6]:

$$\lambda_n = O\left\{\frac{1}{n}\exp\,[-D\; n\; \ln\, n]\right\}, \quad (D = \text{const.}) \tag{6}$$

Moreover, the eigenvalues λ_n are also the normalization constants of the eigenfunctions ψ_n on the interval [-1,+1], i.e.

$$\int_{-1}^{1} \left|\psi_n\,(k,x)\right|^2 dx = \lambda_n \quad (n = 0,1,2,...) \tag{7}$$

Therefore the functions $u_n(x) = \frac{1}{\sqrt{\lambda_n}}\, \psi_n\,(k,x)$ form a basis in $L^2\,(-1,+1)$.

On the other hand, the functions $\psi_n(k,x)$, which are called in literature prolate spheroidal wave functions (P.S.W.F.), can also be viewed as the continuous solutions, on the closed interval [-1,+1], of the following differential equations [5,6]:

$$-\left[(1-x^2)\,\psi_n'\,(x)\right]' + k^2x^2\,\psi_n(x) = \chi_n\psi_n(x) \quad (n = 0,1,2,..) \tag{8}$$

where the eigenvalues χ_n can be ordered as follows: $0 < \chi_0 < \chi_1 < \chi_2 < ....$ Their asymptotic behaviour, when $n \rightarrow \infty$, is given by [5,6]:

$$\chi_n = n\,(n+1) + \frac{1}{2}\,k^2 + O\left(\frac{1}{n^2}\right) \tag{9}$$

The functions $\psi_n\,(k,x)$ can be uniquely extended to entire analytic functions, and they can be normalized as follows:

$$\int_{-\infty}^{+\infty} \left|\psi_n\,(k,x)\right|^2 dx = 1 \quad (n = 0,1,2,...) \tag{10}$$

2 The Regularization of the Antenna Synthesis Problem

As we remarked in the Introduction, even if the function g is specified analytically, nevertheless it will have to be discretized and handled by a computer. Now in view of the behaviour of the eigenvalues λ_n (formula (6)), the numerical errors in the computation of g are sufficient to push the data function outside the range of the operator A. The antenna synthesis problem is numerically unstable. But we have at our disposal two bounds. One is on the solutions $f \in X$ (X denotes the solution space which in our case is given by $L^2(-1,+1)$): i.e.

$$\|Bf\|_Z \leq E \tag{11}$$

where E is a fixed prescribed constant, which does not depend on f; Z and B: X→ Z, denote respectively the constraint space and the constraint operator.

The other bound is on the data, and it can be written as follows:

$$\|Af - g^\epsilon\|_Y \leq \epsilon \tag{12}$$

where ϵ is a constant, which characterizes the uncertainty which we have on the data; g^ϵ is the data function actually given, and Y denotes the data space, which in our case is $L^2(-1,+1)$.

Let us now consider the following differential operator:

$$(B^*Bf)\ (x) = -\left[(1-x^2)f'(x)\right]' + k^2x^2f(x) \tag{13}$$

In view of the equations (8) and taking for the constraint space Z an L^2-norm, the bound (11) reads:

$$\sum_{n=0}^{\infty} \chi_n |f_n|^2 \leq E \tag{14}$$

where $f_n = (f, u_n)_{L^2(-1,+1)}$ and $u_n = \frac{1}{\sqrt{\lambda_n}}\ \psi_n$. Next we recall that the ball $\sum_{n=0}^{\infty} (1+n^2)|f_n|^2 \leq E$ is compact in $L^2(-1,+1)$. Therefore, the set of functions f which satisfy the bound (14) form a compact set in $L^2(-1,+1)$ (see formula (9)). We may conclude that Z is a compact subset of the solution space $X{=}L^2(-1,+1)$.

Now we recall the following Lemma [7]: let $\mathcal{H}_*$ be a compact subset of the Hilbert space $\mathcal{H}$, and let us suppose that the linear, continuous operator A: $\mathcal{H} \to \mathcal{K}$, restricted to $\mathcal{H}_*$, has an inverse. Then the inverse operator is continuous.

In view of this Lemma, and the compactness of Z, we can conclude that the continuous dependence of the solutions on the data is restored if we restrict the admitted solutions to the constraint space Z. We are thus naturally led to look for the minimization of the following functional (see [8,9,10]):

$$\Phi(f) = \|Af - g^{\epsilon}\|^2_{L^2(-1,+1)} + \left(\frac{\epsilon}{E}\right)^2 \|Bf\|^2_{L^2(-1,+1)} \tag{15}$$

Now we introduce the following operator:

$$C = A^*A + \left(\frac{\epsilon}{E}\right)^2 B^*B \tag{16}$$

It is a positive definite operator, it is self-adjoint and it has a continuous inverse. The last property allows us to introduce the function:

$$f_* = C^{-1} A^* g^{(\epsilon)} \tag{17}$$

Then the functional (15) may be rewritten as follows [10]:

$$\Phi(f) = \Big(C[f - f_*],\ [f - f_*]\Big)_{L^2(-1,+1)} + \|g^{(\epsilon)}\|^2_{L^2(-1,+1)} -$$

$$-(g^{(\epsilon)}, Af_*)_{L^2(-1,+1)} \tag{18}$$

It is clear that the function f_* minimizes the functional (18) and that

$$\Phi\left(f_*\right) = \|g^{(\epsilon)}\|^2_{L^2(-1,+1)} - \left(g^{(\epsilon)}, Af_*\right)_{L^2(-1,+1)} \geq 0 \tag{19}$$

Furthermore, the components $(f_*,\ u_n)$ read:

$$(f_*,\ u_n) = \frac{\lambda_n\ g_n^{(\epsilon)}}{\lambda_n^2 + (\frac{\epsilon}{E})^2\ \chi_n};\quad g_n^{(\epsilon)} = (g^{(\epsilon)},\ u_n) \tag{20}$$

and therefore we have:

$$f_*(y) = \sum_{n=0}^{\infty} \frac{\lambda_n\ g_n^{(\epsilon)}}{\lambda_n^2 + (\frac{\epsilon}{E})^2\ \chi_n}\ u_n(y);\ (y \in [-1,+1]) \tag{21}$$

Finally we can prove that the following "a posteriori - compatibility check" [8,9,10].

Proposition: - for any function f which satisfies the bounds (11) and (12), the following inequalities hold:

$$\|A(f - f_*)\|_Y \leq \sqrt{2}\epsilon \tag{22}$$

$$\|B(f - f_*)\|_Z \leq \sqrt{2}E \tag{23}$$

Proof: - If f satisfies the bounds (11) and (12), then the following inequality holds true:

$$\Phi(f) = \Big(C\,[f - f_*],\, [f - f_*]\Big)_{L^2(-1,+1)} + \|g^{(\epsilon)}\|^2_{L^2(-1,+1)} - $$
$$-(g^{(\epsilon)},\, Af_*)_{L^2(-1,+1)} \leq 2\epsilon^2 \qquad (24)$$

In view of the inequality (19) it follows that:

$$\Big(C[f - f_*],\, [f - f_*]\Big)_{L^2(-1,+1)} = \|A(f - f_*)\|^2_Y + (\frac{\epsilon}{E})^2\|B(f - f_*)\|^2_Z \leq 2\epsilon^2 \quad (25)$$

From inequality (25) the inequalities (22) and (23) follow.

REFERENCES

1. G. A. Deschamps and H. S. Cabayan, Antenna synthesis and solution of inverse problems by regularization methods, *IEEE Trans. on Antennas and Propagation, AP-20:* N.3, 268 (1972).

2. C. Pucci, Sui problemi di Cauchy non ben posti, *Atti Acc. Naz. Lincei, 18:* 473 (1955).

3. D. Fox and C. Pucci, The Dirichlet problem for the wave equation, *Annali Mat. Pura Appl., 46:* 155 (1958).

4. N. Magnoli and G. A. Viano, The source identification problem in electromagnetic theory (in preparation).

5. D. Slepian and H. O. Pollak, Prolate spheroidal wave functions, Fourier analysis and uncertainty-I, *Bell syst. techn. J. 40:* 43 (1961).

6. C. Flammer, Spheroidal wave functions, *Stanford University Press* (1957).

7. A. N. Tikhonov and U. Y. Arsenin, Solutions of ill-posed problems, *Wiston/Wiley*, Washington (1977).

8. K. Miller, Least squares methods for ill-posed problems with a prescribed bound, *SIAM J. Math. Anal. 1:* 52 (1970).

9. K. Miller and G. A. Viano, On the necessity of nearly-best-possible methods for analytic continuation of scattering data, *Journal of Math. Phys. 14:*, 1037 (1973).

10. M. Bertero, C. De Mol and G. A. Viano, The stability of inverse problems in "Inverse scattering problems in optics", *Topics in Current Physics 20:*, Berlin Springer, p. 161 (1980).

31. The Problem of Packaging

Piero Villaggio Istituto di Scienza delle Costruzioni, Università di Pisa, Pisa, Italy

1. THE TECHNICAL PROBLEM

Let us consider the problem of having to wrap a parcel, of prescribed size and shape, containing a very delicate substance such as, for instance, an aggregrate of small crystal globules or of organic cells. In order to transport the material, we must apply some forces to its exterior envelope, but, since the material is weak, we must stiffen the envelope so that the stresses inside remain small.

To formulate the problem in more precise terms, let us assume that the material occupies a region D of the Cartesian space, with boundary S, as indicated in the Fig.1. On S, a given distribution of tractions $\vec{p}$ is prescribed, whose magnitude and direction vary from point to point of the boundary. Let us assume, for simplicity, that the weight of the material is negligible. If the material were linearly elastic, and, in particular, homogeneous and isotropic, the stresses in D, arising from the surface tractions, would be determined by solving a boundary value problem in classical elasticity. However, this stress state is, in general, inacceptable because it is accompanied by local stress peaks that the material cannot constitutively sustain. It is thus necessary to apply the device of "packing." This consists in the interposing of an elastic membrane between S and the external loads for the purpose of smoothing out the stresses generated by these tractions. We shall assume the membrane to be elastic and perfectly flexible, that is to say able to trasmit stresses tangential to its mean surface, but not transversal to it. In addition, we can arbitrarily vary the thickness in order to modify the internal stresses so that they do not exceed the desired values. This operation is instinctively done by sellers of fruits or pastries when, in preparing the package, they thicken the paper under the thread in order to avoid the pressure exerted by it squeezing a ripe fruit or a soft tea cake.

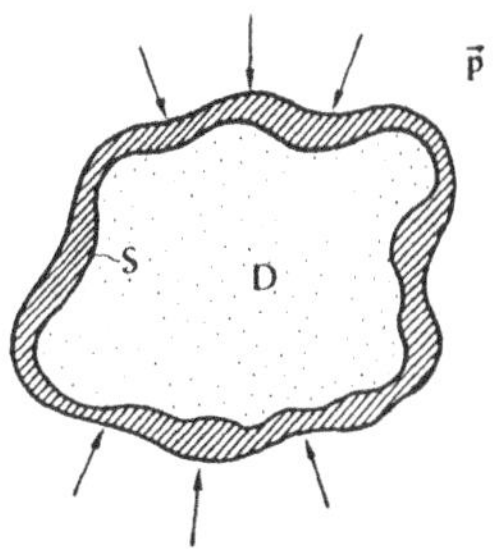

Fig. 1

Formulated in this way, this is a typical inverse problem in elasticity, according to

Keller's definition. More precisely, it is a problem of reinforcement, but not according to the usual meaning of the term, because, in this case, it is not important to minimize reinforcing material, but rather to arrange it in the most suitable manner to control the distribution of a stress field. A certain curiosity has recently arisen for problems of this kind, but with different purposes. In general, problems relating to exterior domains have been treated: for example, the problem of finding the amount of a reinforcement to be placed on the boundary of a hole in order to maintain a prescribed stress state in an infinite plate (Mansfied (1953)); or of finding the shape of the boundary of a hole which does not alter the sum of the principal stresses (Bjorkmann (1976)). The case considered there is a variant of these, which may be useful in the transport of delicate substances.

2. THE ELASTIC PROBLEM

In order to consider the problem of packaging in detail, let us study the two-dimensional case, that is to say when D is a plane domain and S is a simple, regular, closed curve. In this case the reinforcement is a perfectly flexible thread the thickness of which can be adjusted in such a manner that, for any given boundary tractions, the sum of the principal stresses at any point has a given constant value. An analogous problem, but formulated for an exterior domain, have been studied by Dhir (1979).

Let us assume that T is the magnitude of the axial force, varying point to point, acting along the reinforcement. An element of reinforcement, of length ds, will be subjected, beside the force T, to surface tractions, and to contact forces exchanged with the plate. If we call p_u, p_v the radial and tangential components of the exterior tractions, and q_u, q_v the analogous components of the contact forces (*cf.* Fig. 2), the equilibrium of the element ds requires the following equations to be satisfied

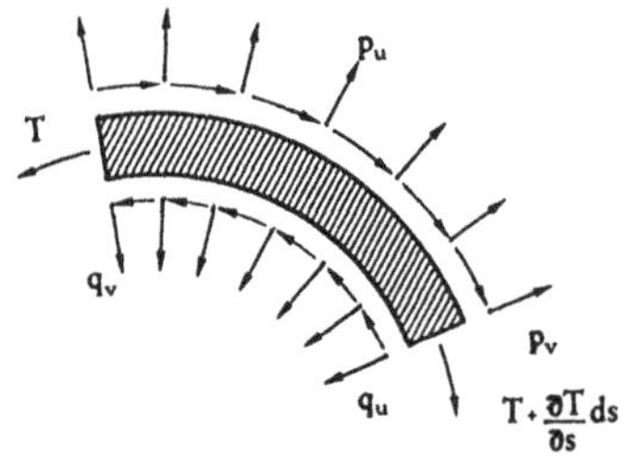

Fig. 2

$$\left.\begin{aligned} &\frac{\partial T}{\partial s} ds + (p_v - q_v) t ds = 0, \\ &- T d\psi + (p_u - q_u) t ds = 0 \end{aligned}\right\}, \tag{2.1}$$

t being the thickness of the plate, and ψ the angle that the unit tangent to the element at its left end makes with the x-axis. In writing equations (2.1), the p_v-components have been taken as positive if oriented in the sense in which the arc s increases, and the p_u-components as positive if directed along outer normal to the boundary (Fig. 2). The q_v- and q_u-components of the contact forces, which act on the opposite side of the element ds are then oriented in opposite senses. The q_v- and q_u-components are also related to the stresses acting tangentially and perpendicular to the boundary of the plate by the equations

$$\tau_{uv} = q_v \quad , \quad \sigma_u = q_u. \tag{2.2}$$

In equations (2.1) the exterior tractions p_u, p_v are given, while the axial force T and the contact forces q_v, q_u are unknown. In order to render the problem determinate, it is necessary to add a further equation which imposes the equality of the tangential stains at the interface between the reinforcement and the plate. If we denote the cross section of the reinforcement by A, variable with s, this condition reads

$$\frac{T}{A} = \sigma_v - \nu\sigma_u, \tag{2.3}$$

where we have assumed, without loss of generality, that the reinforcement is made of the same material as the plate, and ν is Poisson's ratio of the material.

Since we are dealing with a plane problem it is convenient to introduce a stress function $F(x,y)$ such that the Cartesian components of the stresses in D can be expressed as

$$\sigma_x = \frac{\partial^2 F}{\partial y^2}, \quad \sigma_y = \frac{\partial^2 F}{\partial x^2}, \quad \tau_{xy} = -\frac{\partial^2 F}{\partial y \partial x}, \tag{2.4}$$

where, in addition, $F(x,y)$ is a biharmonic function in D in order to guarantee that the stress components defined by (2.4) also satisfy, beside equilibrium, the compatibility equations. By using (2.4), we can give (2.1) a more manageable form (Dihr (1979)). The Cartesian components of T are

$$T_x = T\cos\psi \quad , \quad T_y = T\sin\psi, \tag{2.5}$$

because ψ is the slope of the boundary with the x-axis. Let us denote the Cartesian components of the exterior tractions by (p_x, p_y). The components of the resultant of these forces, evaluated along an arc of boundary of length s, measured from a fixed point, are

$$P_x = \int_0^s p_x ds \ , \ P_y = \int_0^s p_y ds. \tag{2.6}$$

The components of the contact forces, exchanged between the reinforcement and the plate, are

$$Q_x = \int_0^s q_x ds \ , \ Q_y = \int_0^s q_y ds. \tag{2.7}$$

These letter can be expressed in terms of the Cartesian stress components through the relations

$$\sigma_x \ell + \tau_{xy} m = q_x \ , \ \tau_{xy} \ell + \sigma_y m = q_y, \tag{2.8}$$

where ℓ, m denote the components of the unit outer normal to S. Since ℓ, m, can be written as $\ell = -\frac{\partial y}{\partial s}$, $m = \frac{\partial x}{\partial s}$, where x, y are the coordinates of a generic point of the boundary, the relations (2.8), after integration from 0 to s, yield

$$Q_x = -\left[\frac{\partial F}{\partial y}\right]_0^s \ , \ Q_y = \left[\frac{\partial F}{\partial x}\right]_0^s. \tag{2.9}$$

Then the equilibrium equations at the boundary can be written in the more compact form

$$-\frac{\partial F}{\partial y} + i\frac{\partial F}{\partial x} = \frac{T}{t}e^{i\psi} + P_x + iPy + C, \tag{2.10}$$

where i is the imaginary unit and C an arbitrary complex constant.

If we introduce the complex variable $z = x + iy$ and the complex resultant $P = P_x + iPy$, equation (2.10) becomes

$$2i\frac{\partial F}{\partial \bar{z}} = \frac{T}{t}e^{i\psi} + P + C, \tag{2.11}$$

or, by taking the complex conjugate,

$$-2i\frac{\partial F}{\partial z} = \frac{T}{t}e^{-i\psi} + \bar{P} + \bar{C}. \tag{2.12}$$

In absence of reinforcement, T is zero and equation (2.12) becomes the well known boundary condition of the second boundary value problem of plane elasticity (*cf.* Neuber (1985)).

Let now $z = z(w)$ be a conformal mapping which transforms the unit circle of the w–plane ($w = e^{u+iv}$) into the region D of the z–plane. If we call σ the value of w on the boundary of the circle, where $u = 0$, the angle ψ is given by

$$e^{i\psi} = i\sigma\sqrt{\frac{\dot{z}(\sigma)}{\dot{\bar{z}}(\bar{\sigma})}}, \tag{2.13}$$

where, as usual, $\dot{z}(w) = \frac{dz(w)}{dw}$, $\dot{\bar{z}}(\bar{w}) = \frac{d\bar{z}(\bar{w})}{d\bar{w}}$. By substituting (2.13) into (2.12), we write

$$\left.\frac{\partial F}{\partial z}\right|_{z=z(\sigma)} = \frac{\bar{\sigma}}{2}\frac{T}{t}\sqrt{\frac{\dot{\bar{z}}(\bar{\sigma})}{\dot{z}(\sigma)}} - i\frac{\bar{P}}{2} + C_1, \tag{2.14}$$

C_1 being another complex constant.

3. THE ANALYTICAL PROBLEM

Since F is biharmonic, it can be represented through the formula (of Goursat)

$$F = z\bar{f}_1(\bar{z}) + \bar{z}f_1(z) + f_2(z) + \bar{f}_2(\bar{z}), \tag{3.1}$$

where $f,(z), f_2(z)$ are two arbitrary analytic functions of z. Putting $z = z(w)$ into (3.1) we express F in function of the curvilinear coordinates u, v. Then the stress components in the u, v–system can be represented in the form

$$\sigma_u + \sigma_v = 4\left(\frac{\dot{f}_1}{\dot{z}} + \frac{\dot{\bar{f}}_1}{\dot{\bar{z}}}\right), \tag{3.2}$$

$$\sigma_v - \sigma_u + 2i\tau_{uv} = 4\left[\frac{\bar{z}}{\dot{\bar{z}}}\overline{\left(\frac{\dot{f}_1}{\dot{z}}\right)^{\cdot}} + \frac{1}{\dot{\bar{z}}}\overline{\left(\frac{\dot{f}_2}{\dot{z}}\right)^{\cdot}}\right]. \tag{3.3}$$

These formulae are due to Muskhelishvili (1953).

By using Goursat's formula we may write (2.14) as

$$\frac{1}{\dot{z}}\frac{\partial F}{\partial w}\bigg|_{w=\sigma} = \bar{f}_1(\bar{\sigma}) + \frac{\bar{z}(\bar{\sigma})}{\dot{z}(\sigma)}\dot{f}_1(\sigma) + \frac{1}{\dot{z}(\sigma)}\dot{f}_2(\sigma)$$

$$= \frac{\bar{\sigma}}{2}\frac{T}{t}\sqrt{\frac{\dot{\bar{z}}(\bar{\sigma})}{\dot{z}(\sigma)}} - i\frac{\bar{P}}{2} + C_1. \tag{3.4}$$

This equation differs from the classical one of the second boundary value problem of eleasticity for the reason that the unknowns are not f_1, f_2, but the axial force T, and therefore, by virtue of (2.3), the cross–section A of the reinforcement. The magnitude of T must be such that the sum of principal stresses $(\sigma_u + \sigma_v)$ at any point has a fixed value, say S_0. The consequence is that, at any point of D, the stresses must satisfy the condition

$$\sigma_u + \sigma_v = 4\left(\frac{\dot{f}_1}{\dot{z}} + \frac{\dot{\bar{f}}_1}{\dot{\bar{z}}}\right) = 4S_0, \tag{3.5}$$

where S_0 is a prescribed constant. This implies that $f_1(w)$ has the form $f_1(w) = \frac{1}{2}S_0 z(w)$, so that equation (3.4), after multiplication by $\dot{z}(\sigma)$, becomes

$$S_0\bar{z}(\bar{\sigma})\dot{z}(\sigma) + \dot{f}_2(\sigma) = \frac{\bar{\sigma}}{2}\frac{T(\sigma)}{t}\sqrt{\dot{\bar{z}}(\bar{\sigma})\dot{z}(\sigma)} - \frac{i}{2}\bar{P}(\sigma)\dot{z}(\sigma) + C_1\dot{z}(\sigma). \tag{3.6}$$

In order to solve this equation we observe that, on the unit circle, σ and $\bar{\sigma}$ are related by the condition $\bar{\sigma} = e^{-iv} = \frac{1}{\sigma}$, therefore (3.6) assumes the form

$$S_0 \dot{\bar{z}}(\frac{1}{\sigma})\dot{z}(\sigma) + \dot{f}_2(\sigma) = \frac{1}{2\sigma}\frac{T(\sigma)}{t}\sqrt{\dot{\bar{z}}(\frac{1}{\sigma})\dot{z}(\sigma)} - \frac{i}{2}\bar{P}(\sigma)\dot{z}(\sigma) + C_1\dot{z}(\sigma). \tag{3.7}$$

Let us suppose that the boundary S is regular, so that $z(\sigma)$ is an analytic function with $\dot{z}(\sigma) \neq 0$, despite the fact that this assumption can be removed (*cf.* Weinberger (1965, §52)). Let us assume that $\dot{f}_2(w)$ can be represented in the form of a series like

$$\dot{f}_2(w) = \sum_{n=0}^{\infty} b_n w^n, \tag{3.8}$$

and that this series converges, not only for $|w| < 1$, but also at the boundary $|w| = 1$ of the circle. Let us expand the exterior load in a Fourier series of the type

$$P(\sigma) = \sum_{n=-\infty}^{+\infty} A_n e^{inv} = \sum_{n=-\infty}^{+\infty} A_n \sigma^n, \tag{3.9}$$

where A_n are known coefficients. We finally assume that the solutions $T(\sigma)$ admits the representations (*cf.* Dihr (1979))

$$\frac{T(\sigma)}{t}\sqrt{\dot{\bar{z}}(\frac{1}{\sigma})\dot{z}(\sigma)} = \sum_{n=0}^{\infty}\left(c_n\sigma^n + \frac{d_n}{\sigma^n}\right), \tag{3.10}$$

where c_n and d_n are coefficients such that $c_n = \bar{d}_n$ in order to ensure that (3.10) is a real quantity.

By substituting all these expansions into (3.7) and equating the terms of the same power, we can obtain c_n, d_n, and hence $T(\sigma)$. In order to determine A, it is sufficient to combine (2.3) with (2.1), (2.2) to obtain

$$\frac{A}{t} = \frac{\frac{T}{t}\frac{ds}{d\psi}}{4S_0\frac{ds}{d\psi} - (1-\nu)\left(\frac{T}{t} - p_u\frac{ds}{d\psi}\right)}, \tag{3.11}$$

where

$$\frac{ds}{d\psi} = \frac{(\dot{z}\dot{\bar{z}})^{3/2}}{\dot{z}\dot{\bar{z}} + \frac{\dot{z}\ddot{\bar{z}}}{2\sigma} + \frac{\sigma}{2}\dot{\bar{z}}\ddot{z}}, \tag{3.12}$$

is the radius of curvature of the boundary. It is easily seen that $\frac{A}{t}$, which represents the ratio between two essentially positive quantities can never be negative. Thus, if this occurs, the problem does not admit solution.

4. THE CIRCULAR DISK

The solution, as summarized in equations (3.7), (3.9), can be illustrated in some simple cases, which avoid unessential calculations. For example, let us assume that D is a circle of radius R. Then the conformal mapping which takes the unit circle of the w–plane into the circle of radius R of the z–plane is simply

$$z = Rw. \tag{4.1}$$

If we replace (4.1) into (3.7) we obtain

$$2S_0\frac{R^2}{\sigma} + 2\dot{f}_2(\sigma) = \frac{T(\sigma)}{\sigma t} - i\bar{P}(\sigma)R + C_1R. \tag{4.2}$$

Let us first suppose that the boundary tractions consists of a condtant radial pressure $p_u = p_0$ =constant. In this case we have

$$p_x = p_0 \cos v \ , \ p_y = p_0 \sin v,$$

and therefore, from (2.6), we derive

$$P_x = R\int_0^v p_0 \cos v dv = p_0 R \sin v, \quad P_y = R\int_0^v p_0 \sin v dv = p_0 R(1-\cos v), \tag{4.3}$$

and consequently, since $e^{iv} = \sigma$, (4.2) becomes

$$2S_0\frac{R^2}{\sigma} + 2\sum_{n=0}^{\infty} b_n\sigma^n = \sum_{n=0}^{\infty}\left(c_n\sigma^{n-1} + \frac{d_n}{\sigma^{n+1}}\right) - p_0R^2(1-\frac{1}{\sigma}) + C_1R. \tag{4.4}$$

After comparison of the coefficients we obtain the relations

$$2b_0 = c_1 - p_0R^2 + C_1R, \quad 2b_n = c_{n+1} \text{ for } n \geq 1,$$

$$2S_0R^2 = (c_0 + d_0) + p_0R^2, \quad d_n = 0 \text{ for } n \geq 1.$$

In addition, we know that $c_n = \bar{d}_n$, and hence $c_1 = c_2 = 0$ in the first two equations, while the third yields $2b_n = c_{n+1} = 0$ $(n \geq 1)$. Since we may choose $C_1 = 0$, we find that the only non–vanishing coefficients are

$$2b_0 = p_0 R^2 \ , \ c_0 = d_0 = S_0 R^2 - \frac{1}{2} p_0 R^2, \tag{4.5}$$

whence, by using (3.11) and recalling that now $\frac{ds}{d\psi} = R$, we find

$$\frac{A}{t} = \frac{2S_0 R - p_0 R}{2(1-\nu)S_0 + 2(1+\nu)p_0}. \tag{4.6}$$

The result is that $\frac{A}{t}$ is a constant, but the problem does not admit solution for any value of S_0 and p_0, because $\frac{A}{t}$ cannot become negative. For $p_0 = 2S_0$, that is when the radial pressure is equal to the desired value of the internal pressure, $\frac{A}{t}$ is of course zero.

Let us instead assume that the disk is subject to a radial pressure $p_u = p_1 \cos 2v$, where p_1 is a constant. Using again (2.6) we calculate

$$P_x = p_1 R \int_0^v \cos 2v \cos v dv = p_1 R \left(\frac{\sin 3v}{6} + \frac{\sin v}{2} \right),$$

$$P_y = p_1 R \int_0^v \cos 2v \sin v dv = p_1 R \left(\frac{2}{3} - \frac{\cos 3v}{6} - \frac{\cos 2v}{2} \right),$$

and hence, recalling that $e^{iv} = \sigma$, we can write

$$-i\bar{P} = -i(P_x - iP_y) = \frac{p_1 R}{2} \left(-\frac{4}{3} + \frac{1}{\sigma} + \frac{1}{3\sigma^3} \right). \tag{4.7}$$

The equation determining the coefficients c_n, d_n is again (4.4) but where, now, the term $-pR^2(1 - \frac{1}{\sigma})$ is replaced by the right hand side of (4.7) multiplied by R. The comparison of the coefficients of the terms of equal order yields

$$2b_0 = -\frac{2}{3} p_1 R^2 \ , \ 2b_1 = c_2 = d_2 = -\frac{p_1 R^2}{6}$$

$$c_0 = d_0 = S_0 R^2 - \frac{1}{4} p_1 R^2, \tag{4.8}$$

the other coefficients being zero. It follows that $\frac{A}{t}$ has the expression

$$\frac{A}{t} = \frac{2S_0 R - p_1 R(\frac{1}{2} + \frac{1}{3}\cos\sigma)}{2(1-\nu)S_0 + (1+\nu)p_1(\frac{1}{2} + \frac{4}{3}\cos\sigma)}. \tag{4.9}$$

Also here S_0 and p_1 must be such so as to ensure $\frac{A}{t}$ to be positive or zero.

To have an idea of the distribution of the reinforcement, let us take $p_1 = 2S_0$ and $\nu = 0$. Thus (4.9) becomes

$$\frac{A}{t} = R\frac{(3 - \cos 2v)}{(9 - 8\cos 2v)}.$$

This last formula shows that $\frac{A}{t}$, considered as a function of v is periodic with period π and assumes the values

$$\frac{A}{t}(0) = 2R \ , \quad \frac{A}{t}(\frac{\pi}{4}) = 0.333\,R \ , \quad \frac{A}{t}(\frac{\pi}{2}) = 0.237\,R,$$

which confirm the intuitive expectation that the cross section is the thickest at $v = 0$, where the external pressure is the highest.

REFERENCES

1. Bjorkman, G.S.,Jr., and Richards R.: "Harmonic Holes – An Inverse Problem in Elasticity". Journ. Appl. Mech. Vol. 43, pp. 414–418 (1976).

2. Dihr S.K.: "A Hybrid Problem in Plane Elasticity". Journ. Appl. Mech. Vol. 46, pp. 714–716 (1979).

3. Mansfield E.H.: "Neutral Holes in Plane Sheet–Reinforced Holes which are Elastically Equivalent to the Uncut Sheet". Quart. J. Mech. Appl. Math. Vol. VI, pp. 370–378 (1953).

4. Muskhelishvili, N.L: *Some Basic Problems of the Mathematical Theory of Elasticity.* Groningen: Noordhoff (1953)

5. Neuber H.: *Kerbspannungslehre.* Berlin/Heidelberg/New York: Springer (1985).

6. Weinberger H.F.: *A First course in Partial Differential Equations.* Waltam: Blaisdell (1965).

32. The First Digit Problem and Scale-Invariance

Aljoša Volčič Dipartimento di Matematicà, Università di Trieste, Trieste, Italy

Abstract. If we classify an extensive collection of numerical data expressed in decimal form according to the first significant digit, the nine resulting classes are not usually of the same size. The American astronomer Simon Newcomb was the first to write about this phenomenon in 1881. He observed that the pages in well-used tables of logarithms tend to get quite dirty in the front, whereas the last pages stay relatively clean.

F. Benford provided in 1938 for empirical evidence and the law (also known to Newcomb)

$$p_k = \log_{10} \frac{k+1}{k},$$

$1 \leq k \leq 9$, is named after him.

1. Introduction.

Suppose we have a collection of physical measures like distances between visible stars, consumption of water or electricity in April 1995 by users living in New York, areas of the largest 1000 lakes in Finland, areas of the counties in United States, production of oil in various years and states in Middle East, and so on.

1991 *Mathematics Subject Classification.* 60A10.

Key words and phrases. First-digit problem, scale-invariance.

If we pick "at random" a number from such a collection, we would expect that all the digits have the same frequency. In particular, if we disregard the position of decimal point, we would expect that all the integers between 1 and 9 are equaly likely to be in the first place. Surprisingly enough, real data show that this expectation is wrong.

This has been noticed first by the astronomer S. Newcomb [**N**] who observed that library books of logarithm tables used by students were dirtier towards the beginning than towards the end. This is not surprising for a bad novel, but it seems curious that anybody is more interested in his calculations in numbers with leading digit 1 rather than in those with leading digit 2 or more.

Much later F. Benford [**B**] rediscovered the phenomenon and gave an empirical evidence for the expected distribution of the integers as first digits. This distribution is now called Benford's Law:

$$p_k = \log_{10} \frac{k+1}{k},$$

$1 \leq k \leq 9$.

This fact is so surprising that it attracted the attention of many authors. The memoir [**R2**] is an excellent rewiev of the problem.

This distribution, as other laws in probability, has various explanations and it would be misleading to mention just one (as it would be misleading to see the normal distribution only as an appropriate limit of binomial distributions).

One explanation can be found in discrete and continuous density arguments and summability methods (see for instance [**F**], [**R2**] and [**FL**]). This approach brings in the appealing methods offered by finitely additive measures as in [**R1**] and [**S**].

More recently, J. Boyle [**Bo**] showed that Benford's law follows from a kind of central limit theorem for products of independent continuous random variables.

After the preparation of this paper the author learned about two very recent papers by T. P. Hill on the subject. One of the papers proposes a new explanation for Benford's law, namely the base-independence [**H2**]. In [**H1**] the same author proposes a new approach to the scale-invariance hypothesis. The relation to our results will be discussed at the end of this paper.

R. S. Pinkham [**P**] has been the first to deduce Benford's law from the scale-invariance hypothesis. Raimi [**R1**] and Knuth [**K**] showed however that there is a flaw in Pinkham's argument.

We will show how to avoid that flaw by an appropriate choice of the measure space.

Our main result is Theorem 1, which implies that a scale-invariant random variable is absolutely continuous. Benford's law follows then from the explicit expression of the distribution function and the density given in Theorem 1 and its Corollary.

It should be noted also that in proving Theorem 1 we do not require the continuity of the distribution function.

2. Scale-invariance is not compatible with uniform distribution.

Scale-invariance is very nicely motivated by Raimi. In [**R2**] he wrote: "If a table of surface areas of a set of nations or lakes is rewritten in another system of units of measurements, acres for hectars for instance, the result will be a rescaled table whose every entry is the same multiple of the corresponding entry of the original table. If the first digits of all the tables in the universe obey some fixed distribution law, that law must surely be independent of the system of units chosen, since God is not known to favour either the metric system or the English system."

With rather elementary considerations we will show in this Section that if we accept this point of view, uniform distribution has to be excluded. We will find for the nine unknowns p_k, $1 \leq k \leq 9$, seven independent linear equations. This system has infinitely many solutions. The values given by Benford's law provide one of the solutions. It can be seen however that any solution q_k of the system for which $q_k > 0$ for all k necessarily satisfies the inequality $q_1 > q_k$ for every $2 \leq k \leq 9$.

Suppose we have a collection of distances expressed in yards and consider a second collection expressing the same distances, but this time in feet. Let us denote by p'_k and p''_k, $1 \leq k \leq 9$, the frequencies of the first significant digits in the first and in the second collection, respectively.

Any measure x in yards belonging to $[1,2)$ has 1 as first significant digit and is expressed in feet with the number $y = 3x \in [3,6)$, whose first significant digit is either 3, 4 or 5. The viceversa is also true: any measure $y \in [3,6)$ expressed in feet in the second collection corresponds to a measure $x = \frac{y}{3} \in [1,2)$ in the first collection. We have therefore that

$$p'_1 = p''_3 + p''_4 + p''_5 .$$

Similar considerations involving measures in yards $x \in [2,3)$ and in feet $y \in [6,9)$ show that

$$p'_2 = p''_6 + p''_7 + p''_8 .$$

Any statistical regularity present in collections of data of this sort should be present in both tables and in every other collection obtained by expressing the distances in lines, inches, feet, yards, fathoms, rods, chains, furlongs or miles, not to mention the meters. But the metric system is of course much less exciting.

If we assume that the distribution p_k, $1 \leq k \leq 9$, of the first significant digits does not depend on the measure unit we use, we can therefore conclude that

$$p_1 = p_3 + p_4 + p_5$$

and

$$p_2 = p_6 + p_7 + p_8 .$$

Different equations of the same kind can be obtained comparing other tables of distances. For instance, if we take as measure units yards and fathoms (1 fathom = 2 yards), and if we use the same argument as above, we get, for $1 \leq i \leq 4$,

$$p_i = p_{2i} + p_{2i+1} .$$

All the six linear equations we found so far are compatible with Benford's law. They, together with the seventh's obvious equation

$$\sum_{k=1}^{9} p_k = 1,$$

are independent. We could go on, considering other measure units which differ by an integer multiple. For instance, since one chain is four rods, any measure $z \in [1,2)$ in chains is expressed in rods by a number $w \in [4,8)$. The argument given above says that

$$p_1 = p_4 + p_5 + p_6 + p_7 .$$

This equation is not independent from the previous ones, since we found already that $p_1 = p_3 + p_4 + p_5$ and $p_3 = p_6 + p_7$.

On the other hand, the seven equations are not sufficient to derive Benford's law. A little bit of linear algebra shows that there exists a solution of the system such that $q_7 = q_9 = 0$.

In fact, if we assume that Benford's law is true, there cannot exist nine independent linear equations of the sort we found, since they all have integer coefficients and Cramer's rule would then give rational values for every p_k. This is in contrast to the fact that the frequencies given by Benford's law are not rational.

Proposition. *If q_k is a solution of the linear system obtained above and if we assume $q_k > 0$ for all k, then*

$$q_1 > q_k \tag{1}$$

for $2 \leq k \leq 9$.

Proof. Since $q_1 = q_2 + q_3$, $q_1 > q_2$ and $q_1 > q_3$. From $q_1 = q_3 + q_4 + q_5$, we see that (1) holds for all $2 \leq k \leq 5$. From $q_1 > q_2$ and the equation $q_2 = q_6 + q_7 + q_8$, we conclude that (1) is valid also for $6 \leq k \leq 8$. Finally we get the conclusion from $q_4 = q_8 + q_9$ and $q_1 > q_4$. □

Therefore if we assume the scale-invariance hypothesis, 1 has to be the most frequent first significant digit and we cannot expect that the first significant digits are uniformly distributed.

3. Benford's law for scale-invariant random variables.

Let us point out the flaw in Pinkham's approach.

The set of all positive real numbers whose first digit is at most k is

$$D_k = \bigcup_{n=-\infty}^{\infty} [10^n, (k+1)10^n) .$$

Pinkham assumed that there exists a distribution function F on the reals such that $F(0) = 0$ and such that if we denote by μ the measure induced by F on the positive reals by $\mu([a,b)) = F(b) - F(a)$, it satisfies for every positive constant c the equation

$$\mu(D_k) = \mu(cD_k) , \tag{2}$$

where

$$cD_k = \bigcup_{n=-\infty}^{\infty} [c10^n, c(k+1)10^n) .$$

He showed that the validity of equation (2) for every $c > 0$ implies Benford's law.

The problem with this approach is that if we assume that there is a scale-invariant measure on the σ-algebra $\mathcal{A}$ of all the Borel subsets of the positive real line, the equation (2) should be valid not only for the nine sets D_k, but for all $A \in \mathcal{A}$. This would imply in particular that for every $c > 0$

$$\mu((0,1)) = \mu((0,c)) .$$

It is easy to see however that there is no such countably additive probability measure on the Borel sets of the positive real line!

We will overcome this flaw by taking an appropriate measure space and defining the scale invariance in an appropriate way. We need first some definitions.

Every positive real number x can be written as

$$x = a.b_1 b_2 \ldots \ 10^c,$$

where a, b and c are integers, $1 \leq a \leq 9$ and $0 \leq b_i \leq 9$ for every i. To avoid ambiguity, we exclude the representations in which $b_i = 9$ for all sufficiently large i's.

Define, for $x > 0$,

$$N(x) = a.b_1 b_2 \ldots$$

and

$$S(x) = a .$$

The mapping N maps the positive reals onto $[1, 10)$ while S maps the same set onto $\{1, 2, \ldots, 9\}$. The number $S(x)$ is called the *first significant digit* of x.

Note that $N(x) = N(y)$ if and only if $x = y \ 10^c$ for some integer c. Moreover we can define an equivalence relation on the positive reals putting xNy if and only if $N(x) = N(y)$. The set $[1, 10)$ represents all the equivalence classes.

If we represent the data in decimal form and we are looking at the first significant digit, it is natural to consider the positive real numbers modulo the equivalence relation N. The set $[1, 10)$ will be taken as the sample space, with the Borel σ-algebra on it.

We will define, for any Borel set $B \subset [1, 10)$ and each $c > 0$, the product of c with the set B in the following way:

$$cB = \{y : y = cx \bmod [1, 10) , x \in B\} .$$

Let us define now the scale-invariance for a random variable taking its values in $[1, 10)$.

Definition. *We say that a random variable X with values in $[1, 10)$ is scale-invariant, if for all $x \in [1, 10)$ and each $y > 0$ we have*

$$P(1 \leq X < x) = P(N(y) \leq X < xN(y)),$$

if $xN(y) < 10$ and

$$P(1 \leq X < x) = P(N(y) \leq X < 10) + P(1 \leq X < xN(y)10^{-1}),$$

if $xN(y) \geq 10$.

Remark. *It is clear that X is scale-invariant in the above sense, if the condition holds just for all $y \in [1, 10)$.*

The two conditions above can be therefore rewritten as

$$P(1 \leq X < x) = P(y \leq X < xy),$$

when $xy < 10$ and

$$P(1 \leq X < x) = P(y \leq X < 10) + P(1 \leq X < xy\ 10^{-1}),$$

if $xy \geq 10$.

Theorem 1. *A random variable X with values in $[1, 10)$ is scale-invariant if and only if its distribution function F is given, for $x \in [1, 10]$, by*

$$F(x) = \log_{10} x.$$

Proof. Suppose that the distribution function F of X is given by the expression given above. Then clearly X takes its values in $[1, 10)$. Let us check that X is scale-invariant. Let $x \in [1, 10)$ and suppose first that $y \in [1, 10)$ is such that $xy < 10$. Let us evaluate

$$P(y \leq X < xy) = F(xy) - F(y) = \log_{10} xy - \log_{10} y = \log_{10} x = P(1 \leq X < x).$$

If now $xy \geq 10$, we have to evaluate

$$P(y \leq X < 10) + P(1 \leq X < xy\,10^{-1}) = F(10) - F(y) + F(xy\,10^{-1}) - F(1) =$$

$$\log_{10} 10 - \log_{10} y + \log_{10}(xy10^{-1}) = \log_{10} x = P(1 \leq X < x).$$

Let us prove now the viceversa. Assume that X takes its values in $[1, 10)$ and is scale-invariant. If $xy < 10$, the scale-invariance can be rewritten in terms of the distribution function F of X as

$$F(x) + F(y) = F(xy).$$

Letting $x = 10^u$ and $y = 10^v$, we have $u, v \in [0, 1)$, $uv < 1$ and if we define $g(u) = F(10^u)$, we get for g the Cauchy equation

$$g(u) + g(v) = g(u + v)\,. \tag{3}$$

Note that g is the composition of two non decreasing functions, so g itself is nondecreasing. Moreover, $g(0) = 0$ and $g(1) = 1$, since $F(1) = 0$ and $F(10) = 1$. It is well known that a monotone solution of the Cauchy equation is linear but in this case we have to take care of the restrictions on u and v, since g satisfies the equation only for u an v in $[0, 1]$, with the additional condition that $uv \le 1$. We shall check however that even this weaker condition is sufficient to conclude that g is linear.

From (3) we see that $g(2u) = g(u + u) = g(u) + g(u) = 2g(u)$ for all $u \in [0, \frac{1}{2}]$. By induction we conclude that for any positive integer n

$$g(nu) = ng(u) \tag{4}$$

for all $u \in [0, \frac{1}{n}]$. For any $v \in [0, 1]$, apply (4) to $u = \frac{v}{n}$ to get

$$g\left(\frac{v}{n}\right) = \frac{1}{n}g(v) \tag{5}$$

for every $v \in [0, 1]$. Let now m and n be positive integers with $m \le n$ and let $u \in [0, 1]$. Applying first equation (4) and then equation (5), we get

$$g\left(\frac{m}{n}u\right) = mg\left(\frac{u}{n}\right) = \frac{m}{n}g(u)\,.$$

From $g(1) = 1$ it follows immediately that $g(q) = q$ for allrational numbers $q \in [0, 1]$. But since g is non decreasing, we can conclude that $g(u) = u$ for all $u \in [0, 1]$. Substituting now $u = \log_{10} x$ in $g(u) = F(10^u)$, we see that

$$F(x) = \log_{10} x$$

for all $x \in [1, 10]$.

It is easy to verify now that this distribution function F satisfies also the scale-invariance condition if $xy > 10$. In fact,

$$P(y \le X < 10) + P(1 \le X < xy\,10^{-1}) =$$

$$\log_{10} 10 - \log_{10} y + \log_{10} x + \log_{10} y - \log_{10} 10 = P(1 \le X < x)\,.$$

□

From Theorem 1 we can immediately deduce the following important consequence.

Corollary. *A random variable X with values in $[1, 10)$ is scale-invariant if and only if X is absolutely continuous and its density is*

$$f(x) = \frac{\log_{10} e}{x}$$

for $x \in [1, 10]$ and 0 otherwise.

Let us show now that Benford's law holds for a scale-invariant random variable.

Theorem 2. *If X is a scale-invariant random variable taking its values in $[1,10)$, then the probability p_k that the first significant digit is k, for $1 \le k \le 9$, is given by*

$$p_k = \log_{10} \frac{k+1}{k}.$$

Proof. It follows from Theorem 1 that

$$p_k = P(S(x)=k) = P(k \le X < k+1) = F(k+1) - F(k) = \log_{10} \frac{k+1}{k}.$$

□

For a scale-invariant random variable we can also find the formulas for the distributions of the second, third and more in general n-th significant digit. They are known in the literature and go back to Newcomb. We can deduce them from Theorem 1.

Theorem 3. *If X is a scale-invariant random variable taking its values in $[1,10)$, then the probability p_k^n that the n-th digit is k, $0 \le k \le 9$, is given by*

$$p_k^n = \log_{10} \prod_{i=10^{n-1}}^{10^n-1} \frac{10\,i+k+1}{10\,i+k}. \tag{6}$$

Moreover, $p_0^n > p_1^n > \cdots > p_9^n$.

Proof. Applying again Theorem 1 we get

$$p_k^n = \sum_{i=10^{n-1}}^{10^n-1} P(i\,10^{-n+1} + k\,10^{-n} \le X < i\,10^{-n+1} + (k+1)\,10^{-n}) \tag{7}$$

$$= \log_{10} \prod_{i=10^{n-1}}^{10^n-1} \frac{10\,i+k+1}{10\,i+k},$$

therefore (6) is proved. Since the density $f(x)$ of X is decreasing,

$$P(i\,10^{-n+1} + k\,10^{-n} \le X < i\,10^{-n+1} + (k+1)\,10^{-n}) \le$$
$$P(i\,10^{-n+1} + (k-1)\,10^{-n} \le X < i\,10^{-n+1} + k\,10^{-n})$$

for each $1 \le k \le 8$ and every $10^{n-1} \le i \le 10^n - 1$. The conclusion follows now from (7). □

Theorem 4. *If the probabilities p_k^n are defined as above,*

$$\lim_{n\to\infty} p_k^n = \frac{1}{10}.$$

Proof. Since $p_k^n > p_{k-1}^n$ for all $1 \le k \le 9$ and $\sum_{k=0}^9 p_k^n = 1$ for all n, we will prove the claim if we show that

$$\lim_{n\to\infty} p_0^n = \frac{1}{10}.$$

But we have

$$\frac{1}{10} < p_0^n = \sum_{i=10^{n-1}}^{10^n-1} \log_{10}(1+\frac{1}{10i}) < \frac{1}{10}\sum_{i=10^{n-1}}^{10^n-1} \frac{1}{i} <$$

$$\frac{1}{10}\int_{10^{n-1}-1}^{10^n-2} \frac{1}{x}\,dx = \frac{1}{10}\log_{10}\frac{10^n-2}{10^{n-1}-1},$$

and the conclusion follows. □

Explicit computation gives $p_0^2 - p_9^2 = 0.0289$, $p_0^3 - p_9^3 = 0.00352$, $p_0^4 - p_9^4 = 0.00035$ and $p_0^5 - p_9^5 = 0.000037$.

The next theorem gives the generalized Benford's law for the scale-invariant random variable giving the probability that the first n digits are $k_1, k_1, \ldots, k_n$, respectively.

Theorem 5. *If X is a scale-invariant random variable taking its values in $[1,10)$, then the probability $p_{k_1,k_1,\ldots,k_n}$ that the first n digits are $k_1, k_1, \ldots, k_n$, respectively, is*

$$p_{k_1,k_1,\ldots,k_n} = \log_{10}\frac{k_1 10^{n-1} + k_2 10^{n-2} + \cdots + k_n + 1}{k_1 10^{n-1} + k_2 10^{n-2} + \cdots + k_n}.$$

Proof. The proof follows immediately from Theorem 1. □

The distribution function F defines a measure ν on the Borel subsets of $[1,10)$ by

$$\nu(B) = \int_B \frac{\log_{10} e}{x}\,dx.$$

The next theorem shows that this measure is scale-invariant.

Theorem 6. *For any Borel set $B \subset [1,10)$ and each $c > 0$, it is*

$$\nu(cB) = \nu(B). \tag{8}$$

Proof. Since all the Borel sets can be approximated by finite unions of disjoint intervals, it is enough to prove (8) for all intervals $B = [a,b) \subset [1,10)$. We may also assume that $c \in [1,10)$.

If $bc < 10$, then $cB = [ac, bc)$ and therefore

$$\nu(cB) = \log_{10} bc - \log_{10} ac = \nu(B).$$

If $bc \geq 10$ and $ac < 10$, then $cB = [ac, 10) \cup [1, bc\,10^{-1})$ and therefore

$$\nu(cB) = \log_{10} 10 - \log_{10} ac + \log_{10} bc - \log_{10} 10 = \nu(B).$$

Finally, if $ac \geq 10$, then $cB = [ac\,10^{-1}, bc\,10^{-1})$ and we have again

$$\nu(cB) = \log_{10} bc - \log_{10} 10 - \log_{10} ac + \log_{10} 10 = \nu(B)\,.$$

□

Other bases.

There exists an extensive literature on Benford's law for data written in bases b other than 10.

The results of this paper extend easily to that case. We define a random variable X with values in $[1, b)$ to be scale-invariant if for every $y \in [1, b)$ we have

$$P(1 \leq X < x) = P(y \leq X < xy)$$

if $xy < b$ and

$$P(1 \leq X < x) = P(y \leq X < b) + P(1 \leq X < xyb^{-1})$$

if $xy \geq b$.

Following the arguments of Theorem 1 we find that such a random variable is absolutely continuous and that its distribution function is

$$F_b(x) = \log_b x$$

for $x \in [1, b]$. The density of X is in this case $f_b(x) = \frac{\log_b e}{x}$ for $x \in [1, b]$ and 0 otherwise.

Benford's law in base b is then

$$p_{b,k} = \log_b \frac{k+1}{k}$$

for $1 \leq k \leq b$. In the same way we can prove the distributions of the n-th digit and the generalized Benford's law.

4. Final remarks.

Since we were interested in scale-invariance, we mentioned only data for which it makes sense to change the measure unit and which can, at least in principle, be described by positive real numbers.

It is reasonable to pose the question whether in practice the digits which follow the third or the forth are still reliable. It is quite possible that the measurements cannot be made sufficiently precise. For instance, the hieght of a mountain is expressed in meters or feet. The most recent determination of the heigth of Mount Everest [**PMB**] is 8,848.65 meters above the geoid, with a possible error of ±0.41 centimeters. Therefore the fourth digit could as well be 9 and any digit expressing fractions of meter would be completely meaningless. The measure expressed in feet is 29,031.00 ±1.35 and the fourth digit could as well be 2, the fifth digit could be 9, 0 or 2, while the sixth digit could be anything.

In some cases the precision of the measurements can be made very precise, but nobody cares to achieve it. For instance, there is no reason to measure the consumption of electricity if this does not affect the consumer's bill.

Another example of this sort are the distances between towns. They are expressed in kilometers or miles. The fractions have almost no interest since it is even not clear what do we measure: do we measure the distance between the two city halls, between the main railway stations or between the city limits?

Precise measurements are frequently rounded off in order to make them easier to remember and compare. For instance in the monthly report [**I**] on statistical indicators published by ISTAT, the Italian national statistical institute, the investments in industry and the italian export trade are indicated in bilions of liras with five or six digits, railway trafic is indicated in milions of passengers-kilometers with four digits, the consumption of oil products is indicated in thousands of quintals with four digits.

There are statistical data of different nature, which also apparently obey Benford's law, like street numbers, population of regions, number of towns in various states having at least thousand inhabitants and so on. The scale-invariance has nothing to do with this examples, since the "natural" measure unit is used. The most convincing explanation for that phenomenon is offered by the discrete summability methods first considered by [**F**].

Surprisingly enough, there are also "deterministic" tables of numbers which obey quite accurately Benford's law, like powers of integers [**D**]. This observation goes back to Benford.

5. Addendum. After the preparation of this paper, the author learned about the two papers by T. P. Hill [**H1**] and [**H2**].

There is a close relation between his results in [**H1**] and the results obtained in this paper. Since there is no scale-invariant measure on the Borel subsets of the positive real line, Hill considers a smaller σ-algebra $\mathcal{F}$ and shows that there is a scale-invariant measure φ on it. As noted in [**H2**] (last Lemma of Section 2), $\mathcal{F}$ is nothing else than (in our notation) $\mathcal{F} = \{F : F = N^{-1}(B); B \in \mathcal{B}\}$, where $\mathcal{B}$ denotes the σ-algebra of the Borel subsets of $[1,10)$. The measure φ on $\mathcal{F}$ is related to our measure ν by the relation

$$\varphi(N^{-1}(B)) = \nu(B).$$

Considering the probability space $([1,10), \mathcal{B}, \nu)$ instead of $(\mathbb{R}^+, \mathcal{F}, \varphi)$ has the advantage that the problem can be put and solved in the most classical setting by finding that the scale-invariant random variable is absolutely continuous and calculating its density and its distribution function.

References

[B] F. Benford, *The law of anomalous numbers*, Proc. Amer. Phil. Soc. **78** (1938), 551-572.

[D] R. Durrett, *Probability: Theory and examples*, Wadsworth, Belmont, 1991.

[F] B. J. Flechinger, *On the probability that a random number has initial digit A*, Amer. Math. Monthly **73** (1966), 1056-1061.

[FL] A. Fuchs, G. Letta, *Sur le problème du premier ciffre décimal*, Bollettino U.M.I. **(6) 3-B** (1984), 451-461.

[H1] T. P. Hill, *The significant-digit problem*, Amer. Math. Monthly **102** (1995), 322-327.

[H2] T. P. Hill, *Base-invariance implies Benford's law*, Proc. Amer. Math. Soc **123** (1995), 887-895.

[I] ISTAT, Indicatori Mensili **35** (December 1994).

[K] D. Knuth, *The Art of Computer Programming*, vol. 2, Addison-Wesley, 1969, pp. 219-229.

[N] S. Newcomb, *Note on the frequency of use of the different digits in natural numbers*, Amer. J. Math. **4** (1881), 39-40.

[P] R. S. Pinkham, *On the distribution of first significant digits*, Ann. Mat. Statist. **32** (1961), 1223-1230.

[PMB] G. Poretti, C. Marchesini, A. Beinat, *GPS Surveys: Mount Everest*, GPS World (October 1994), 33-44.

[R1] R. A. Raimi, *On the distribution of first significant figures*, Amer. Math. Monthly **76** (1969), 342-348.

[R2] R. A. Raimi, *The first digit problem*, Amer. Math. Monthly **83** (1976), 521-538.

[S] R. Scozzafava, *Un esempio concreto di probabilità non σ-additiva: la distribuzione della prima cifra significativa dei dati statistici*, Bollettino U.M.I. **18-A** (1981), 403-410.

33. Change of Variable in the *SL*-Integral

Rudolf Výborný Mathematics Department, University of Queensland, Queensland, Victoria, Australia

1 INTRODUCTION.

In this paper we shall discuss the change of variables formula

$$\int_{\varphi(A)}^{\varphi(B)} f = \int_A^B (f \circ \varphi)\varphi' \tag{1}$$

for the SL integral which was introduced by Lee Peng Yee and R. Výborný (1993). This integral is equivalent to the Kurzweil-Henstock integral which in turn is equivalent to the Perron or Denjoy integral. For the latter powerful theorems on change of variables exist, due to K. Karták (1956), K. Krzyżewski (1962a), (1962b) and G. Goodman (1978a), (1978b). Our aim is different, we wish to stay within the framework of SL integration and integrate over infinite intervals. Simple examples show that if f is defined only a.e. than the integrand on the right hand side of (1) may fail to be defined on a set of positive measure (even on all of $[A, B]$), therefore we make the following convention: *If $f(\varphi(t))$ is not defined and $\varphi'(t) = 0$ then we postulate that $f(\varphi(t))\varphi'(t) = 0$.* This will be sufficient (under the assumtions specified later) to make (1) meaningful. For convenience we assume throughout the paper that neither A nor B belong to an interval on which φ is constant, clearly such intervals can be eliminated from (1).

2 NOTATION, PRELIMINARIES.

The symbols $\mathbb{R}_+$, $\mathbb{R}$ and $\overline{\mathbb{R}}$ denote in order the positive reals, reals and extended reals. We employ the usual convention regarding arithmetic operations and order for ∞ and

$-\infty$. An interval $[A,\ B]$ could be a part of $\overline{\mathbb{R}}$ but the symbol $[x,\ y]$ with small letters is reserved for a compact interval, degenerate intervals consisting of one point are allowed. All functions in this article are real valued. In the entire paper φ denotes a function which is continuous throughout and differentiable a.e. on $[A,\ B]$.

A *tagged partial division* of an interval $[A,\ B]$ is a finite set of couples $(t_k, [u_k,\ v_k])$ such that the points $t_k \in [u_k, , v_k]$, the compact intervals $[u_k, , v_k]$ are without common interior points and $[u_k,\ v_k] \subset [A,\ B]$. The qualification "of $[A,\ B]$" will often be omitted. We shall call the point t_k the *tag* of $[u_k, , v_k]$. Small letters of the Greek alphabet, particularly π and ω, will be used, possibly with subscript, to denote tagged partial divisions. A tagged partial division

$$\pi \equiv \{(t_i, [u_i,\ v_i]) : i = 1, 2, \ldots n\} \tag{2}$$

shall be often abbreviated to $\{(t, [u,\ v])\}$ if the range of subscripts i is clear from the context. If the tags of π lie in S then we say that π is a tagged partial division on S. A function is called *an open gauge* if it is posititive, it is called *a tight gauge* if it is positive a.e.. If δ is a gauge defined on S then a tagged partial division π on S, given by (2), for which

$$t_i - \delta(t) < u_i \le t_i \le v_i < t_i + \delta(t) \tag{3}$$

for all i with $1 \le i \le n$ is called a δ-fine, in symbols $\pi \ll \delta$.

Given a function f, a tagged partial division (2) associates with it a Riemann sum $\sum_\pi f(t)(v-u)$ which is given by $\sum_{i=1}^n f(t_i)(v_i - u_i)$. We extend our short-hand also to other similar sums, e.g. we shall denote by $\sum_\pi f(u,\ v)$ the sum $\sum_{i=1}^n [f(v_i) - f(u_i)]$ or

$$\sum_\pi (F(u,v) - f(t)(v-u)) = \sum_{i=1}^n (F(u_i) - F(v_i) - f(t_i)(v_i - u_i))\,.$$

The letter e is reserved for a positive continuous function defined on $[A,\ B]$ for which there is a open gauge δ_e such that for any tagged partial division π of $\overline{\mathbb{R}}$ which is δ_e-fine we have $\sum_\pi e(t)(v-u) < \varepsilon$. A function is said to satisfy *the Strong Luzin condition,* or briefly SL, on a set $S \subset \mathbb{R}$ if for every set E of measure zero and every positive ε there exists $\gamma : S \to \mathbb{R}_+$ such that for any γ-fine tagged partial division π on $E \cap S$ we have

$$\sum_\pi |F(u,v)| < \varepsilon. \tag{4}$$

A function f is said to satisfy SL on an interval $[A,\ B] \subset \overline{\mathbb{R}}$ if it satisfies SL on (A, B) and is continuous on $[A,\ B]$. It is obvious that condition SL implies continuity. It can be shown by using Vitali's covering theorem (or by the Covering Lemma of McLeod (1980) p.143) that an SL function necessarily satisfies the condition N of Luzin. Conversely any continuous function of bounded variation satisfying the N condition is SL.

DEFINITION 1 (Definition of the SL integral.) *A function f is said to be SL-integrable on $[A,\ B] \subset \overline{\mathbb{R}}$ if there exists a SL function F with the property that for every positive ε there is a tight gauge δ such that*

$$\sum_\pi |f(t)(v-u) - F(u,v)| < \varepsilon \tag{5}$$

whenever π is a δ-fine tagged partial division of $[A,\ B]$. The number $F(B) - F(A)$ is then the SL-integral of f. The function F itself is called the SL-primitive of f, or simply the primitive and throughout the paper F will denote the primitive of f.

We shall use the notation usual in integration for the SL-integral. It can be shown that the definition is meaningfull, i.e. F is determined uniquely up to a constant, the space of SL-integrable functions is a linear space on which the SL-integral is a linear functional which preserves inequalities. If F is the SL-primitive of f then $F' = f$ a.e.. Conversely, if $H' = f$ a.e. and H is SL then H is an SL primitive of f. If f is SL-integrable then there exists an *open* gauge δ_o such that (5) holds for every tagged partial division which is δ_o-fine.

3 MONOTONE φ.

If φ is monotone it is possible to show that existence of one integral in equation (1) implies the existence of the other. For sake of definitness let φ be increasing. There is at most a countable number of closed intervals K_n such that φ is constant on each K_n and one-to-one on $U = [A, B] - K$, $K = \cup K_n$. There is a correspondence between tagged partial divisions of $[A, B]$ and $[\varphi(A(, \varphi(B)]$. In one direction we have π and $\varphi \circ \pi$ given by $\pi \equiv \{(t, [u, v])\}$ and $\varphi \circ \pi \equiv \{(\varphi(t), [\varphi(u), \varphi(v)])\}$, respectively. In the other direction we associate with every tagged partial division ω on U a tagged partial division $\varphi_{-1} \circ \omega$ of $[A, B]$ in the following way: We set

$$\overline{\varphi}_{-1}(x) = \sup\{t;\, x = \varphi(t)\} \quad \text{and} \quad \underline{\varphi}_{-1}(x) = \inf\{t;\, x = \varphi(t)\}.$$

Then $(x, [w, z]) \in \omega$ corresponds to

$$(\varphi_{-1}(x), [\overline{\varphi}_{-1}(w), \underline{\varphi}_{-1}(z)]) \in \varphi_{-1} \circ \omega.$$

If $D \subset [\varphi(A), \varphi(B)]$ and $\eta : D \to \mathbb{R}_+$ it is possible using continuity of φ to find δ_η such that $\pi \ll \delta_\eta$ implies $\varphi \circ \pi \ll \eta$. Also conversely, given $\delta : C \to \mathbb{R}_+$ with $C \subset U$ there is a positive η_δ such that $\omega \ll \eta_\delta$ implies $\varphi_{-1} \circ \omega \ll \delta$, η_δ is given at $x = \varphi(t)$ by

$$\eta_\delta(x) = \min[\varphi(t) - \varphi(t - \delta(t)), \varphi(t + \delta(t)) - \varphi(t)]. \tag{6}$$

We need the following

LEMMA 1 *If φ is differentiable almost everywhere on $[A, B]$, if $Z \subset \operatorname{rng} \varphi$ is of measure zero then $\varphi' = 0$ almost everywhere on $\varphi_{-1}(Z)$.*

This lemma is due to Krzyżewski (1962a), a reasonably simple proof can be found in Serrin and Varberg (1969) (Theorem 1.) This lemma is a special case of (1) with f the characteristic function of Z and is easy to prove for a monotone φ.

Now we can proceed with

THEOREM 1 *If $[A, B] \subset \overline{\mathbb{R}}$, the function φ satisfies the condition N and is monotone on $[A, B]$ then if one side of equation (1) exists then so does the other and the equality holds.*

Proof: Obviously it is sufficient to prove the theorem for an increasing φ. Let $\varepsilon > 0$ be given. Assume that f is integrable, F its SL-primitive and $\mathcal{N}_F$ is the set where $F' = \pm\infty$ or F' does not exist or does not equal f[1]. Set $\mathcal{M} = \varphi_{-1}(\mathcal{N}_F)$. There is an

[1]$\mathcal{N}_F$ includes all points where f is not defined. The meaning of $\mathcal{N}_F$ is retained throughout the paper.

open gauge η such that if $\omega_N \equiv \{(x, [w, z])\}$ is a tagged partial division on $\mathcal{N}$ with $\omega_N \ll \eta$ then $\sum_{\omega_N} |F(w,z)| < \varepsilon$. For any $t \in \mathcal{M}$ there is by continuity a positive $\delta(t)$ such that $|\varphi(\tau) - \varphi(t)| < \eta(\varphi(t))$ for $|t - \tau| < \delta(t)$. Consequently, if π_M is a tagged partial division on $\mathcal{M}$ and $\pi_M \ll \delta$ then

$$\sum_{\pi_M} |(F \circ \varphi)(u,v)| < \varepsilon. \tag{7}$$

Let $\delta(t) = 0$ if $\varphi'(t) > 0$ and $t \in \mathcal{M}$ or $\varphi'(t)$ does not exist. For the rest of (A, B) there is a positive $\delta(t) \leq \delta_e(t)$ such that that

$$|(F \circ \varphi)(u,v) - f(\varphi(t))\varphi'(t)(v-u)| < e(t)(v-u), \tag{8}$$

whenever $t - \delta(t) < u \leq t \leq v < t + \delta(t)$. Let π be δ-fine, $\pi_M \subset \pi$ with tags in M and $\pi_c = \pi \setminus \pi_M$. Then by inequalities (7) and (8)

$$\begin{aligned} &\sum_{\pi} |(F \circ \varphi)(u,v) - f(\varphi(t))\varphi'(t)(v-u)| \leq \\ \leq\ &\sum_{\pi_c} |(F \circ \varphi)(u,v) - f(\varphi(t))\varphi'(t)(v-u)| + \sum_{\pi_M} |(F \circ \varphi)(u,v)| \\ \leq\ &\sum_{\pi_c} e(t)(v-u) + \varepsilon < 2\varepsilon. \end{aligned}$$

It is easy to see that $F \circ \varphi$ is SL because F is and φ is an increasing N function. That proves the change of variable formula "from left to right".

Going in the other direction let H be the SL primitive of $(f \circ \varphi)\varphi'$, denote $F = H \circ \underline{\varphi}_{-1}$. Since H is constant on each K_n we have also $F = H \circ \overline{\varphi}_{-1}$ and that F is continuous. There are an open gauge $\delta_1 : [A, B] \to \mathbb{R}_+$ and a tight gauge δ_2 such

$$\sum_{\pi_1} |H(u,v) - f(\varphi(t))\varphi'(t)(v-u)| < \varepsilon \tag{9}$$

and

$$\sum_{\pi_2} |f(\varphi(t))\varphi(u,v) - f(\varphi(t))\varphi'(t)(v-u)| < \varepsilon. \tag{10}$$

whenever $\pi_1 \ll \delta_1$ or $\pi_2 \ll \delta_2$, respectively. The existence of δ_1 is obvious from the fact that H is the SL-primitive, δ_2 is defined at every point of differentiability of φ by requiring

$$|\varphi(u,v) - \varphi'(t)(v-u)| < \frac{e(t)(v-u)}{1 + |f(\varphi(t))|} \tag{11}$$

for $t - \delta_2(t) < u \leq t \leq v < t + \delta_2(t)$ and $\delta_2(t) \leq \delta_e(t)$. On the rest of $[A, B]$ we define $\delta_2 = 0$. Define $\delta = \min(\delta_1, \delta_2)$ and denote $Z_\delta = \{t;\, \delta(t) = 0\}$. For $x \notin \varphi(Z_\delta) \cup K$ we obtain η_δ according to (6), for $x \in \varphi(Z_\delta) \cup K$ we set $\eta_\delta = 0$. The function η_δ is a tight gauge and if $\omega \ll \eta_\delta$ and $\pi = \varphi_{-1} \circ \omega$ then

$$\begin{aligned} \sum_{\omega} |F(w,z) - f(x)(z-w)| &\leq \sum_{\pi} |H(u,v) - f(\varphi(t))\varphi'(t)(v-u)| \\ &+ \sum_{\pi} |f(\varphi(t))\varphi(u,v) - f(\varphi(t))\varphi'(t)(v-u)|. \end{aligned} \tag{12}$$

and by inequalities (9), (10)

$$\sum_{\omega} |F(w,z) - f(x)(v-u)| < 2\varepsilon. \tag{13}$$

Since F is continuous it is SL on the countable set $\varphi(K)$. To complete the proof it suffices to show that F is SL on $\varphi(U)$. Let $Z \subset \varphi(U)$ be of measure zero, $T = \varphi_{-1}(Z)$ and $S \subset T$ where $\varphi' = 0$. The set $T \setminus S$ is of measure zero, hence there is $\gamma : T \setminus S \to \mathbb{R}_+$ such that

$$\sum_\pi |H(u,v)| < \varepsilon \tag{14}$$

whenever $\pi \ll \gamma$ and the tags of π lie in $T \setminus S$. For $t \in S$ we can let $\gamma(t) = \delta_1(t)$. If $\omega \ll \eta_\gamma$ and $\pi = \varphi_{-1} \circ \omega$ then by (9) and (14)

$$\sum_\omega |F(w,z)| = \sum_\pi |H(u,v)| \quad = \quad \sum_{t \in T \setminus S} |H(u,v)| \tag{15}$$

$$+ \sum_{t \in S} |H(u,v) - f(\varphi(t))\varphi'(t)(v-u)| \quad < \quad \varepsilon + \varepsilon. \qquad \text{QED} \tag{16}$$

REMARK 1 The condition that φ satisfies N is essential, the theorem fails without it even if $f = 1$.

4 THE GENERAL CASE.

For a general φ we can only prove the formula from left to right.

THEOREM 2 *If f is integrable on the range of φ then $f \circ \varphi$ is integrable and formula (1) holds if and only if $F \circ \varphi$ is SL and differentiable almost everywhere.*

Proof: The only if part is obvious. For the if part it sufices to show that

$$(F \circ \varphi)' = (f \circ \varphi)\varphi' \quad \text{a.e. on} \quad [A,\ B]. \tag{17}$$

Equation (17) is obvious on $[A,\ B] \setminus \mathcal{M}$. Since $(F \circ \varphi)(\mathcal{M})$ is of measure zero the derivative of $F \circ \varphi$ is a.e. zero on $\mathcal{M}$ by Lemma 1. So is φ' by the same Lemma. QED

If $\varphi' \neq 0$ a.e. then $F \circ \varphi$ is differentiable a.e. Indeed it is differentiable a.e. outside $\mathcal{M}$ and $\varphi' = 0$ on $\mathcal{M}$ by Lemma 1, hence $\mathcal{M}$ itself must be of measure zero. Let us assume that φ satisfies the condition SL. Then, clearly, $F \circ \varphi$ is SL if F is Lipschitz which happens if f is bounded a.e.. It is also SL on the set where F has bounded Dini derivates. A function which is SL on each set of a countable family is also SL on the union. Consequently if F has finite Dini derivates everywhere on the range of φ except possibly a countable set then $F \circ \varphi$ is SL.

Theorem 2 is a counterpart of Theorem 3 in Serrin and Varberg (1969) for the Lebesgue integral and Theorem 1 in Krzyżewski (1962a) for the Denjoy integral.

REFERENCES.

Goodman, G. (1978) N-functions and integation by substitution, *Rend. Sem. Mat. Fis. Milano; 47, 124–134.*

Goodman, G. (1978) Integration by substitution, *Proc. Amer.Math. Soc. 89–91.*

Krzyżewski, K. (1962a) On change of variable in the Denjoy-Perron integral I, *Coll. Math (9) 99–104.*

Krzyżewski, K. (1962b) On change of variable in the Denjoy-Perron integral II, *Coll. Math (9) 317–323.* Lee, P. Y. and Výborný. R. (1993) Kurzweil-Henstock integration and the Strong Luzin condition, *Bolletino U.M.I. (7), 761-773.*

McLeod, R. (1980). The Generalized Riemann integral. The Carus Math. Monograph published by MAA.

Karták, K. (1956) Věta o substituci pro Denjoyovy integrály. *Časopis pro pěstování mat. (81), 410–419.*

Serrin, J and Varberg. D.E. (1969) A general chain rule for derivatives and the change of variable formula for the Lebesgue integral, *Amer. Math. Monthly 76, 514–519.*

34. The Minimum Energy Configuration of a Mixed-Material Column

Hans F. Weinberger School of Mathematics, University of Minnesota, Minneapolis, Minnesota

To my friend Carlo Pucci.

INTRODUCTION

In this work we consider a vertical cylindrical column of prescribed cross section which consist of a mixture of two constituent materials with a prescribed volume fraction of each. We suppose that the mean vertical displacements of the top and bottom are prescribed, that the sides of the column are free of tractions, and that the tractions on the ends are vertical and constant.

It was shown in two recent papers [1,2] that among all such columns for which the mixture in each horizontal cross section is statistically homogeneous, the material arrangement which minimizes the potential energy places all of one of the materials in a single horizontal layer. This fact is of interest if one assumes that the materials in the column are able to diffuse toward the configuration of minimum potential energy.

The above result makes essential use of the hypothesis of statistical homogeneity. Because this hypothesis is difficult to define and even more difficult to verify for a particular mixture of materials, it is interesting to know whether one can find the same result without this assumption.

We shall show that at least in the special case in which the two materials are isotropic and have the same mass density and the same value of the ratio of Young's modulus to Poisson's ratio, it is true that the arrangement of the materials which minimizes the potential energy places all of one of the materials in a single horizontal layer.

2 THE MINIMUM ENERGY CONFIGURATION

We suppose that a cylindrical column is composed of a mixture of two materials which happen to have the same density ρ. We introduce a coordinate system with the origin on the top of the cylinder and with the z-axis oriented downward. Then the top of the cylinder is at $z = 0$, while the the bottom is is at $z = L$, where L is the length of the cylinder.

We label the two materials with the labels 0 and 1. We assume that the elastic behavior of both materials is governed by the usual equations of (not necessarily isotropic) linear elasticity. That is, in material α the stress σ_{ij} is related to the infinitesimal strain by a relation of the form

$$\sigma_{ij} = \sum_{k,\ell=1}^{3} C_{ijk\ell}^{(\alpha)} e_{k\ell},$$

where the four-tensor $C_{ijk\ell}^{(\alpha)}$ represents a positive definite symmetric linear transformation of the space of symmetric 3×3 matrices. The inverse of this transformation is represented by a four-tensor $T_{ijk\ell}^{(\alpha)}$ which gives the strain-stress relation

$$e_{ij} = \sum_{k,\ell=1}^{3} T_{ijk\ell}^{(\alpha)} \sigma_{k\ell}.$$

We are concerned with the equilibrium deformation which is produced in the cylinder by gravitational forces when the top is fixed at 0 while the bottom of the cylinder is given a vertical displacement Δ. No force is applied to the lateral boundaries of the cylinder, and only vertical forces are applied at the ends.

Because it is known [4, §86] that even in the case of a single material, uniaxial vertical stress produces a warping of the horizontal cross sections, we shall assume that the boundary conditions are only imposed in the mean sense

$$\begin{aligned} \frac{1}{A}\int u_3(x, y, 0)\,dx\,dy &= 0 \\ \frac{1}{A}\int u_3(x, y, L)\,dx\,dy &= \Delta, \end{aligned} \tag{1}$$

where A is the area of the cross sections of the column. The potential energy of this problem is defined to be

$$\mathcal{E} = \int \left\{ \frac{1}{2} \sum_{i,j,k,\ell} C_{ijk\ell}^{(\alpha)} e_{ij} e_{k\ell} - g\rho u_3 \right\} dx\,dy\,dz.$$

Because the cylinder is a mixture of the two materials, the integer α is a function of (x, y, z) in this and all subsequent integrals.

The boundary value problem is obtained by stating that the actual potential energy is the minimum of this integral among all displacement fields which satisfy the end conditions (1). This variational principle implies that the stress satisfies the equilibrium conditions and the natural boundary conditions that the traction vanishes on the lateral boundaries and is a constant vertical vector on each of the ends.

The resulting boundary value problem produces unique strains. The displacements are only determined to within an arbitrary horizontal translation and infinitesimal rotation. Because this nonuniqueness of the displacements is easily removed by introducing some simple normalizing conditions and because it does not affect the potential energy, we shall ignore it.

Castigliano's principle states that the potential energy can also be characterized as the maximum of the complementary energy

$$\hat{\mathcal{E}}[\tilde{\sigma}] = \Delta \int \tilde{\sigma}_{33}(x, y, L)dxdy - \frac{1}{2}\int \sum_{i,j,k,\ell} T^{(\alpha)}_{ijk\ell}\tilde{\sigma}_{ij}\tilde{\sigma}_{k\ell}dxdydz \tag{2}$$

among symmetric matrix fields $\tilde{\sigma}_{ij}$ which give zero traction on the lateral boundaries of the cylinder and constant vertical tractions on the top and bottom, and which satisfy the equilibrium condition

$$\sum_{j=1}^{3} \partial\tilde{\sigma}_{ij}/\partial x_j = -g\rho,$$

where g is the gravitational constant.

We shall confine our attention to matrix fields of the form

$$\tilde{\sigma}_{ij} = s(x, y, z)\delta_{i3}\delta_{j3}.$$

Then the only surviving component of the matrix $T^{(\alpha)}_{ijk\ell}$ in (2) is $T^{(\alpha)}_{3333}$. When the material α is isotropic, this component is just the reciprocal of Young's modulus. Whether the material is or is not not isotropic, we define

$$T^{(\alpha)} := \max\left\{ \sum_{i,j,k,\ell=1}^{3} T^{(\alpha)}_{ijk\ell}z_i z_j z_k z_\ell \ : \ \sum_{i=1}^{3} z_i^2 = 1 \right\}.$$

It is easily seen that when the material α is oriented in such a way that the maximizing vector $\mathbf{z}$ points downward, then $T^{(\alpha)}_{3333} = T^{(\alpha)}$, and $T^{(\alpha)}_{1333} = T^{(\alpha)}_{2333} = 0$. We now make the additional assumption that one of the two materials can then be rotated about the z-axis in such a way that the matrix $T^{(\alpha)}_{\mu\nu 33}$ in which μ and ν take the values 1 and 2 is the same for the two materials. When the materials are isotropic, this simply means that Poisson's ratio divided by Young's modulus has the same value in both materials.

This additional condition is easily seen to be necessary if a vertical uniaxial stress field is to be compatible with a smooth displacement field near a horizontal interface of the two materials. The condition is also sufficient to force the stress field for a column in which the two materials are arranged in horizontal layers to be uniaxial.

Castigliano's principle now gives the lower bound

$$\mathcal{E} \geq \Delta \int s(x,y,L)dxdy - \frac{1}{2}\int T^{(\alpha)}s^2 dxdydz, \tag{3}$$

provided the function s satisfies the conditions

$$\frac{\partial s}{\partial z} = -g\rho,$$
$$s(x,y,0) \text{ and } s(x,y,L) \text{ constant.}$$

By hypothesis the right-hand side is the same for both materials. Thus

$$s(x,y,z) = s(x,y,L) + g\rho(L-z).$$

Because $s(x,y,L)$ is a constant, s a function of z only, which we denote by $s(z)$. We are free to choose the constant $s(L)$.

We now ask for a lower bound for the potential energy which depends upon the prescribed volume fractions of the two materials, but not upon the particular arrangement of the materials. We do this by using a rearrangement argument which was introduced by M. G. Krein [3]. We first fix the constant $s(L)$, and hence the function s.

Suppose that

$$T^{(1)} < T^{(0)},$$

and let R be the volume fraction of material 1 in the cylinder. Because $s(z)$ is strictly decreasing, there is a unique subinterval I of length RL of the interval [0,L] such that

$$\sup_{z\in I} s^2 = \inf_{z\notin I} s^2. \tag{4}$$

Let

$$\hat{\alpha}(x,y,z) = \begin{cases} 1 & \text{for } z \in I \\ 0 & \text{for } z \notin I. \end{cases}$$

Then

$$\int T^{(\hat{\alpha})}dxdydz = \int T^{(\alpha)}dxdydz.$$

Therefore,

$$\int T^{(\hat{\alpha})}s^2 dxdydz - \int T^{(\alpha)}s^2 dxdydz = \int \{T^{(\hat{\alpha})} - T^{(\alpha)}\}\{s^2 - \sup_{z\in I} s^2\}dxdydz.$$

The first factor on the right is nonpositive in I and nonnegative outside I. By (4), the same is true of the second factor. We conclude that the first integral on the left is an upper bound for the second.

It now follows from (3) that

$$\mathcal{E} \geq \Delta A s(L) - \frac{A}{2}\int T^{(\hat{\alpha})}s^2 dz, \tag{5}$$

which is independent of the arrangement of the materials in the column. We note that this lower bound corresponds to the rearrangement of the material in such a way that all the material 1 with $T^{(1)} < T^{(0)}$ is in a single horizontal layer of height RL. For this arrangement with the materials oriented in such a way that $T^{(\alpha)}_{3333} = T^{(\alpha)}$ and that the matrices $T^{(\alpha)}_{\mu\nu 33}$ with $\mu, \nu = 1, 2$ agree, the correct stress has the form $s\delta_{3i}\delta_{3j}$ with s a function of z only. Therefore maximizing the right-hand side of (5) with respect to the choice of $s(L)$ will give the potential energy for the column with this arrangement of materials.

We have thus shown that under the above conditions on the densities and the strain-stress tensors of the two materials, **among all arrangements of the materials with a prescribed volume fraction the potential energy is minimized by a unique arrangement, and in this arrangement the material with the smaller value of $T^{(\alpha)}$ is in a single horizontal layer.**

That is, in this special case the result of [1,2] is valid without any assumption about homogenization.

REFERENCES

1. R. L. Fosdick and G. F. Royer, Equilibrium of a two-phase solid mixture in a stressed elastic bar, Int. J. of Solids and Structures, in print.
2. Roger L. Fosdick, Gianni F. Royer-Carfagni, and Hans F. Weinberger, A note on the optimal state of a binary solid mixture in a stressed elastic bar, Meccanica, in print.
3. M. G. Krein, On certain problems on the maximum and minimum of characteristic values and on the Lyapounov zones of stability, Prik. Mat. Mech. 15 (1951), pp. 323-348, Amer. Math. Soc. Transl. (2) 1 (1955), pp. 163-187.
4. A. E. H. Love, A Treatise on the Mathematical Theory of Elasticity, Dover (1944).

35. Convergence of Regularized Solutions of Nonlinear Ill-Posed Problems with Monotone Operators

Fengshan Liu Department of Mathematics, Delaware State University, Dover, Delaware

M. Zuhair Nashed Department of Mathematics, University of Delaware, Newark, Delaware

In honor of Professor Carlo Pucci on the occasion of his seventieth birthday.

Abstract We consider convergence of regularized solutions of the ill-posed problem $b \in Au$, where A is a nonlinear monotone set-valued mapping on a Hilbert space. The regularized solutions u_ε are solutions of the perturbed inclusion $b \in Au + \varepsilon Bu$, where B plays the role of a regularizer and $\varepsilon > 0$. We also provide a convergence rate in the case of single-valued mappings.

1 INTRODUCTION

Let F be a mapping from a metric space X into a metric space Y. At the begining of this century, Hadamard formulated the concept of well-posedness of the operator equation

$$F(x) = y \tag{1}$$

as follows: The problem (1) is said to be *well-posed* if for each $y \in Y$,

(a) a solution exists;

(b) the solution is unique; and

(c) the solution depends continuously on the data.

A problem which is not well-posed is called *ill-posed*, and a method for finding stable approximate solutions of an ill-posed problem is called a regularization method.

Ill-posed problems arise frequently in many areas of applied mathematics, geophysics, technology and medicine. Thus the study of regularization methods has been, and continues to be, an important area of research in mathematics, numerical analysis and the applied sciences.

When the operator in (1) is linear, regularization methods of solving (1) have been initiated by Tikhonov since the 1940s and studied extensively since the early 1960s by Soviet mathematicians and others including Tikhonov, Lavrentjev, Morozov and Ivanov. For further information, see for example [14], [15], [16], [8], [12], [7], [10], [11] , [3], [9], and the references cited therein. Ill-posed problems in partial differential equations, and integral equations of the first kind have also been extensively studied by many mathematicians (see Payne [12] and references cited therein).

In this paper we consider the problem of finding $u \in H$ such that

$$b \in Au, \tag{2}$$

where A is a nonlinear monotone set-valued mapping from Hilbert space H to 2^H. The graph of A is denoted by

$$G(A) = \{(u,v) \in H \times H : u \in \mathrm{dom}(A),\ v \in Au\}.$$

Instead of $(u,v) \in G(A)$ we briefly write $v \in Au$.

We recall a few basic definitions.

Definition 1 *A mapping $A : H \to 2^H$ is said to be*

(i) monotone if

$$\langle u^* - v^*, u - v\rangle \ \geq 0, \tag{3}$$

for all $u^* \in Au$ and $v^* \in Av$.

(ii) maximal monotone if A is monotone and has no proper monotone extension.

(iii) strongly monotone if there exists a positive constant λ such that

$$\langle u^* - v^*, u - v\rangle \ \geq \lambda \parallel u - v \parallel^2, \tag{4}$$

for all $u^* \in Au$ and $v^* \in Av$.

In particular, an operator $B :\ H \to\ H$ is strongly monotone if

$$\langle Bu - Bv, u - v\rangle \ \geq \lambda \parallel u - v \parallel^2, \tag{5}$$

for all $u,\ v \in \mathrm{dom}B \subset H$.

Definition 2 *An operator $Z : H \to 2^H$ is called coercive with respect to a given element $b \in H$, if there exist $r > 0$ and $u_0 \in domZ$ such that $\langle u^* - b, u - u_0\rangle > 0$ for all $(u, u^*) \in Z$ with $\parallel u \parallel > r$.*

Definition 3 *An operator A is hemicontinuous if and only if the real function*

$$t \mapsto \langle A(u + tv), w\rangle$$

is continuous on [0,1] for all $u, v, w \in H$.

The study of monotone operators and nonlinear operator equations with monotone operators was initiated by Minty and Browder. Minty proved that if $F : H \to H$ is a continuous monotone operator on a Hilbert space H into H, then the operator equation $x + F(x) = y$ has a unique solution for each $y \in H$ and the solution depends continuously on y. Browder reformulated this result and gave a different proof: If $G : H \to H$ is a strongly monotone, continuous operator, then G is one-to-one, onto and its inverse is continuous. These results provide sufficient conditions under which the operator equation is globally well-posed. Many inportant extensions and generalizations of these results have been given during 1960s and 1970s by Browder, Minty, Stampacchia and others (see Vol.3 of Zeidler [17]). In particular, Browder [6] gave the following fundamental theorem for monotone mappings:

Theorem 1 *Let $b \in H$ be given and assume that the mapping A is maximal monotone, and the mapping $B : H \to H$ is monotone, hemicontinuous, bounded, and coercive with respect to each element $b \in H$. Then $A + B : H \to 2^H$ is surjective.*

As a Corollary of the above theorem, the following theorem holds (for the proof see, e.g. [17], p.868):

Theorem 2 *Let mapping $A : H \to 2^H$ be maximal monotone and coercive with respect to each element $b \in H$. Then A is surjective, i.e., inclusion (1) has a solution for each $b \in H$.*

Furthermore if the operator $A : H \to H$ in the Existence Theorem 2 is strongly monotone, then A is one-to-one, onto and its inverse is continuous; hence the problem is well-posed. However, the problem is ill-posed in general for $A : H \to 2^H$.

Several regularization methods have been considered for nonlinear ill-posed problems. Here we consider regularization via the perturbation method and seek solutions of the inclusion (2).

2 MAIN RESULTS

We study the problem (2) where b is in the range of A. For a fixed b, we denote the solution set of (2) by Φ.

We study the convergece and convergence rate of the solutions of the inclusions

$$b \in Au + \varepsilon Bu, \; u \in H, \tag{6}$$

where $\varepsilon > 0$ is a small parameter and $B : H \to H$ is a monotone (not necessary linear) operator.

In 1967, Browder and Ton [4], [5] established the strong convergence when the operator A is a demicontinuous monotone mapping , and also satisfies the coercivity property

$$\frac{(Av, v)}{\| v \|} \to \infty, \text{ as } \| v \| \to \infty.$$

In 1975, Alber [1] improved the results of strong convergence for single-valued monotone operators without requiring coercivity. Another convergence result for the regularization method can be found in Zeidler's book (see Theorem 32.K. in [17]), where the convergence of the solutions of the following regularized inclusions is given,

$$b \in Au_n + \varepsilon_n Bu_n,\ u_n \in H,\ n = 1, 2, ...,$$

where $\varepsilon_n > 0$ for all n and $\varepsilon_n \to 0$ as $n \to \infty$, the operator $B : H \to H$ is a strictly monotone, hemicontinuous, bounded, and $B(0) = 0$. In 1992, Bakushinskii [2] established an iterative method for the nonlinear ill-posed problem $Au = f$.

We study the convergence of solutions of (6) for set-valued mappings without the coerciveness condition. Furthermore, we give a convergence rate for the solutions of (6) when the operator A is single-valued. To our knowledge, this is the first time that the convergence rate of the perturbation inclusion (6) is given.

Theorem 3 *Let $A : H \to 2^H$ be a maximal monotone operator and $B : H \to H$ be a strongly monotone, hemicontinuous, bounded (i.e., there exists a constant C such that $\| Bu \| \leq C \| u \|$ for all $u \in H$), and $B(0) = 0$. Then,*

(a) the solution set Φ of (2) is convex and closed;

(b) for each $\varepsilon > 0$, (6) has a unique solution u_ε and the solution u_ε satisfies

$$\| u_\varepsilon \| \leq C \min_{u \in \Phi} \| u \|, \tag{7}$$

where C is a constant depending on λ(the λ is same as in (4));

(c) as $\varepsilon \to 0, \{u_\varepsilon\}$ converges in H to the unique solution $u_0 \in \Phi$ of the variational inequality

$$\langle Bu, v - u\rangle \geq 0, \tag{8}$$

for all $v \in \Phi$.

Proof: (a) (see[17]) $u \in \Phi$ iff
$\langle b - v^*, u - v\rangle \geq 0$ for all $v \in H$ and $v^* \in Av$, so Φ is convex and closed in H.

(b) For each fixed $\varepsilon > 0$, by Theorem 32.I in [17], we have that $A + \varepsilon B$ is maximal monotone. Hence, by Theorem 1, the inclusion (6) has a solution u_ε for each $\varepsilon > 0$. Furthermore, since B is strongly monotone, $A + \varepsilon B$ is also strongly monotone and the solution of (6) must be unique.

Now, we want to show that

$$\| u_\varepsilon \| \leq C \min_{u \in \Phi} \| u \| .$$

For each $u \in \Phi$, we have that $b \in Au_\varepsilon + \varepsilon Bu_\varepsilon$, and

$$\varepsilon Bu + b \in Au + \varepsilon Bu. \tag{9}$$

Therefore $-\varepsilon Bu \in Au_\varepsilon - Au + \varepsilon(Bu_\varepsilon - Bu)$, or $-\varepsilon Bu = u_\varepsilon^* - u^* + \varepsilon(Bu_\varepsilon - Bu)$ for some $u_\varepsilon^* \in Au_\varepsilon$ and $u^* \in Au$.

Taking inner products with $u_\varepsilon - u$, we get

$$\langle u_\varepsilon^* - u^*, u_\varepsilon - u\rangle + \varepsilon\langle Bu_\varepsilon - Bu, u_\varepsilon - u\rangle = -\varepsilon\langle Bu, u_\varepsilon - u\rangle. \tag{10}$$

Since A is monotone, this implies

$$\langle Bu_\varepsilon - Bu, u_\varepsilon - u\rangle \leq -\langle Bu, u_\varepsilon - u\rangle.$$

Therefore $\lambda \| u_\varepsilon - u \|^2 \leq \| Bu \| \| u_\varepsilon - u \|$, or

$$\| u_\varepsilon - u \| \leq \frac{1}{\lambda} \| Bu \| \leq C \| u \|, \tag{11}$$

or $\| u_\varepsilon \| \leq C \| u \|$. Hence, there exists a constant $C > 0$ such that

$$\| u_\varepsilon \| \leq C \min_{u\in\Phi} \| u \| .$$

(c) We know that the variational inequality (8) has at most one solution (see the proof of Theorem 32.K in [17]). From (b), we get that u_ε is bounded.

Let $\{\varepsilon_n\}$ be a sequence of positive numbers converges to zero as $n \to \infty$. From (b) we know that $\{u_{\varepsilon_n}\}$ is bounded. So there exists a subsequence, again denoted by $\{u_{\varepsilon_n}\}$, such that

$$u_{\varepsilon_n} \rightharpoonup u \quad as \quad n \to \infty. \tag{12}$$

We want to show that $u \in \Phi$. Using the monotonicity of A, we obtain, for each $v \in \mathrm{dom}A$ and for some $v^* \in Av$, $u_{\varepsilon_n}^* \in Au_{\varepsilon_n}$,

$$\begin{aligned} 0 &\leq \lim_{n\to\infty} \langle u_{\varepsilon_n}^* - v^*, u_{\varepsilon_n} - v\rangle = \lim_{n\to\infty} \langle b - \varepsilon_n Bu_{\varepsilon_n} - v^*, u_{\varepsilon_n} - v\rangle \\ &= \langle b - v^*, u - v\rangle. \end{aligned}$$

Because of the maximal monotonicity of A, we get $b \in Au$, i.e., $u \in \Phi$.

We want to prove that u is a solution of the variational inequality (8). To this end, we set $u_t = u + t(v - u)$ for $0 < t \leq 1$ and $v \in \Phi$, where t and v are fixed. Since Φ is convex, we obtain $u_t \in \Phi$ and hence $b \in Au_t$. From the regularized equation (6) it follows that for all n

$$\begin{aligned} 0 &= \langle u_{\varepsilon n}^* - u_t^*, u_{\varepsilon n} - u_t\rangle + \varepsilon_n\langle Bu_{\varepsilon_n} - b, u_{\varepsilon_n} - u_t\rangle \\ &\geq \varepsilon_n\langle Bu_{\varepsilon_n} - b, u_{\varepsilon_n} - u_t\rangle. \end{aligned}$$

Hence

$$\begin{aligned} 0 &\geq \langle Bu_{\varepsilon_n} - b, u_{\varepsilon_n} - u_t\rangle \\ &= \langle Bu_{\varepsilon_n} - Bu_t, u_{\varepsilon_n} - u_t\rangle + \langle Bu_t - b, u_{\varepsilon_n} - u_t\rangle \\ &\geq \langle Bu_t - b, u_{\varepsilon_n} - u_t\rangle. \end{aligned}$$

Letting $n \to \infty$, we get

$$0 \geq \langle Bu_t, u - u_t\rangle = t\langle Bu_t, u - v\rangle.$$

Dividing by t and letting $t \to 0^+$, we obtain

$\langle Bu, v-u\rangle \geq 0 \quad$ for all $\ v \in \Phi,$

i.e., u satisfies (8).

According to our consideration above, each sequence of u_{ε_n} has, in turn, a subsequence which converges weakly to the unique solution u of (8). So, the entire family of u_ε must converge weakly to u as $\varepsilon \to 0$.

Now we want to prove that u_ε converges in norm. From equation (6) and $b \in Au$, we obtain

$$\begin{aligned} 0 &= \langle u_\varepsilon^* - b + \varepsilon Bu_\varepsilon, u_\varepsilon - u\rangle \\ &= \langle u_\varepsilon^* - u^*, u_\varepsilon - u\rangle + \langle \varepsilon Bu_\varepsilon, u_\varepsilon - u\rangle \\ &\geq \varepsilon\langle Bu_\varepsilon, u_\varepsilon - u\rangle \end{aligned}$$

and hence

$$\begin{aligned} 0 &\geq \langle Bu_\varepsilon, u_\varepsilon - u\rangle \\ &= \langle Bu_\varepsilon - Bu, u_\varepsilon - u\rangle + \langle Bu, u_\varepsilon - u\rangle \end{aligned}$$

Noting that $u_\varepsilon \rightharpoonup u$, as $\varepsilon \to 0$, this implies

$$0 \leq \langle Bu_\varepsilon - Bu, u_\varepsilon - u\rangle \to 0 \ \ as \ \ \varepsilon \to 0.$$

That is $u_\varepsilon \to u$ as $\varepsilon \to 0$.

□

Theorem 3 does not take into consideration of the contamination in the data b. Usually we are given b_δ with $\| b_\delta - b \| < \delta$. Then we consider the operator equations

$$b_\delta \in Au + \varepsilon Bu. \tag{13}$$

Theorem 4 *Under the assumptions of Theorem 3, we have:*

(b′) For each $\varepsilon > 0$ and $\delta > 0$, the equation (13) has a unique solution u_ε and the solution u_ε satisfies that

$$\| u_{\varepsilon,\delta} \| \leq C \min_{u\in\Phi} \| u \| + \delta/\varepsilon, \tag{14}$$

where C is a constant depending only on λ.

(c′) As $\varepsilon \to 0$, if $\delta = o(\varepsilon)$, then $\{u_\varepsilon\}$ converges in H to a solution $u_0 \in \Phi$, where u_0 is the unique solution of the variational inequality

$$\langle Bu, v-u\rangle \ \geq 0, \ for \ all \ v \in \Phi. \tag{15}$$

Proof: The proof is similar to the proof of Theorem 3.

□

Now, we study the rate of convergence of the regularized solutions.

Assumptions:

Let $u_0 \in \Phi$ be the solution of (2) satisfying the variational inequality (8), and $A : H \to H$ be an operator that satisfies the following assumptions:

(a1) A is monotone and hemicontinuous.
(a2) A is Fréchet differentiable, and there exists a constant $L > 0$, such that
$\| A'(u_0) - A'(u) \| \leq L \| u_0 - u \|$
for all $u \in H$.
(a3) There exists an element $\omega \in H$, such that
$u_0 = A'(u_0)^* \omega$.
(a4) $L \| \omega \| < 2$.

First, we consider the rate of the convergence of the solutions u_ε of

$$Au + \varepsilon u = b. \tag{16}$$

Theorem 5 *Suppose that the assumptions (a1), (a2), (a3), and (a4) hold. Then the equation (16) has a unique solution $u_\varepsilon \in H$ which converges to u_0 as $\varepsilon \to 0$, and the following convergence rate holds:*

$$\| u_\varepsilon - u_0 \| = O(\varepsilon^{1/2}).$$

Proof: From $Au_0 = b$ and $Au_\varepsilon + \varepsilon u_\varepsilon = b$, we obtain

$$\begin{aligned} & \langle Au_\varepsilon - Au_0, u_\varepsilon - u_0 \rangle + \varepsilon \langle u_\varepsilon - u_0, u_\varepsilon - u_0 \rangle \\ = \ & -\varepsilon \langle u_0, u_\varepsilon - u_0 \rangle, \end{aligned} \tag{17}$$

and

$$\begin{aligned} & \langle Au_\varepsilon - Au_0, Au_\varepsilon - Au_0 \rangle + \varepsilon \langle u_\varepsilon - u_0, Au_\varepsilon - Au_0 \rangle \\ = \ & -\varepsilon \langle u_0, Au_\varepsilon - Au_0 \rangle. \end{aligned} \tag{18}$$

From the above two equations, we obtain

$$\| Au_\varepsilon - Au_0 \| \leq \varepsilon \| u_0 \|, \tag{19}$$

and $\| u_\varepsilon - u_0 \|^2 \leq -\langle u_0, u_\varepsilon - u_0 \rangle$. By using (a3), we have

$$\| u_\varepsilon - u_0 \|^2 \leq -\langle \omega, A'(u_0)(u_\varepsilon - u_0) \rangle. \tag{20}$$

From assumption (a2) and the mean-value theorem, we get

$$Au_\varepsilon = Au_0 + A'(u_0)(u_\varepsilon - u_0) + r_\varepsilon,$$

where

$$\| r_\varepsilon \| \leq L/2 \| u_\varepsilon - u_0 \|^2 .$$

From (19), (20) and the above equation, we have

$$\begin{aligned} \| u_\varepsilon - u_0 \|^2 & \leq \langle \omega, Au_\varepsilon - Au_0 - r_\varepsilon \rangle \leq \| \omega \| (\| Au_\varepsilon - Au_0 \| + \| r_\varepsilon \|) \\ & \leq \| \omega \| (\varepsilon \| u_0 \| + L/2 \| u_\varepsilon - u_0 \|^2). \end{aligned}$$

Thus $(1- \| \omega \| L/2) \| u_\varepsilon - u_0 \|^2 \leq \| \omega \| \| u_0 \|$. Since $\| \omega \| L < 2(see(a4))$, we have

$$\| u_\varepsilon - u_0 \| \leq \left(\frac{\| \omega \| \| u_0 \|}{1- \| \omega \| L/2}\right)^{1/2} \varepsilon^{1/2} = O(\varepsilon^{1/2}).$$

□

Now we conclude with a result on the convergence rate of the solutions $u_{\varepsilon,\delta}$ of

$$Au + \varepsilon u = b_\delta. \tag{21}$$

Theorem 6 *With the assumptions (a1), (a3), (a4), and (a5), if* $\delta = o(\varepsilon)$, *then*
(1) the equation (21) has a unique solution $u_{\varepsilon,\delta} \in H$ *and* $u_{\varepsilon,\delta}$ *converges to* u_0;
(2) $\| u_{\varepsilon,\delta} - u_0 \| = O(\varepsilon^{1/2})$.

Proof: The proof is similar to the proof of Theorem 5.

□

References

[1] Alber, Ya.I., The Solution of Nonlinear Equations with Monotone Operators in Banach Spaces, *Siberian Math. J.*, 16 (1975), pp. 1-8.

[2] Bakushinskii, A.B., Iterative Method for The Solution of Nonlinear Ill-Posed Problems and Their Applications, *Ill-Posed Problems in Natural Sciences,* (A.N. Tikhonov et. al., Eds.), VSP, Utrecht/TVP Science Publishers, Moscow, 1992, pp. 13-17.

[3] Bakushinskii, A.B. and Goncharskii, A.G., *Ill-Posed Problems: Theory and Applications*, Kluwer Academic Publishers, Dordrecht, 1994.

[4] Browder, F.E. and Ton, B.A., Nonlinear Functional Equations in Banach Spaces and Elliptic Super-regularization, *Math. Zeitschr.*, 105 (1968), pp. 177-195.

[5] Browder, F.E. and Ton, B.A., Convergence of Approximants by Regularization for Solutions of Nonlinear Functions in Banach Spaces, *Math. Zeitschr.*, 106 (1968), pp. 1-16.

[6] Browder, F.E., Nonlinear Operators and Nonlinear Equations of Evolution in Banach Spaces, *Proc. Sympos. Pure Math.*, 18, Part 2, 1968.

[7] Groetsch, C.W., *The Theory of Tikhonov Regularization for Fredholm Integral Equations of the First Kind*, Pitman, Boston, 1984.

[8] Ivanov, V.K., Vasin, V.V. and Tanna, V.P., *Theory of Ill-Posed Linear Problems and Its Applications*, Nauka, Moscow, 1978.

[9] Lavrentiev, M., Romanov, V. and Shishatskii, S., *Ill-Posed Problems of Mathematical Physics and Analysis*, Amer. Math. Soc., Providence, 1987.

[10] Morozov, V.A., *Methods for Solving Incorrectly Posed Problems*, Springer-Verlag, Berlin, Heidelberg, New York, 1984.

[11] Morozov, V.A., *Regularization Methods for Ill-Posed Problems*, CRC Press, Inc., Florida 1993.

[12] Payne, L.E., Improperly Posed Problems in Partial Differential Equations, *Regional Conference Series in Applied Mathematics*, SIAM, 1975.

[13] Rjazantseva, I.P., Variational Inequalities with Monotone Operators on Sets Which Are Specified Approximately, *USSR Comput. Maths. Math. Phys.*, 24 (1984), pp. 194-197.

[14] Tikhonov, A.N., On Stability of Inverse Problems, *Dkad. Nauk.* USSR, 39 (1994), pp. 195-198.

[15] Tikhonov, A.N., On The Solution of Ill-Posed Problems, *Ibid.*, 153 (1963), pp. 49-52.

[16] Tikhonov, A.N. and Arsenin, U.Y., *Solutions of Ill-Posed Problems,* V.H. Winston and Sons, Washington, D.C., John Wiley and Sons, New York, 1977.

[17] Zeidler, E., *Nonlinear Functional Analysis and its Applications*, Volume 2, Springer-Verlag, 1989.

Index